T0281602

Paving the Way for Apollo 11

David M. Harland

Paving the Way for Apollo 11

 Springer

Published in association with
Praxis Publishing
Chichester, UK

 PRAXIS

David M. Harland
Space Historian
Kelvinbridge
Glasgow
UK

SPRINGER–PRAXIS BOOKS IN SPACE EXPLORATION
SUBJECT *ADVISORY EDITOR*: John Mason M.B.E., B.Sc., M.Sc., Ph.D.

ISBN 978-0-387-68131-3 Springer Berlin Heidelberg New York

Springer is a part of Springer Science + Business Media (*springer.com*)

Library of Congress Control Number: 2009925770

Apart from any fair dealing for the purposes of research or private study, or criticism or review, as permitted under the Copyright, Designs and Patents Act 1988, this publication may only be reproduced, stored or transmitted, in any form or by any means, with the prior permission in writing of the publishers, or in the case of reprographic reproduction in accordance with the terms of licences issued by the Copyright Licensing Agency. Enquiries concerning reproduction outside those terms should be sent to the publishers.

© Copyright, 2009 Praxis Publishing Ltd.

The use of general descriptive names, registered names, trademarks, etc. in this publication does not imply, even in the absence of a specific statement, that such names are exempt from the relevant protective laws and regulations and therefore free for general use.

Cover design: Jim Wilkie
Typesetting: BookEns Ltd, Royston, Herts., UK

Printed in Germany on acid-free paper

To Doris Spence Harland

Contents

Illustrations

Chapter 13

Chapter 14

Tables

Foreword

The historic flight of Apollo 11 in July 1969 did not sprout from the head of Zeus, fully grown like Athena. The events from when men first gazed at the Moon up in the night sky until the launch of humanity's first voyage to another world is a broad and complex topic and is the subject of this book.

Readers of David Harland's previous books on the exploration of the Moon are well acquainted with his detailed research, lucid style and ability to summarise complex events. But on this present topic, David has truly surpassed himself. The story of the detailed study, preparation and flight of robotic space missions that prepared us to send people to the Moon is a long and complicated one, with many simultaneous events occurring in different parts of the world and on the Moon. His account clearly details how we advanced our understanding of the Moon, from a beacon in the night sky to a neighboring world with its own history and processes.

Before people could land on the Moon, we needed to know what awaited these explorers – one can only surmise so much by peering through a telescope from a range of over 400,000 km. The robotic precursor program that blazed a trail to the Moon involved crash landers, orbiters and soft landers. This progression now seems logical and well considered, but in fact at that time it was one that grew piecemeal in response to geopolitical, budgetary and bureaucratic pressures. Understanding this complicated and sometimes confusing story is necessary in order to appreciate fully the accomplishment of the Apollo program.

Paving the Way for Apollo 11 holds many lessons for our return to the Moon, some depressingly familiar. Different factions within NASA had differing agendas and desires which were usually worked out for the best, but not always. Science and engineering in the space program were in constant tension then, and have been ever since. The pressing need to know what the Soviets were up to on the Moon led to accelerated schedules and simultaneous exploration programs and missions. Some investigators predicted that disaster awaited us on the Moon, with spacecraft likely to be swallowed whole by giant bowls of choking dust. The information from these robotic probes disproved some of the wilder speculations on lunar surface conditions and allowed us to plan and execute the Apollo missions in a near flawless manner.

In this book David Harland recounts this fascinating story with clarity and verve, re-creating the excitement of the Apollo days when we were not merely going to the

Moon, but racing there. People of that time didn't know how events would unfold, yet they made many excellent decisions, and from these efforts we gained a new and more complete understanding of the Moon, the Earth and their intimate relationship and history. The exploration of the Moon revolutionised science in ways we are still trying to understand. And now, the Moon beckons us to return and capitalise on that wonderful legacy.

Paul D. Spudis
Lunar and Planetary Institute
Houston, Texas

Author's preface

For millennia human beings have peered at the Moon in the sky and wondered what it might be. Within months of its establishment on 1 October 1958, the National Aeronautics and Space Administration set out to develop a program of robotic lunar exploration. In 1961 President John F. Kennedy raised the stakes by challenging his nation to land a man on the Moon within that decade. The resulting Apollo program dominated the agency's activities throughout the 1960s and into the early 1970s.

It is impractical to cover all the strands of this effort in a single volume in equal detail. Nor can any given strand be properly appreciated in isolation. My approach is therefore to write a series of books, each of which applies a magnifying glass to a certain number of strands and glosses over others. This book focuses on what was known about the Moon at the dawn of the space age and details the robotic projects that paved the way for the first Apollo lunar landing, in particular the Surveyors that soft-landed to investigate the physical and chemical nature of the lunar surface and the Lunar Orbiters sent to reconnoitre possible landing sites.

As such, this book complements: *Apollo – The Definitive Sourcebook*, which was compiled with Richard W. Orloff and supplements an account of how the Apollo program was organised with the minutiae of each flight; *How NASA Learned to Fly in Space – An Exciting Account of the Gemini Missions*, which explains the key contribution that the Gemini crews made to the success of Apollo; and *The First Men on the Moon – The Story of Apollo 11*, which covers that mission from start to finish. In *Exploring the Moon – The Apollo Expeditions*, which I recently reissued in enlarged format, I detailed what the astronauts of each mission did whilst on the lunar surface. It also complements the excellent *To a Rocky Moon – A Geologist's History of Lunar Exploration* by Donald E. Wilhelms, and the *International Atlas of Lunar Exploration* by Philip J. Stooke.

I used the mission reports as my primary source of information – there are many thousands of pages available on the NASA Technical Report Server. Millions of dollars were spent developing and flying the vehicles used to take close-up pictures of the Moon and, like the mission reports, until recently they remained in archives. I have assembled some of the contiguous photographic sequences taken by the Lunar Orbiters to illustrate the process by which the site for the first Apollo landing was selected. To my knowledge, they have never previously been made available to the public in this form. I have also freely intermixed units of measure, largely following the choice of the appropriate mission reports. Unless stated otherwise, all times are GMT in 24-hour format. Launch, parking orbit, midcourse and terminal phase times are usually specified to the nearest second, but for a Surveyor spacecraft's powered descent the event times are specified to several decimal places.

In the 1960s NASA was a young and aggressive agency which embodied the 'can do' spirit of America at that time in tackling audacious engineering challenges with a tremendous sense of urgency – motivated by the desire to be the first to explore a new world. This is an account of a strand of that story that is often reduced to a few paragraphs in popular histories.

David M. Harland
Kelvinbridge, Glasgow
January 2009

Acknowledgements

I must thank, in no particular order, Patrick Moore, Mike Gentry, Harald Kucharek, Marc Rayman, W.D. Woods, Philip J. Stooke, Roland Suhr, Ken MacTaggart, Chris Gamble, Paul Spudis and, of course, Clive Horwood of Praxis.

Acronyms

ALS	Apollo Landing Site
ASPO	Apollo Spacecraft Project Office
CSM	Command and Service Module of the Apollo spacecraft
IU	Instrument Unit of the Saturn launch vehicle family
JPL	Jet Propulsion Laboratory
LEM	Lunar Excursion Module (later contracted to LM)
LM	Lunar Module of the Apollo spacecraft
LMP	LM Pilot
LTA	LM Test Article
NASA	National Aeronautics and Space Administration
PI	Principal Investigator
RADVS	Radar Altimeter and Doppler Velocity Sensor
S-IB	First stage of the Saturn IB launch vehicle
S-IC	First stage of the Saturn V launch vehicle
S-II	Second stage of the Saturn V launch vehicle
S-IVB	Second stage of the Saturn IB or third stage of the Saturn V
SLA	Spacecraft/LM Adapter
TLI	Translunar Injection
VAB	Vehicle Assembly Building

1

Astronomers' Moon

CLASSICAL PHILOSOPHERS

Greek astronomy began with Thales, who was born shortly before 600 BC and lived in Miletus, a city of Ionia, which was a state on the western coast of what is now Turkey. As a philosopher he is regarded as one of the Seven Sages of Greece, and is considered to be the 'father of science'. He set the seasons of the year and divided the year into 365 days. He also predicted a solar eclipse that occurred in 585 BC. It had been believed that the Moon was self-luminous, but he suggested that it shone by reflecting sunlight. Anaximander, a student of Thales, went to Italy in 518 BC. He opined that Earth floated in space – the prevailing view was that it was in some way supported on pillars through with the Sun passed during the night.

Pythagoras was born about 575 BC on Samos, an island off the coast of Ionia that was a crossroads between Asia, Africa and Europe. In his youth he reputedly visited Thales. Pythagoras considered the Moon to mark a fundamental boundary, in that it and everything 'above' was 'perfect', while Earth was subject to change and thus to decay. When critics argued that the markings on the face of the Moon indicated that it, too, was imperfect, it was suggested that the Moon was a mirror and the markings it displayed were really on Earth.

Around 450 BC Anaxagoras of Athens decided that Thales was correct in saying the Moon shone by reflecting sunlight. He realised that the Moon was spherical, and used this to explain its monthly cycle of 'phases'. A generation later, Democritus, who travelled widely in ancient Greece, reasoned that the Moon was a world in its own right with a rugged surface, and he speculated that it might be an abode of life.

In the early fourth century BC, Plato, a student of Socrates, founded the Academy in Athens as the first institution of higher learning. Eudoxus briefly studied under Plato. After learning astronomy, he devised an explanation for the manner in which the constellations on view change with the seasons. He imagined the stars to be on a sphere that was centred on Earth, and the Sun to be on a slightly smaller concentric sphere made of transparent crystal which allowed the stars to be seen through it. The solar sphere turned around Earth on a daily basis, as did that with the stars, but there was a slight differential in their rates that took a year to complete. Aristotle, another student of Plato, seized on this idea of 'crystal spheres' by proposing that there was one for each object that had an independent motion in the sky, and that their rotation was due to the action of angels. Although Eudoxus had envisaged crystal spheres only as a means of exposition, Aristotle believed them to be real and his views would come to dominate natural philosophy.

The points of light in the sky which move against the background of stars were called 'planets', meaning 'wanderers'. In the third century BC Aristarchus of Samos suggested that the Sun might be located at the centre of the 'planetary system', with Earth being a sphere, rotating daily on its axis, and travelling around the Sun on an annual basis; but the idea attracted little support and was soon forgotten. Aristarchus also reasoned that because the Moon occults the Sun at a solar eclipse, the Sun must be further away – in fact, *much* further away. He also inferred that the stars must be considerably further away than the Sun, because they show no parallax when viewed from opposite sides of Earth's path around the Sun. However, his reasoning on these matters was ignored. He interpreted a lunar eclipse as the Moon's passage through the shadow cast by Earth, and made a fair estimate of the distance between the Moon and Earth in relation to the diameter of Earth. His contemporary, Eratosthenes of Cyrene, made the first realistic estimate of the Earth's true diameter, thereby providing a scale to Aristarchus's calculations.

At the end of the third century BC, Apollonius of Perga on the southern coast of modern Turkey was a Greek geometer with an interest in conic sections, and it was he who introduced the names to the ellipse, parabola and hyperbola. Although it was inconceivable that celestial objects should be less than perfect, detailed observations had shown their motions to be anomalous. Apollonius devised a geometrical scheme in which a celestial body would trace a small circle whose central point travelled in a circle around Earth; the small circle was termed the 'epicycle', and its centre was the 'deferent'. This allowed the Moon to appear at times to lead and at other times to trail its perfect position. Furthermore, this accounted for why the size of the Moon appeared to vary in a cyclical manner. And of course, because the scheme involved only circles it restored purity.

Hipparchus, a Greek living in Alexandria, Egypt, in the second century BC, was the greatest of the classical Greek astronomers. His legacy was a star catalogue, but he also used a solar eclipse to estimate the relative distances of the Sun and Moon to a greater accuracy than had Aristarchus. He reasoned that although the Moon must orbit the Earth's centre, the location of observers on the Earth's surface provided the basis for parallax. On scrutinising records of eclipses that had been observed from both Alexandria and Nicaea, which lie on the same meridian but are some distance apart, he used the extents to which the Moon had masked the Sun's disk to calculate the distance to the Moon relative to the Earth's diameter. In fact, he calculated the distance of the Moon to within a few thousand kilometres and its diameter to within several hundred kilometres – although obviously he didn't use kilometres as a unit of measure. Hipparchus also used measurements of the Moon's orbit to assess Apollonius's suggestion of deferents and epicycles, found it satisfactory, and provided measurements of the sizes of the epicycles.

   In 80 AD the Greek historian Plutarch, who became a citizen of Rome, wrote the philosophical treatise *Faces in Orbe Lunare* in which he discussed the motion of the Moon across the sky, and how it maintained one face towards Earth as it turned on its axis. He thought that it was a world similar to Earth, and suggested it might be inhabited. A generation later, this latter point led the Greek storyteller Lucien of Samosata to write *Vera Historia* describing how a whirlwind lifted a ship from the sea and deposited it onto the Moon, where there was a battle in progress between the local inhabitants and invaders from the Sun. The story was a satire on the wars raged by the Greeks.

   Claudius Ptolemaeus was born around 85 AD, probably in Alexandria, which was at that time under Hellenistic control. The Royal Library of Alexandria was founded at the start of the third century BC. Over the centuries it had built up an unrivalled catalogue, because whenever a ship docked in the harbour the authorities ordered copies made of any books that were on board. Ptolemy (as he is known in English) used his own observations of the stars and the resources of the library to refine the work of Hipparchus, and wrote up his findings in a book of his own. The library was sacked several times and eventually destroyed, but when this occurred is disputed. Although Ptolemy's book was lost, an Arabic translation survived as the *Almagest*. He accepted Earth to be centrally located, celestial objects to be travelling in circles, Aristotle's belief in the reality of concentric celestial spheres, and also Hipparchus's endorsement of the deferents and epicycles as the reason for the anomalous motions. The Church of Rome accepted Aristotle's philosophy, and so, despite its contrived nature, the 'Ptolemaic system' – as it became known, even although Ptolemy had not invented it – survived for over 1,000 years.

THE RENAISSANCE IN ASTRONOMY

In 1330 AD the Italian scholar Francesco Petrarca coined the term 'dark ages' for the centuries of cultural decline in Europe after the fall of Rome in the fifth century. Intellectual development did not resume until the start of the Italian Renaissance in the fourteenth century. During this interregnum, the works of classical Greece and Rome were available only in Arabic translation. On being 'rediscovered', they were translated from Arabic into Latin.

   In 1505 Leonardo da Vinci, who had exceptional eyesight, drew an impression of the face of the Moon. He interpreted the brighter part to be water, the dark areas as land, and believed that there were clouds. He was the first to explain the old-Moon-in-the-new-Moon's-arms effect that occurs when the Moon is a narrow crescent. At such times the majority of the Earth's disk in the lunar sky must be illuminated, and the dark part of the remainder of the lunar disk is dimly lit by sunlight reflecting off Earth. Late in the 13th century, it had been realised that light was bent by passing through a glass lens. The term 'refraction' was not invented until some time later. In 1490 da Vinci had speculated upon whether lenses could be used in combination to make an enlarged view of a distant object. In 1504 he conducted experiments, and by 1510 had the optical principle of the telescope.

After further experiments, three years later he described how a concave mirror could produce a magnified image.

As the Renaissance progressed, some of the ancient beliefs were questioned. By the Ptolemaic system, all celestial bodies travelled around Earth on a daily basis, but Nicolaus Copernicus, a Polish canon, realised that this was not entirely true. In his book *De Revolutionibus Orbium Coelestium* he revived the heliocentric system of Aristarchus of Samos. Copernicus said only the Moon travels around Earth, but he retained circular orbits, deferents and epicycles. The planets, including Earth, are in orbit of the Sun. But knowing that the Church of Rome would construe this to be heresy, he kept silent, and his book was not released until after he died in 1543. His caution was justified, as in 1600 Giordano Bruno was burned at the stake in Rome for arguing in favour of the heliocentric hypothesis.

Johann Kepler was born near Stuttgart in Germany in 1571. He went to Prague in 1600 to assist the Danish astronomer Tycho Brahe, who held the title of Imperial Mathematician to the Holy Roman Emperor Rudolph II. Over a period of 20 years Brahe had compiled a highly accurate catalogue of planetary motions. When Brahe died in 1601, Kepler inherited the title of Imperial Mathematician, together with the archive of observations, which he set about analysing – something that Brahe had never attempted. Brahe was convinced of the view that Earth was central, but Kepler found otherwise. In his book *Astronomica Nova*, published in 1609, he announced that a planet pursues an ellipse with the Sun at one focus and the other focus vacant. The same applies to the Moon, but with Earth at one of the foci instead of the Sun. Whilst this rendered obsolete the Ptolemaic system with its circular orbits, deferents and epicycles, the Church was reluctant to concede the point.

In fact, Kepler also realised that the speed of a body in its orbit is proportional to its distance from its primary. In the case of the Moon, with Earth at one focus of its orbit, it travels more rapidly at perigee than at apogee. As a result, whilst the rate at which the Moon turns on its axis is fixed and is synchronised with its orbital period, the Moon is sometimes leading and sometimes trailing the mean position of its orbit, at which times we can see a portion of the otherwise hidden hemisphere around first one equatorial limb and then the other. Similarly since the Moon's orbit is inclined to the Earth's equator, when the Moon is in the southern sky we can observe slightly beyond its north pole at a time when that is illuminated, and when the Moon is in the northern sky we can see beyond its south pole when that is illuminated. This effect is known as libration. As for the Moon as a body, Kepler introduced the terms 'terrae' and 'maria' to describe the light and dark areas respectively.

EARLY TELESCOPIC IMPRESSIONS OF THE MOON

In 1600 William Gilbert, one of Queen Elizabeth I's physicians and a keen natural philosopher, published the book *De Magnete* in which he proposed that the planets circled the Sun as a result of some attractive force – and as his great interest was the Earth's magnetic field he suggested that this force was magnetism. In regard to the Moon, Gilbert shared Leonardo Da Vinci's view that the bright areas were seas, and after making a sketch of the face of the Moon he assigned names to the dark areas. The drawing was not published until 1651, long after his death.

Hans Lippershey was born in Germany in 1570, but became a citizen of Flemish Zeeland in 1602 and lived in Middleberg, earning his living as a spectacle maker. He is usually credited with discovering in 1608 that using lenses in combination could provide a magnified view of a remote object, but such instruments were apparently developed several times in different places in preceding decades – the telescope was evidently a device whose time had come. On 2 October 1608 Lippershey applied to The Hague for a patent. Several weeks later so did optical instrument maker, Jacob Metius. Both applications were refused. After an account of Lippershey's telescope was included in a widely circulated diplomatic dispatch, there was a proliferation of telescopes across Europe.

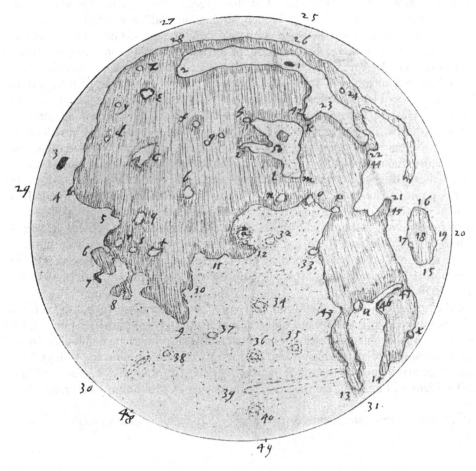

Thomas Harriot began to study the Moon telescopically in the summer of 1609 and in 1611 compiled a map based on his observations.

Thomas Harriot graduated in mathematics from Oxford in England. He tutored Sir Walter Raleigh on navigational issues, in particular the 'longitude problem', and on several occasions sailed with him. By 1603 Harriot was wealthy, living in London and pursuing his interest in optics. The appearance of a comet in 1607 prompted his interest in astronomy. In 1608 he obtained from Holland a crude telescope that had a magnification factor of six, and on 26 July 1609 aimed it at the Moon and sketched what he saw. After further observations, in 1611 he compiled a whole-disk map. His work was never published, and it remained unknown until discovered in 1965 by E. Strout of the Institute of the History of Science in the Soviet Union.

Galileo Galilei was born in 1564 in Pisa in Tuscany. He was a mathematician and experimentalist. In 1589 he was made professor of mathematics at the University of Pisa, but three years later took a similar position at the University of Padua. During a visit to Venice in July 1609 he heard of the invention of the telescope via a letter written to one of his friends by a French nobleman. Galileo promptly set out to make one. Whereas Lippershey had used two convex lenses, Galileo combined a convex lens with a concave one to obtain an upright image. It had a magnification factor of ten. On 25 August he displayed it to the Venetian Senate, pointing out that it would enable an inbound ship to be identified several hours earlier than would otherwise be possible – knowledge which would be commercially valuable in a city of merchants. He was rewarded with an increase in salary. In October he went to Florence to show the telescope to his former pupil, Cosimo de Medici, now Grand Duke of Florence. On returning to Padua, Galileo made one with double the magnification factor. On 30 November turned this to the Moon, and again five times over the next 18 nights as the illumination phase changed. As he had a leaning towards painting, his depictions of what he saw were more representational than technical – with the result that few of the features he drew are recognisable. He never drew a full disk to consolidate his observations. After this burst of activity, he seems to have paid little attention to the Moon – but, to be fair, he was busy making discoveries about other celestial bodies. He wrote an account of his observations in the pamphlet *Sidereus Nuncius*, which he dedicated to de Medici and published on 12 March 1610.

Of the 'imperfections' on the Moon he wrote, "We could perceive that the surface of the Moon is neither smooth nor uniform, nor accurately spherical, [...] but that it is uneven, rough, replete with cavities and packed with protruding eminences, in no other wise than the Earth, which is also characterised by mountains and valleys." He was particularly struck by the shadows, which appeared totally black – there was no detail evident within them. He used the shadows cast by mountains to estimate their heights.[1]

But Galileo's most significant discovery was that Venus displays phases similar to

---

[1]   He did *not* infer from the absence of detail in the shadows that the Moon was airless, nor did he suggest the presence of open water.

Representations of the Moon at two illumination phases made by Galileo Galilei in late 1609 and published in 1610.

the Moon, which was proof that this planet orbits the Sun, not Earth. Although the Church was not overly concerned about imperfections on the Moon, it was firmly of the belief that Aristotle was correct in saying that Earth was located at the centre of a system of crystal spheres. However, the phases of Venus meant that Copernicus was correct, which was a serious matter. After being hounded by the Roman Inquisition, in 1633 Galileo was obliged to publicly "curse and detest" the false opinion that the Sun held the central position. He was then placed under house arrest, and that is how he lived until his death in 1642.

Although Galileo may not have been the first to aim a telescope at the heavens, he was the first to publish, and the rapid distribution of his pamphlet prompted a great many people to obtain instruments to look for themselves.

Francesco Fontana, a Neapolitan lawyer, began to observe the Moon in 1630, and in 1646 published *Novae Coelestium, Terrestriumque Rerum Observationes*, featuring wood-cut engravings of two of his drawings made at different illumination phases.

Although Kepler knew that Galileo's findings confirmed the Sun to be centrally located, he was wary of saying so explicitly. Since the Moon was evidently a world in its own right, he wrote an allegorical fantasy, *Somnium*, in which he related how 'demons' transferred his hero character to the Moon at the time of a lunar eclipse by sending him down the Earth's shadow. The fact that the explanatory footnotes were longer than the text of the story established it to be a technical treatise disguised as a work of fiction. Even so, it was not published until 1634, four years after his death.

Jeremiah Horrocks in England observed in 1637 the dark limb of the Moon occult the stars of the Pleiades cluster, one by one. If the Moon possessed an atmosphere, the starlight would have flickered and faded as it was attenuated and refracted by the gas, but in each case the star's disappearance was instantaneous.

In 1638 John Wilkins, an English clergyman, published *Discovery of a World in the Moone; Or a Discourse tending to prove that 'tis probable there may be another Habitable World in that Planet.* He took a serious look at how a voyage to the Moon might be attempted utilising some form of 'engine'. Wilkins was so enthusiastic that after he helped to establish the Royal Society of London in 1660 he had this petition the government to undertake such a venture with the objective of claiming the Moon for the British Empire!

MAPPING THE MOON

Jacob Floris van Langren founded a business in Amsterdam in 1586 which made globes, and as Dutch explorers reported discoveries he could barely keep up with the demand for updates. In 1627 his grandson, Michel van Langren, observed the Moon and made a sketch. After moving to Madrid as Court Astronomer to King Félipe IV of Spain in 1630, the grandson convinced the King that tables listing the sunrise and sunset times of specific lunar features would enable the time at the observing site to be determined, which would in turn solve the 'longitude problem'. The prerequisite was a map of the Moon. In 1643, having made 30 sketches, van Langren realised he had competitors, so in 1645 he issued a whole-disk map 34 cm in diameter on which he named 325 features after prominent philosophers, mathematicians, astronomers, explorers, religious figures and (recognising his sponsor) members of the Spanish royal family. However, at that time the revolt of Protestantism which would later be called the Thirty Years War was well underway, and a nomenclature drawn from Catholic Europe was sure to be contentious.[2]

In 1637 Pierre Gassendi, a mathematician in Paris, also came to the conclusion that it should be possible to use observations of the Moon to determine the time and thereby resolve the 'longitude problem'. After he had made some drawings, he heard that Johann Hevelius, whom he had once met, was starting to make a map, and upon seeing the quality of the younger man's sketches Gassendi stopped and handed over his own work. A city councillor in Danzig in Poland, Hevelius built an observatory on the roof of his house and installed a telescope with a 5-cm-diameter lens, a focal length of 3.6 metres and a magnification of 50 – in fact, one of the best telescopes of the time. In 1647 he published *Selenographica sive Lunae Descriptio*, with fine drawings and a consolidated map 30 cm in diameter.[3] He named 275 features after terrestrial landforms, including oceans, seas, bays and lakes – although he realised there were no bodies of open water. Like Galileo, Hevelius estimated the heights of the lunar peaks by their shadows, but much more accurately. Being a Protestant, his nomenclature had little in common with that of van Langren. In fact, Hevelius had presumed himself to be the first to name features, and said the task was arduous. He

---

[2]   In fact, one of the few names introduced by van Langren to have survived is Langrenus, by which he honoured his own family.

[3]   Selene was the Greek moon-goddess.

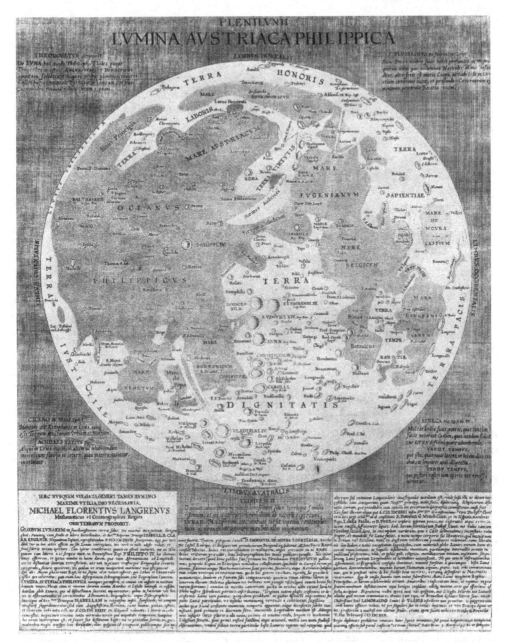

The map of the Moon published by Michel van Langren in 1645 was the first to assign names to features.

A map of the Moon published by Johann Hevelius in 1647.

honoured astronomers and scientists, including Gassendi, but not himself. However, only a dozen of his names have survived.

As a Jesuit professor of astronomy and theology at the University of Bologna, Giovanni Battista Riccioli believed implicitly in the Aristotlean system as written of by Ptolemy. In an effort to counter the growing belief that Earth travels around the Sun, he set out to write an authoritative account of astronomy. But while developing his argument he came to suspect he was wrong! He could never admit this publicly, however. In 1651 he published *Almagestum Novum*, with a whole-disk map of the Moon that was 28 cm in diameter and was based on observations made by his pupil, Francesco Maria Grimaldi. Although the map was little better than that by Hevelius, its historical significance was the nomenclature. This retained oceans, seas and bays for the dark areas, but renamed them for states of mind: e.g. Oceanus Procellarum and Mare Tranquillitatis. Craters were named after astronomers and philosophers, including Riccioli and Grimaldi. The despised Copernicus was assigned a crater in Oceanus Procellarum – the Ocean of Storms. To Helvelius's frustration, soon copies of *his* map were in circulation relabelled with Riccioli's nomenclature! Nearly all of the 200 names introduced by Riccioli and Grimaldi are still in use today.

Giovanni Domenico Cassini was born in 1625 in the Republic of Genoa. After a Jesuit education he was hired by the Marquis Cornelio Malvasia in Bologna, who

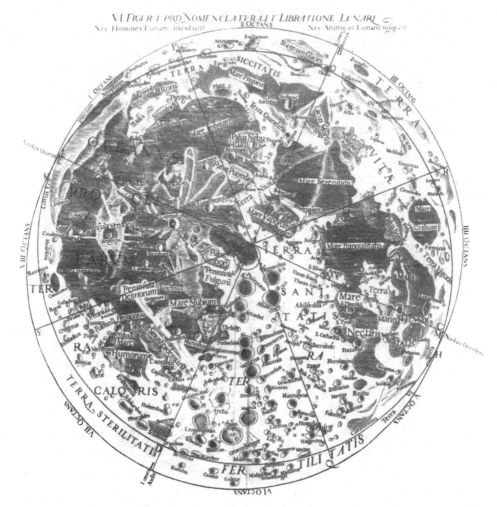

A map of the Moon published by Giovanni Battista Riccioli in 1651.

derived ephemerides for astrological purposes. Utilising the excellent instruments of his employer's observatory, Cassini made observations of exceptional precision and quality, and in 1650 became professor of astronomy at the University of Bologna. In 1666 plans were initiated to establish a national observatory in Paris, and in 1669 Cassini, now with several significant discoveries to his name, was invited by King Louis XIV to become its first director; he accepted and promptly moved to France to oversee the construction of the observatory, which was finished in 1671. In 1679 he published a map of the Moon which, at 52 cm in diameter, was much larger than its predecessors. Although very accurate, so few copies were made that it did not gain the attention which it warranted.

Meanwhile, Isaac Newton at the University of Cambridge in England had made a study of gravity and, contrary to the accepted wisdom that it remained constant

A map of the Moon published by Giovanni Domenico Cassini in 1679.

with distance, realised that its strength declined with the inverse square of distance. His book *Philosophiae Naturalis Principia Mathematica*, published in 1687, provided a basis for the laws of planetary motion that Kepler had derived empirically.

The Moon was essentially ignored for a century, then Tobius Mayer in Germany became interested in its use in relation to the 'longitude problem'. In 1751 he gained the chair of economics and mathematics at the University of Göttingen, where, a few years later, he became superintendent of the observatory. As a skilled draughtsman, he utilised a micrometer to measure the geographical positions of the lunar features. His map was published posthumously in 1775, and although only 20 cm in diameter it was the first to include lines of latitude and longitude. It superseded the map by Hevelius (re-annotated with the nomenclature of Riccioli) which had been standard for almost 150 years, and would itself not be surpassed for half a century.

T. Mayer del.   *J. P. Kaltenhofer sculp. Gottingae.*

A map of the Moon by Tobius Mayer that was published posthumously in 1775.

William Herschel, who discovered the planet Uranus in 1781 and became the first president of the Royal Astronomical Society, did not devote much attention to the Moon. However, he believed it to possess an atmosphere (even though the way stars were occulted contradicted this) and active volcanoes. Although by now the idea of open water had been abandoned, Herschel was "absolutely certain" the Moon was inhabited.

Johann Hieronymus Schröter was born in 1745 in Erfurt in Germany. In 1767 he graduated in law from the University of Göttingen. In 1781 he moved to Lilienthal, near Bremen, to become chief magistrate. Having been inspired by Mayer's map, he built an observatory alongside his house and installed a series of ever more powerful telescopes. Over a period of 30 years he made hundreds of detailed drawings of the Moon, recording individual features under different angles of illumination. In this respect, he founded modern selenography. He paid particular attention to craters and rilles – both of which terms he introduced. His measurements of the heights of lunar peaks were better than those of his predecessors. Although an accurate observer, he was not a skilled draughtsman and utilised a 'schematic' style. In 1791 he published *Selenotopographische Fragmente zur genauern Kenntniss der Mondfläche* in two volumes containing a total of 75 engravings. He did not consolidate his observations into a full-disk map, but did include an enlarged version of Mayer's map. In 1813 the invading Napoleonic army ransacked and destroyed his observatory, and most of his unpublished work was lost. Schröter inferred the Moon to have an atmosphere, but estimated its pressure to be less than that of the best vacuum pump available at that time. Nevertheless, like Herschel, he believed the Moon to be inhabited.

Wilhelm G. Lohrmann was a surveyor in Dresden, and a skilled draughtsman. He had developed a keen interest in astronomy as a boy, and in 1821, aged 25, began to study the Moon. He set out to make a map in 25 sections on a scale at which the disk would be 95 cm in diameter. By 1824 he had released the first four sections, and by 1836 had the drawings for the remaining sections, but then his eyesight failed and he was unable to finish the editing. Nevertheless, in 1838 he published a full-disk map at 40 per cent scale.

Johann von Mädler was born in 1794 and became a teacher in Berlin. One of his students, the wealthy banker Wilhelm Beer, was only a few years younger and they became friends. In 1829 Beer built an observatory at his house and bought a 95-mm-diameter refracting telescope produced by Joseph von Fraunhofer's firm. Beer hired Mädler as observer, and they measured almost 1,000 features to trigonometrically survey the Moon. Between 1834 and 1837 they published *Mappa Selenographica* in four parts, which together made a whole-disk map 95 cm in diameter. Owing to the recent improvement in telescopes, their map surpassed all its predecessors. In 1838 they republished the map with a dissertation in their book *Der Mond*. It became the definitive work on the subject, but their convincing argument for the Moon being an airless and unchanging body prompted a hiatus in observing.

While reading Schröter's 1791 *Selenotopographische Fragmente* as a boy, J.F.J. Schmidt in Germany decided to study the Moon. After acting as assistant at various German observatories, in 1858 he became director of the Athens Observatory and set out to make a full-disk map with a diameter of 180 cm which would show more craters, rilles and mountains than its predecessors. In fact, for many years he was the only observer engaged in systematic lunar work! He completed the observations in 1868, having produced in excess of 1,000 sketches, measured the positions of over 4,000 points, catalogued 278 rilles and used shadow details to measure the depths of craters and the heights of mountains. When issued in 25 sections in 1874, the map specified 33,000 craters and the heights of 3,000 mountains. In 1878 he reprinted it in his book *Charte der Gebirge des Mondes*. Having acquired Lohrmann's files, Schmidt had his predecessor's map engraved at the scale originally intended, and in 1878 published it as *Mondkarte in 25 Sektionen*; it would have been a fine map for its time, but was now obsolete.

In 1864 the British Association established a Lunar Committee, but this achieved little. In 1876 Edmund Neville Nevill (using the surname Neison) published a book, *The Moon*, which included a full-disk map in 22 sections with a diameter of 60 cm. In fact, it was a reworking of Beer and Mädler's chart supplemented with a detailed description of each named feature – some 500 in all – making it a monumental work. It stimulated sufficient interest to prompt the establishment of the Selenographical Society with William Radcliffe Birt, one of the most active of British amateur lunar observers of that time, as president, and Neison as secretary. But it was disbanded in 1883 after the death of Birt in 1881 and the departure of Neison in 1882 to become the first director of the Natal Observatory in Durban in South Africa. Nevertheless, when the British Astronomical Association was established in London in 1890 the former members of the Selenographical Society set up a Lunar Section as a means of coordinating their activities. T.G. Elger, the Director of the Lunar Section, published a book entitled *The Moon* in 1895 to assist new observers.

By 1890 photography had matured sufficiently to facilitate surveys of the Moon. Two photographic atlases were published in 1897. *Atlas Photographique de la Lune* covered the whole face in many small sections using plates taken by Maurice Loewy at the Paris Observatory and text provided by his assistant Pierre Puiseux. The *Lick Observatory Atlas of the Moon* by Edward Singleton Holden comprised 19 sheets of reproduced photographs. After establishing a temporary astronomical outstation in Jamaica, W.H. Pickering of the Harvard College Observatory set himself the task of photographing the lunar disk in several sections at five illumination phases. When *The Moon – A summary of existing knowledge of our satellite, with a complete photographic atlas* was published in 1903 it was the first true atlas, because the pictures were reproduced at the same scale. Although Jamaica had particularly clear skies, Pickering's pictures of the Moon were still blurry and so there remained scope for visual studies, particularly in the limb regions – but as professional astronomers turned their attention to the stars and even more distant objects, they left the Moon, which they regarded as a source of 'light pollution', to their amateur brethren.

On the nights of 12 to 15 September 1919 Francis Pease photographed the Moon while testing the new 100-inch Hooker reflector at the Mount Wilson Observatory in California. Walter Goodacre was born in 1856, lived in London, and developed an interest in astronomy as a boy. In 1910, after making thousands of observations, he compiled a whole-disk map of the Moon almost 200 cm in diameter. In 1931, by which time he had replaced T.G. Elger as Director of the Lunar Section of the British Astronomical Association, he published *The Moon*. It featured a map which combined fine detail obtained by visual observing in good 'seeing', with positional accuracy derived from the Mount Wilson photographs, but few copies were issued.

The International Astronomical Union was established in 1919 to oversee general issues, and it took responsibility for regulating lunar nomenclature.

THE SURFACE MATERIAL

As William Herschel was passing sunlight through a prism in 1800, he found that heat was refracted just beyond the red end of the visible spectrum, so he named this infrared radiation. The Estonian physicist Thomas Johann Seebeck discovered in 1821 that if two wires of different metal are made into a loop by soldering their ends together, then an electric current will flow if the joins are at different temperatures. In 1856 Charles Piazzi Smyth utilised such a thermocouple to detect solar infrared reflecting off the Moon. Laurence Parsons inherited the 72-inch reflecting telescope built by his father at Birr Castle in Ireland. It was the largest telescope in the world at that time. The common view was that since the airless lunar surface was exposed to the intense cold of space, it simply must be covered by ice. In fact, S. Ericsson of Norway had proposed in 1869 that the lunar landscape was shaped by glaciation. In 1870 Parsons equipped his telescope with a thermocouple and found that at lunar noon the temperature of the equatorial zone – where the Sun would pass close to the zenith – exceeded that of the boiling point of water, which indicated that the surface could not·be ice. Measurements of the angle of polarisation of the surface published by M. Landerum in 1890 confirmed that it could not be ice. Despite the measured high temperatures at lunar noon, P.J.H. Fauth in Germany endorsed the idea that the landscape was shaped by glaciation, and in 1913 he and Hans Hörbiger announced the highly unorthodox theory that ice was the essence of the cosmos! However, the vapour pressure of ice would cause it to sublime in the vacuum. If ice were indeed present, it would have to be subterranean. In 1916 Pierre Puiseux in Paris pointed out that if ice were present in the amounts claimed by Fauth, then it should be most evident at high latitudes where the Sun did not rise far above the horizon – yet there were no polar caps. Nevertheless, W.H. Pickering speculated that there might be ice at the summits of lunar peaks. The outcome of these studies was therefore that the majority of the surface was not ice.

In 1930 Edison Pettit and Seth B. Nicholson put a thermocouple on the 100-inch reflector on Mount Wilson, which at that time was the largest telescope in the world, and discovered that the surface temperature in the equatorial zone varied by several hundred degrees during the monthly cycle. At the onset of a lunar eclipse in 1939 they measured the temperature plunge by 120°C in the space of an hour as the Moon entered the Earth's shadow. This implied that the material on the surface was poor at retaining heat. On making more sophisticated measurements, they found that at the equator the temperature was +101°C at noon, fell to –39°C at sunset and –160°C at midnight. In 1948 A.J. Wesselink in Holland inferred from these cooling rates that the Moon could not be exposed solid rock but must be covered by a blanket of loose material.

After the Second World War, the Moon was investigated at radio wavelengths. In 1946 Robert H. Dicke and Robert Beringer in America detected thermal emission from the Moon at a microwave wavelength of 1.25 cm. Using the same wavelength, in 1949 J.H. Piddington and H.C. Minnett in Australia measured the temperature of the whole disk at a variety of phases over three lunations. The variation proved to be less extreme than it was at infrared wavelengths. The fact that the radio temperature lagged behind the optical phase of the Moon by 3.5 days suggested the presence of a thin insulating layer with low thermal conductivity. In 1950 John Conrad Jaeger in Australia matched materials to the microwave observations made by Piddington and Minnett. Agreeing with Wesselink's inference of loose material, Jaeger argued for a layer of 'dust', typically only several millimetres thick, resting on top of a granular material. Observations of lunar eclipses on 29 January 1953 and 18 January 1954 at microwave wavelengths by the US Naval Research Laboratory implied that only the uppermost part of the surface underwent a large variation in temperature. This was consistent with a thin layer of dust on a loose granular material. In 1962 J.F. Denisse in France announced that for wavelengths exceeding 30 cm there was no variation in temperature over the monthly cycle.

Taken together, these investigations indicated that whereas an optical telescope fitted with a thermocouple measured the temperature of the surface itself, the radio temperatures were *average*s for granular material to depths corresponding to several times the wavelength. The constancy at wavelengths greater than 30 cm implied that the material in the uppermost metre or so was such a poor conductor of heat that even when the Sun was at the zenith its heat did not penetrate that far. And at night, although the surface rapidly radiated away the heat it had gained during the day, the poor conductivity of the deeper material served to insulate it. The temperature at a depth of about one metre was estimated to be a constant –40°C. Candidates for the uppermost metre of material were a porous volcanic rock like pumice or a granular conglomerate. A colloquium held in Dallas, Texas, in 1959 concluded that the fine dust that formed the actual surface was probably of meteoritic origin. It was initially believed that the Moon is particularly bright at its 'full' phase due to there being no shadows in view – the objects at the centre of the disk cast no shadows, and objects away from the centre mask their shadows to terrestrial observers. But the absence of appreciable darkening of the limb proved to be a result of the fact that the surface 'scatters' more light back towards its source than it does in other directions. It was inferred from this that the material at the surface was a porous vacuum-sintered dust, and that sunlight which penetrated a 'cavity' was not absorbed but reflected back out towards its source.

In 1955 Thomas Gold, an astronomer with a wide-ranging interest who was then at the Royal Greenwich Observatory in England, proposed that particles of dust on the lunar surface would become electrically charged by the harsh ionising ultraviolet radiation from the Sun, and that in making the grains of dust repel each other this would cause them to flow remorselessly 'down hill' and collect in low-lying areas. Tests using powdered cement in a vacuum had shown that this tended to form fragile 'fairy castle' structures full of voids, which was consistent with the inference that the surface material was porous. Gold claimed that the maria were accumulations of dust, possibly several kilometres thick, and were of low albedo because the dust had been darkened through exposure to radiation. But whilst dust moving down hill could bury craters in low-lying terrain, it could *not* explain the missing 'seaward' wall of a crater such as Le Monnier on the margin of Mare Serenitatis, nor the dark floors of Archimedes sitting on elevated terrain or Plato embedded in the lunar Alps.

A. Deutsch in Leningrad suggested in 1961 that there might be life in the granular material where the temperature was constant, and that it lived off gases leaking from the interior. Expanding on this, Carl Sagan in America speculated that if the granular material were tens of metres deep, then it might contain a considerable amount of ice and organic material.

As the space age dawned, therefore, there were already interesting insights and speculations into the nature of the lunar surface material.

2

Geologists' Moon

EARLY IDEAS ABOUT LUNAR CRATERS

In 1662 Robert Hooke was made curator of the recently formed Royal Society of London. He was charged with devising demonstration experiments. As an extremely skilled technical artist, in 1665 he published *Micrographica*, which was profusely illustrated with his own observations using a telescope and a microscope. Although he included a detailed drawing of the lunar crater Hipparchus, which is at the centre of the Moon's disk, he had no desire to map the Moon. However, he undertook a series of experiments to investigate how craters may have formed. First he dropped heavy balls into tightly packed wet clay, and examined the imprints that they made. He also heated alabaster until it bubbled, and then let it set so that the last bubbles to break the surface produced craters. However, just as Hooke could not imagine where the projectiles could have come from to scar the Moon so intensively, nor could he conceive how the surface could have been sufficiently hot to blister on such a scale.

Following the discovery of the first two asteroids in 1801 and 1802, Marshal von Bieberstein in Germany suggested that lunar craters were created by the impact of such bodies. This was reiterated independently in 1815 by Karl Ehrenbert von Moll. In 1829 Franz von Gruithuisen agreed. However, the idea was rejected by those who supported the anti-catastrophist paradigm of uniformitarianism in terrestrial geology which was developed in the 1830s and 1840s.[1] In 1873 Richard A. Proctor published *The Moon*. Although this book was largely devoted to the motions of the Moon, he revived the idea that the craters marked impacts. But when the second edition of the book was issued in 1878 this section had been deleted. What puzzled the nineteenth century proponents of the impact hypothesis was that the lunar

---

[1] Like Herschel and Schröter, von Gruithuisen believed the Moon to be inhabited, and after using a small telescope he reported in 1824 his discovery of a city in the equatorial zone near the meridian; but this was later shown to be merely a group of shallow ridges that were visible only when the Sun was low on the local horizon.

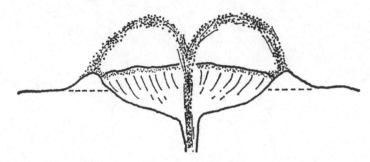

In 1874 James Nasmyth and James Carpenter proposed that volcanic 'fountains'
produced the lunar craters. (Courtesy Patrick Moore, *Survey of the Moon*, 1963)

craters are almost all circular, whereas the majority of bodies must have struck at an
oblique angle and, it was presumed, produced elliptical craters.

In 1874 James Nasmyth and James Carpenter in England published *The Moon*, in
which they attempted to explain how the surface features may have formed. As had
Hooke two centuries earlier, they made model craters in experiments. They came to
the conclusion that the lunar craters were produced by 'fountains' of material. In the
early part of an eruption, when the velocity of the material ejected from the vent was
great, the material would spray out in an umbrella-shaped plume and fall back some
distance away to build up a concentric ring that became the wall of the crater. In
many cases, as the eruption declined the fallout formed a succession of terraces
interior to the wall. They presumed that in some cases the final phase of the eruption
either built up the central peak, or in the case of craters with dark floors and no
peak, switched to fluid lava that was confined to the cavity and buried the vent. In
view of the weak gravity and absence of an atmosphere, it seemed plausible that this
process could have produced very large structures.

Other explanations were offered for the origin of lunar craters. In 1854 the Danish
astronomer Peter Andreas Hansen argued that the Moon bulged towards Earth, that
its centre of gravity was displaced 50 km towards the far-side, and that this had
drawn all the air and water on the surface around to the far-side, to where the
inhabitants had relocated. In 1917 D.P. Beard suggested that the Moon was once
immersed in a deep ocean, that the craters were limestone structures similar to coral
reefs, and they were left exposed when the water flowed to the far-side.

A DIFFERENT POINT OF VIEW

Grove Karl Gilbert was born in Rochester, New York, in 1843. After conducting a
number of surveys as a field geologist, he was made Senior Geologist when the US
Geological Survey was founded in 1879. Over 18 nights during August, September
and October 1892, Gilbert used the 26-inch refractor of the US Naval Observatory in
Washington DC to study the Moon. Pointing out that lunar craters have floors lying
generally below rather than above the level of the adjacent terrain, he rejected the

In 1892 Grove Karl Gilbert drew attention to 'sculpture' radial to Imbrium.

volcanic interpretation and argued that craters must be the result of impacts. He further proposed that the arcuate chains of mountains at the periphery of the 'circular maria' are the walls of craters produced by vast impacts. As evidence, Gilbert cited what he called 'sculpture' as the fall of ejecta thrown out during the formation of Imbrium. He announced his results in *The Moon's Face: A Study of its Origin and its Surface Features*, a paper presented orally to the Philosophical Society of Washington on 10 December to mark his retirement as its president. The paper was published in the *Bulletin* in 1893, but as this was not a publication on the reading list of astronomers his remarkable intrusion into their bailiwick passed unnoticed.

In 1946 Harvard geologist R.A. Daly rejected the endogenic origin of craters and, argued in favour of impact, citing Gilbert's paper. Also in 1946, geologist R.S. Dietz expanded on the subject, listing several criteria that showed how lunar craters differ from terrestrial volcanic craters.

The American geologist J.E. Spurr began to study the Moon in 1937, having been inspired by the photographs taken by Francis Pease using the 100-inch telescope on Mount Wilson. He presumed that the Moon could be described in terrestrial terms, and between 1944 and 1949 wrote up his systematic analysis in four volumes under the general title *Geology Applied to Selenology*, offering volcanic explanations for a wide variety of lunar features. In particular, he said that early in lunar history large calderas left cavities which were later flooded by lava to make the 'irregular maria', and subsequently the better preserved 'circular maria'. Critics of the impact origin of craters pointed out that whilst there were many examples of small craters on the rims of larger ones, there were no cases of large craters overlapping smaller ones. Spurr said craters were volcanic, and were produced with progressively smaller diameters. He interpreted faults and ridges as evidence of lines of weakness in the crust. Since he mapped these up the meridian and around the limb regions, he said they were due to stresses imparted as the Moon's rotation synchronised with its orbital period. He

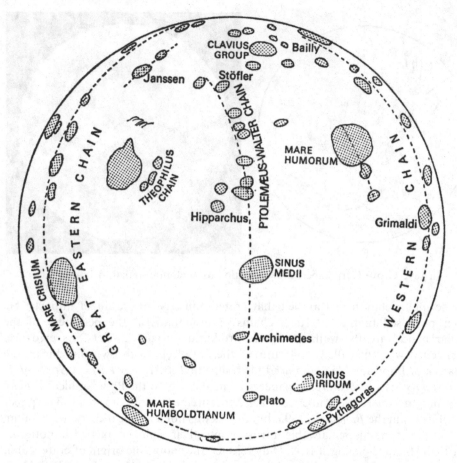

Astronomers who favoured the 'lunar grid system' devised by J.E. Spurr claimed that lines of weakness in the lunar crust, particularly up the meridian and around the limb regions, had prompted eruptions which made 'chains' of large craters whose members were isolated from one other by significant distances. (Courtesy Patrick Moore, *Survey of the Moon*, 1963)

claimed this 'lunar grid system' had significantly controlled the formation of craters. This thesis was eagerly accepted by those who believed volcanism played the main role in shaping the lunar surface. In particular, it was claimed that lines of weakness had prompted eruptions that produced 'chains' of large craters whose members were isolated from one other by significant distances. But critics argued that the lines of weakness were illusory, since relief highlighted by the sunrise or sunset terminator will favour north–south trends and not east–west trends. And, of course, any pair of craters can be said to be related if an observer is so inclined.

It was not until Ralph B. Baldwin made an analysis of bomb craters in the Second World War that the impact origin of lunar craters began to make real headway. As a businessman trained in physics, he developed an interest in the Moon in 1941 during a visit to a planetarium when, in viewing the pictures on display, he independently noticed Imbrium sculpture. On later reading up and finding no explanation (since he did not happen across Gilbert's paper) Baldwin decided to conduct his own study. In an article published in the magazine *Popular Astronomy* in 1942 he argued that the ridges and grooves were "caused by material ejected radially from the point of explosion" by the impact which formed Imbrium – although, like everyone else, he presumed that the impact formed the mare itself. In a follow-up in 1943 he reasoned that the projectile had been "flattened" by the shock and had excavated the cavity in a lateral manner, which was why the nearest sculpture consisted of grooves rather than chains of craters made by plunging debris – the latter occurred further out. He published in a popular outlet because his work was rejected by professional journals – evidently the Moon was not an object for worthwhile study. In his book *The Face of the Moon*, published in 1949 by the University of Chicago, Baldwin reported his observations, experiments and analyses, and included a review of the literature (by now he knew of Gilbert). His own contribution as a physicist drew upon an analysis of bomb craters in which he showed that the greater the deceleration on impact, the greater the energy released. He reasoned that although the weak lunar gravity would enable an explosion to throw ejecta to a greater distance, it would not actually make the crater larger. He logarithmically plotted the relationship between the diameters and depths of explosive craters on Earth, the craters on Earth accepted to have been made by cosmic impacts, and 'fresh-looking' lunar craters (those which had not yet slumped and distorted the ratio that he utilised). He compiled 300 measurements of lunar craters from the literature, and measured several dozen others himself. There was a clear trend.

Significantly, Baldwin realised that although most sculpture could be attributed to Imbrium, there was some which seemed to be associated with other 'circular maria', from which he concluded that they resulted from individual impacts. Furthermore, the mountains peripheral to Serenitatis must have formed prior to the impact that etched the Haemus with Imbrium sculpture, yet before lava flooded the Serenitatis cavity. This established that Mare Serenitatis formed a significant interval after the impact had excavated the cavity in which it resides. Baldwin (as had Gilbert) believed all the maria to have been formed at the same time and to be associated with Imbrium, which at that time was presumed to have been the greatest impact in lunar history. However, whereas Gilbert envisaged the Imbrium impact splashing out liquid ejecta which pooled in low-lying areas to form the various maria, Baldwin saw there had been a significant interval between the formation of the Imbrium cavity and its being filled in. He proposed that the impact raised a vast dome which remained inflated for long enough to be cratered (for example by Archimedes), then collapsed (forming a system of peripheral arcuate faults) and released a pulse of extremely low viscosity lava that not only filled in the cavity but also burst through the containing walls to spread across the surface and fill in other cavities to create the maria. Irrespective of whether the maria were liquid ejecta or erupted lava, it was evident that the large circular cavities were made *by individual impacts* over a period of time and that there was a significant interval before the formation of the maria.

Harold C. Urey, chemist.

Astronomers were not impressed by Baldwin's arguments, however, and for many years continued to associate the maria with the cavities they occupied.

After reading Baldwin's book, Harold C. Urey developed an interest in the Moon. But Urey was not particularly interested in the surface features – as a chemist at the University of Chicago who gained the 1934 Nobel Prize for chemistry, he was more interested in the Moon's composition. He accepted that the craters were impacts and the maria were the by-product of a giant impact, but rejected Baldwin's inference of a significant interval between the Imbrium impact and the formation of the maria. Urey agreed with Gilbert that the maria were splashes of impact melt, and said that because they were molten they could not have preserved sculpture. He also made the remarkable suggestion that the semicircular Sinus Iridum on the northern margin of Mare Imbrium marked the 'entry hole' of the asteroidal body whose impact created the Imbrium cavity.

In 1943 Gerard P. Kuiper began to exploit recent technical developments to make observations of bodies in the solar system. He essentially had the field to himself, at least in professional circles, and was able to make a series of discoveries. In 1953 he turned his attention to the Moon. Although photography was the norm, he mounted a binocular eye-piece on the 82-inch reflector of the McDonald Observatory in Texas to exploit moments of exceptional 'seeing' to discern details of the lunar surface that

Gerard P. Kuiper, astronomer.

would have been blurred in photographs. In his first paper on the subject, in 1954, he argued that in the case of a body of the Moon's size, radiogenic heating would have caused sufficient melting for dense minerals to sink to create a core and lightweight minerals to rise to form a crust. This thermal differentiation would become known as the 'hot Moon' hypothesis. As volcanism is a means of enabling heat to escape from the interior, Kuiper argued that the maria were formed by lava upwelling at various times from deep fractures in the floors of the cavities excavated by major impacts.

In 1891, while studying the desert between Flagstaff and Winslow in Arizona in which the Canyon Diablo meteorites had been recovered, G.K. Gilbert inspected the circular hole known as Coon Butte. It was 1.2 km in diameter, had a rim which rose 45 metres above its surroundings, and a floor lying 200 metres below the rim. If it marked the site of an impact, then, he reasoned, there might be a large iron meteorite beneath its floor. A buried iron mass should be detectable by its magnetic signature, but there was no such indication. He concluded that the hole was a marr, made some 50,000 years ago when magma caused underground ice to flash to steam and blast a hole in the overlying rock. Nevertheless, in 1903 mining engineer D.M. Barringer began to drill in search of the meteorite, to no effect. In 1916 E.J. Öpik realised that a cosmic impact was such a violent event that the projectile would be vaporised, but he published in an Estonian journal and his insight passed unnoticed. In 1924 A.C.

Gifford independently came to the same conclusion and published in a New Zealand journal that had a broader readership. Öpik and Gifford both realised that high-speed impacts always create circular craters because whilst momentum is a vector, energy is not, and as the projectile hits the surface it essentially explodes, liberating energy in a symmetric manner and forming a circular crater. Furthermore, they realised, the crater is always much larger than the projectile. If Coon Butte was an impact crater, then the only relic of the projectile was the field of Canyon Diablo meteorites which littered the surrounding desert.

When Eugene M. Shoemaker joined the US Geological Survey in 1948 he already had an interest in the Moon. In 1949 he made a review of the literature and turned up both Gilbert's paper and Baldwin's recently released book, both of which advocated the impact hypothesis. In 1955 he studied two craters about 100 metres in diameter created by underground nuclear tests at the Nevada Test Site to investigate how such explosions shocked and dispersed rock. He was impressed by their resemblance to lunar craters. In 1957, with Gilbert's analysis in mind, he began a study of Coon Butte. He had already done the field work for his PhD thesis on salt structures, but had never gotten around to writing it up. After hearing Shoemaker give a seminar on his study of Coon Butte, his advisor at Princeton, Harry Hess, suggested that he use that as the basis of his thesis. Shoemaker put in some more field work, wrote it up, and, despite it being rather on the short side for the purpose, submitted it in 1959. The fact that the crater was recent and in a desert environment made the manner in which it was excavated readily evident. In particular, the strike had not just penetrated the surface and pushed the rock aside, as Gilbert imagined; the process, as Öpik and Gifford had inferred, was explosive. Significantly, Shoemaker found that *two* shock waves were involved: one vaporised the projectile, and the other propagated into the 'target rock', compressing it so thoroughly that the rock reacted just as if an explosion had occurred *beneath* it. In his field work Shoemaker methodically traced how the rim and the ejecta blanket formed by the stratigraphy being flipped into an inverted sequence around a circular 'hinge', in the process making a hole much wider than the projectile.[2]

Early in 1960 L.R. Stieff of the US Geological Survey in Washington DC set out to obtain NASA funding for an investigation of lunar geology. NASA deliberated. In 1953 Loring Coes had produced a new very dense mineral using a hydraulic press to squeeze quartz. This 'shocked quartz' was named coesite. In 1956 H.H. Nininger had suggested searching for coesite at Coon Butte, but this was not done. In 1960 Stieff obtained samples of rock taken from Coon Butte which were in the archive of the Smithsonian Institution in Washington DC to enable him to say to NASA that the Survey was *already* at work on craters. Ed Chao identified coesite using X-ray diffraction, which proved Coon Butte to be an impact crater. A press statement was released to this effect on 20 June. When Chao wrote the scientific paper, the Survey

---

[2] For over half a century, geologists had argued about how the Coon Butte crater formed – and this was for a structure that was accessible to in-situ examination. Could there be any hope of resolving the issue of the lunar craters, which could only be peered at from afar!?

Eugene M. Shoemaker, astrogeologist.

added Shoemaker and his assistant Beth Madsen as co-authors to imply that it had a *team* of specialists at work. When the paper was published in *Science* in July 1960, Shoemaker was on his way to present a paper about Coon Butte to the Geological Congress in Copenhagen. The Rieskessel in Bavaria is a 24-km-diameter structure with the town of Nördlingen at its centre. Although widely believed to be volcanic, a study in 1904 had suggested that the circular structure might mark an impact, and its characteristics had led Baldwin to classify it as such. Shoemaker examined samples of quartz from a quarry. To his trained eye, using no more than a hand-lens, the rock showed evidence of shock. The next day he airmailed samples to Chao, who called straight back to confirm that coesite was present. In giving his paper in Copenhagen about Coon Butte, Shoemaker announced the Rieskessel finding.

With the two structures having been shown to be impacts, NASA finally released the funding to enable the Survey to undertake its lunar studies, and on 25 August the Astrogeologic Studies Group was established at the Menlo Park office, south of San Francisco, with Shoemaker in charge. A year later, in 1961, it became the Branch of Astrogeology. In March 1962 Shoemaker decided to move his team to Flagstaff. The move began in December, but some people refused to relocate and were allowed to remain at Menlo Park.

STRATIGRAPHIC MAPPING

In April 1957 the National Academy of Sciences awarded Kuiper the funding to start work on a new lunar atlas, and supplementary money was provided later in the year by the Air Force. The resulting *Photographic Lunar Atlas* was published in 1960. The best available photographs were printed on a scale at which the lunar disk spanned 2.5 metres. It formed a striking contrast to the similarly sized map based on visual observations that was published in 1959 by H.P. Wilkins of the Lunar Section of the British Astronomical Association. Although very different in presentation, the two maps were comparable near the centre of the Moon's disk but even in the best pictures the limb regions were marred by 'seeing', and it was in these areas that the visual observers had the advantage. However, the pictures were able to be projected onto a white globe and rephotographed to eliminate foreshortening and thereby gain a new perspective of the limb regions. This *Rectified Lunar Atlas* was issued in 1963 as a supplement to the 1960 atlas.

In 1959 the Air Force Chart and Information Center in St Louis, Missouri, began to use airbrushing to represent topography on a scale of 1:1,000,000 for a series of Lunar Astronautical Charts. Meanwhile, the Army Map Service issued 'photomaps'. The US Geological Survey wished to map the Moon geologically. The first step was to identify the various distinct *geological units* in terms of their textures, delineate their outlines on a 'base map', and use the principle of superposition (as defined by Nicolas Steno in 1669) to determine the order of their deposition. The objective was to obtain insight into the *history* of the lunar surface. In 1960 Robert Hackman of the Photogeology Branch of the Survey in Washington DC demonstrated that it was possible to apply stratigraphic analysis to the Moon. When issued in 1961, his map of what he referred to as pre-maria, maria and post-maria units marked a significant departure from the astronomers' means of mapping. The superposition relationships suggested to Hackman that the maria were volcanic, not splashes of impact melt. He drew attention to a patch of light-toned material between the Apennine mountains and the crater Archimedes. There was ejecta from Archimedes on this patch, and the dark mare had encroached upon the ejecta. The sequence was clear: the light-toned material was the floor of the cavity created by the Imbrium impact, this had been hit by Archimedes some time later, and the mare had appeared after that. Since the light patch was sufficiently elevated not to be overrun, he named it the Apennine Bench. A factor of two difference in the cratering densities of the bench and the adjacent mare was evidence that a significant interval had elapsed between the Imbrium impact and the appearance of the mare within the cavity.

Meanwhile, Gene Shoemaker had independently made a stratigraphic study of a section of the Moon to demonstrate the technique. Visiting a bookstore shortly after being shown the prototype Lunar Astronautical Chart of the Copernicus area, he had happened across a picture of this area taken by Francis Pease in 1919 while testing the 100-inch telescope at Mount Wilson. It was of sufficient clarity to show craters down to 1 km in diameter, so Shoemaker had it enlarged and set to work. Whereas Hackman had used only pre-maria, maria and post-maria units, Shoemaker mapped seven units, which he named the pre-Imbrian, Imbrian, Procellarian, Eratosthenian and Copernican systems. In essence the Eratosthenian and Copernican corresponded to Hackman's post-maria, but Shoemaker distinguished the Eratosthenian from the Copernican because rays from Copernicus were superimposed on the Eratosthenes ejecta – in effect, the difference was whether a post-mare crater's rays were fresh, or faded. On 17 March 1960 Shoemaker presented a paper showing that whereas much of the material excavated by Copernicus had been 'hinged' to produce the rim and adjacent blanket of ejecta, some of the material was hurled ballistically and fell further out, where its impact made distinctive chains of small secondary craters. The secondary craters were less energetic because, to have fallen back at all, the ejecta could not have exceeded the escape velocity – which is an order of magnitude lower than the typical cosmic velocity of material arriving from space. This study not only established Copernicus to be an impact crater, it also refuted the assertion by the advocates of the volcanic origin of craters that the chains of small craters marked eruptions along fractures in the crust.

At the International Astronomical Union Symposium in December 1960, which was a major event for astronomers, Shoemaker and Hackman presented a joint paper entitled *Stratigraphic Basis for a Lunar Time Scale*. This laid the foundation for how geological units could be recognised on an extraterrestrial surface and placed into a stratigraphic sequence. In the case of Earth the units were identified by studies in the field, but for the Moon they would have to be inferred from overhead imagery – at least until expeditions were made to the lunar surface.

Having established that the maria were formed after the Imbrium impact, it was expected that all maria would be able to be assigned to the Procellarian system, but in late 1963, when patches of mare were found to be stratigraphically younger than craters attributed to the Eratosthenian system, the Procellarian system was dismissed and each mare unit was assigned to the system implied by its particular stratigraphy.

Also in late 1963, the scheme was refined by the introduction of *formation* names for the geological units. The reason for the change was that a formation name was objective, and did not imply a specific physical process. Also, because a formation defined a terrain type by its texture, it did not require to be contiguous. This was the case for the hummocky material peripheral to Imbrium. It had just been mapped by Richard Eggleton, who had transferred to Shoemaker's team from the Engineering Geology Branch. It was labelled the Fra Mauro Formation, after a prominent crater within it. Although there was little doubt that it was Imbrium ejecta, to have labelled it as such would have been subjective and would have set a poor precedent.

In September 1961 Gerard Kuiper convinced the Air Force Chart and Information Center to exploit visual observations in compiling the Lunar Astronautical Charts, since in moments of good 'seeing' the eye can resolve finer detail than is able to be recorded during a photographic exposure. The pictures were to provide the basis for mapping and the visual observations would provide the detail. On joining the team, each 'astrogeologist' was assigned a quadrangle to map geologically, in addition to his principal task. As one of the first such recruits, Eggleton provided training for those who followed. Observing time was allotted when the terminator was near the assigned area, to emphasise subtle topography. Those in Arizona used the 24-inch refractor of the Lowell Observatory in Flagstaff, and those in Menlo Park used the 36-inch refractor of the Lick Observatory on Mount Hamilton near San Jose.

In February 1960 the University of Arizona in Tucson established the Lunar and Planetary Laboratory, and made Gerard Kuiper its head. When William Hartmann joined in mid-1961, he assisted the team which was producing the *Rectified Lunar Atlas*. A major finding was the existence of systems of concentric rings. These had not been recognised from Earth owing to foreshortening, but when viewed from an 'overhead' perspective they stood out clearly. The most spectacular case surrounded a small dark patch which was itself only glimpsed at times of favourable libration and had been named Mare Orientale for the reason that it was on the eastern limb – a rationale rendered obsolete by the decision of the International Astronomical Union in August 1961 to switch the east and west limbs! On realising that the multiple-ring structures were a distinct class of geological feature, Hartmann introduced the term 'basin'. He wrote up the discovery with Kuiper and published in-house on 20 June 1962 in the paper *Concentric Structures Surrounding Lunar Basins*. Soon, similar patterns were identified in degraded states around a dozen 'circular maria'. This insight revealed the true violence of a basin-forming impact. Namely, a vast impact excavated a cavity, forming one or more concentric rings of mountains composed of individually faulted blocks with their steep 'fronts' facing inwards, whilst also piling up material in blankets immediately beyond and etching sculpture as ballistic ejecta fell further out – all of which occurred literally *in an instant*. Some time later, and perhaps after a considerable interval, lava rose through deep fractures in the cavity of the basin to flood it, often to a depth sufficient to submerge the inner rings. As a result, a basin consisted not only of the cavity, but also the concentric rings, the inner blankets of ejecta and the outer sculpture. The clear fact that a basin was distinct from the mare that formed later was highlighted by the discovery of concentric rings around large craters which had not been fill with mare. Since multiple-ring structures were *not* of volcanic origin, this lent support to the case for smaller craters also being of impact origin. In fact, although it was recognised early on that sculpture was gouged by the fall of material thrown out on shallow-angle trajectories, it was a while before it was realised that a lot of basin ejecta must have struck at a high angle and, consequently, many well-known sizeable craters are probably not primary impacts but secondaries from basin-forming events. By 1963, photogeologists were working to determine the order in which the dozen or so recognised basins were formed.

In just a few years, therefore, an examination of the Moon by geologists applying standard mapping methods had provided insights into the history of the lunar surface which had eluded astronomers for centuries.

WHENCE THE MOON?

A number of theories have been suggested over the years to explain the origin of the Moon, which is unique as a planetary satellite in that it has the greatest mass as a fraction of its primary, with the result that its orbital angular momentum exceeds the rotational momentum of the planet.

In 1796 the French mathematician Pierre Simon de Laplace, inspired by the rings of Saturn, proposed that the solar system formed by the gravitational collapse of an enormous cloud of gas which was in a state of rotation. The conservation of angular momentum would have required the rate of rotation to increase, causing material to be shed every so often and making a series of concentric rings in a single plane. The central mass eventually formed the Sun, which was sufficiently hot to become self-luminous. As each ring of material condensed to become a planet, the process would have shed local rings which in turn formed satellites – in Earth's case, the Moon. In Laplace's time, the solar system appeared to comprise the entire celestial realm apart from the stars, and therefore his nebular hypothesis was the first serious attempt at cosmogony. Although accepted for many years, mathematical analysis later showed that it would not work as Laplace had imagined.

In 1878 George H. Darwin posited that the Earth and Moon formed together. The rapidly rotating body of hot liquid became an ellipsoid, rotating about its minor axis in an unstable equilibrium with two forces acting upon it: its own natural period of vibration, and tides raised by the Sun's gravity. Once the forces achieved resonance, the shape became progressively more like a dumbbell until one day the narrow 'neck' collapsed, leaving two masses, the larger becoming Earth and the smaller the Moon. This fission hypothesis was popular for some time, but was later discarded owing to mathematical deficiencies, not least because a rapidly spinning ball of fluid would tend to divide into two more or less comparable masses, whereas the Moon has only 1/81st the mass of Earth.

In *The Planets: Their Origin and Development*, which was based on lectures he gave at Yale University and published in 1952, Harold Urey discussed the Moon in relation to the solar system as a whole. He argued that the Moon condensed from the solar nebula independently, and was later captured by Earth. Furthermore, he said it had never undergone thermal differentiation and that, consequently, its surface had no volcanic structures. This was dubbed the 'cold Moon' hypothesis.

In 1954 Gerard Kuiper proposed that the Earth and Moon formed simultaneously in a common envelope within the solar nebula, and soon became gravitationally bound. He said the preponderance of craters was due to the Moon sweeping up all the debris in the neighbourhood. As the Moon's mass is relatively large as a ratio of its primary, this made the Earth and its Moon essentially a 'double planet'.

Nevertheless, as the space age dawned the origin of the Moon and the state of its interior were contested.

3

The space age dawns

MISSILES AND SPACE

When a team of German rocket experts surrendered to the US Army in May 1945 and General Holger 'Ludy' Toftoy, an artillery officer serving as Chief of Ordnance Technical Intelligence in Europe, set out to arrange their relocation to the USA, the V-2 missile was seen as an important military technology. However, this perception changed with the introduction of the atomic bomb in August against Japan. In the immediate post-war years the US military felt that strategic aircraft carrying atomic bombs would enable it to defeat *any* enemy. In this context, a ballistic missile which could fly only several hundred kilometres to deliver about 1,000 kg of conventional explosive was insignificant. Consequently, upon being settled in El Paso, Texas, the German team led by Wernher von Braun found themselves with little to do.

Although the ballistic missile had seemingly become obsolete as a weapon, it held out the prospect of serving a more benign role, and in November 1945 the US Navy recommended the development of a satellite. The Army Air Force agreed. However, each service felt that *it* alone should be assigned this task.

In 1946 the RAND Corporation, created as a 'think tank' for the Army Air Force, said: "The achievement of a satellite craft by the United States would inflame the imagination of mankind, and would probably produce repercussions in the world comparable to the explosion of the atomic bomb. [...] Since mastery of the elements is a reliable index of material progress, the nation which first makes significant achievements in space travel will be acknowledged as the world leader in both military and scientific techniques. To visualise the impact on the world, one can imagine the consternation and admiration that would be felt here if the US were to discover suddenly that some other nation had already put up a successful satellite."

Meanwhile, von Braun was showing the Army how to assemble, prepare and fire V-2 missiles at the White Sands Proving Grounds in New Mexico. They were made from parts either recovered from Germany or manufactured to his specifications in America. In 1948, while in Texas, von Braun wrote a book, *Das Marsprojekt*, in which he outlined how an expedition to explore Mars might be undertaken. It was a 'grand design' which left the details to be developed in due course. He set out "more or less to project the technology that existed then" to motivate young engineers. He argued that a mission would be feasible "in 15 to 20 years" if a nuclear-powered ion engine could be created. The expedition would involve ten space ships with a crew totalling around 70 people. The ships were to be assembled in Earth orbit, with three carrying 'landing boats' for Mars. Later in 1948, von Braun's team was relocated to the Redstone Arsenal of the Army Ordnance Corps in Huntsville, Alabama. It was a new establishment on the site of facilities used by the Chemical Corps in the Second World War, and was to undertake research and development of rockets and missiles.

In September 1949 the Soviets exploded an atomic bomb – at least 3 years earlier than the US had expected. Although the Soviet bomb was not yet a weapon, it was evident that America would soon lose its monopoly. In early 1950 President Harry S. Truman authorised the hydrogen bomb. In 1951 funding was made available for preliminary work for what would become the Atlas intercontinental-range ballistic missile. The hydrogen bomb test at Eniwetok Atoll on 1 November 1952 was not a viable weapon, because it weighed 60 tonnes. But as the bomb's weight was reduced for carriage by aircraft it was realised that if it were to prove possible to make the device even smaller, it might become feasible to develop a ballistic missile capable of delivering it. The Air Force (which had gained its independence from the Army in 1947) created a committee chaired by physicist John von Neumann. This was asked to predict the trend in weight-to-yield ratio of hydrogen bomb development, estimate the warhead that a ballistic missile might deliver over intercontinental range by the end of the decade, and assess whether the probable accuracy would make a warhead of that yield a viable weapon. In February 1954 the committee reported that progress with warheads would make missiles viable. The RAND Corporation endorsed this conclusion. Although the Air Force responded by assigning the development of an intercontinental-range ballistic missile 'top priority', Secretary of Defense Charles E. Wilson, who was in tune with the 'economic conservatism' of the administration of President Dwight D. Eisenhower deliberated on the matter for over 12 months until informed in 1955 that a recently established radar intelligence station in Turkey that was operated by the US had discovered that the Soviets were well advanced in the development of their own intercontinental-range ballistic missile – test flights were launched from a site east of the Black Sea and passed across Soviet territory to fall near the Kamchatka Peninsula. America had felt safe because the USSR had no strategic bombers, but a ballistic missile would be able to circumvent America's air defences. The risk was that when the Soviet missile entered service with a nuclear warhead it would be able to wipe out the US bomber bases in a 'first strike' which would prevent retaliation against the Soviet Union. The US therefore simply had to have its own fleet of missiles.

SATELLITE SHOCK

The first International Polar Year was held between 1882 and 1883 to coordinate meteorological, magnetic and auroral studies. The eruption of Krakatoa in

Indonesia on 20 May 1883 had a temporary but significant effect on the atmosphere. A second International Polar Year was held 50 years later. In 1950 the International Council of Scientific Unions proposed to exploit the technologies developed in the years since the Second World War to undertake geophysical research on a global basis to study the solar–terrestrial relationship. In early 1952 it was agreed that this International Geophysical Year would run from July 1957 to December 1958, a period which was expected to coincide with the time of maximum solar activity in the 11-year cycle of sunspots. In early 1954 the National Security Council said the US "should make a major effort during the International Geophysical Year", and directed the Pentagon to provide "whatever support was necessary to place scientists and their instruments in remote locations" to make observations.

In August 1953 physicist Fred Singer outlined to the International Congress of Astronautics a 45-kg satellite for MOUSE (Minimum Orbital Unmanned Scientific Experiment). He spent the next year promoting it. In October 1954 he canvassed the US delegation to the meeting in Rome, Italy, of the International Geophysical Year's Steering Committee, and as a result a resolution was passed which encouraged participants to investigate the possibility of launching a satellite as the highlight of the program.

In November 1954 Charles Wilson told journalists he did not care if the Soviets were first to put up a satellite. Despite the National Security Council directive for "a major effort" in support of the International Geophysical Year, it was not until 1955 that Wilson endorsed a satellite. In July 1955 Eisenhower announced that the US would put up a satellite for the International Geophysical Year. Eisenhower saw it as a one-off scientific venture. He assigned to the Pentagon the decision for how it should be achieved. There was intense rivalry between the services, because such a spectacle would boost that service's claim to be assigned a greater responsibility for long-range missiles. Shortly before Eisenhower's announcement, Donald Quarles, Chief of Research and Development at the Pentagon, had set up a committee chaired by Homer Joe Stewart, a physicist at the University of California at Los Angeles, to review the capabilities of the services. The National Security Council had stipulated that the satellite must not impede the development of the Atlas missile, which was only now beginning to gear up as a 'crash' national program. This ruled out the Air Force.

The Army proposed Project Orbiter, claiming that if the Redstone missile, which was an improved V-2, were to be fitted with three upper stages, a satellite would be able to be launched by January 1957, which was before the start of the International Geophysical Year. The Navy had Project Vanguard, in which an improved form of the Viking 'sounding' rocket introduced in 1949 for stratospheric research would be augmented with two upper stages. Part of the rationale for the Stewart Committee selecting Vanguard was the perceived greater reliability of requiring only two upper stages, instead of three. In addition, the Committee was impressed by the in-line configuration of the Vanguard stages, as opposed to clustering small solid rockets to form the upper stages of the Redstone launch vehicle. Nevertheless, Stewart himself had supported the Army's proposal. One factor was that whereas the Redstone was a weapon and was classified, the Viking was not classified. Another rationale, added later, was that it would be better to use a 'civilian' rocket for this scientific project. The Committee was not concerned that Vanguard would not deliver as early as the Army claimed for Orbiter – it was simply presumed that the first satellite would be American, and provided that it was launched within the period of the International Geophysical Year it would serve its purpose. On 9 September 1955 the Pentagon endorsed the Committee's recommendation. The spherical Vanguard satellite would weigh about 1.5 kg, and would transmit a radio signal that would allow the study of electrons in the ionosphere and thus make a unique contribution to the International Geophysical Year.

Since the services were only loosely controlled by the Department of Defense, the Army set out to contest the decision, emphasising that the Redstone could launch a satellite without impeding military work. When on the Stewart Committee, Clifford C. Furnas of Buffalo University had sided with the Army. Now at the Pentagon, he advised the Army to have its missile ready as a backup in case Vanguard faltered.

On 1 February 1956 the Army Ballistic Missile Agency was established at the Redstone Arsenal, Major General John B. Medaris commanding. It was to develop an intermediate-range ballistic missile named Jupiter. As the warhead would enter the atmosphere at a faster speed and be subjected to greater heating than that of the short-range Redstone, it was decided to test the new re-entry vehicle by firing it on a 'stretched' Redstone equipped with two upper stages made by clustering small solid rockets. The fact that this 'Jupiter-C' would enable the Army to develop and test a vehicle capable of launching a satellite was, of course, entirely coincidental! When the first test flight on 20 September 1956 reached a peak altitude of 1,000 km and flew 4,800 km down the Air Force's Eastern Test Range from Cape Canaveral, the Pentagon directed Medaris to personally guarantee that Wernher von Braun did not inadvertently place anything into orbit! One criticism of Vanguard was that although its first stage was based on the Viking, the project really involved developing a new integrated vehicle in a period of only 2 years. With Vanguard running late, Medaris sought permission to launch a satellite, but the Secretary of the Army refused – in fact, the Army Ballistic Missile Agency was ordered to destroy the remaining solid rockets obtained for the upper stages. In response, Medaris decided to leave them in storage to 'assess' their shelf life!

In public, Eisenhower maintained that launching a satellite was a one-off venture for the International Geophysical Year. In fact, this was a cunning ruse, because the aim was to use Vanguard to set the precedent of a US satellite passing over foreign territory, and thus preclude a legal challenge when the US began to send up satellites for military functions such as reconnaissance.

Soon after the US announced that it would launch a satellite for the International Geophysical Year, the Soviet Union said it intended to do the same. In mid-1957 the Soviet magazine *Radio* told its readers how to go about 'listening' to this satellite. In late August the *TASS* news agency announced the successful test flight of a "super long range" missile which was capable of striking "any part of the world". When a Soviet delegate at an International Geophysical Year meeting in Washington in late September was asked whether the promised satellite was imminent, he replied: "We won't cackle until we've laid our egg." In other words, wait and see!

On 4 October the 84-kg Sputnik was placed into an orbit which ranged in altitude between 220 and 950 km and transmitted its incessant 'beep, beep, beep' signal.

The news caused a world-wide sensation, but Eisenhower was not concerned. At a press conference on 9 October he dismissed Sputnik as a "small ball in the air" that "does not raise my apprehensions, not one iota". On the other hand, the mass of the satellite showed the capability of the Soviet intercontinental-range ballistic missile, and Eisenhower ordered an end to the administrative difficulties that were impeding funding for the American missile programs.

Lyndon Baines Johnson was not only the senior Democratic senator for Texas, as the Democratic leader in the Senate he essentially controlled majority congressional support for the legislative program: put simply, without his backing, the Republican administration was ineffective. Johnson saw Sputnik in terms of national security – the satellite could well have been an orbital bomb, waiting to be instructed to fall on an American city. He ordered a Congressional investigation into the state of national security preparedness. As a result, the public became aware that there was a "missile gap"; and, almost overnight, 'space' was transformed from a fantasy into something that the US should be leading, since otherwise national prestige would be damaged.[1]

After the launch on 3 November of a heavier Sputnik with a canine passenger, Eisenhower demanded an increase in the pace of Vanguard, which was in trouble, and also authorised the Army Ballistic Missile Agency to prepare a Jupiter-C in case Vanguard should fail. Medaris had the solid rockets for the upper stages retrieved from storage and let von Braun loose.

On 6 December 1957 Vanguard ignited, lifted a few centimetres off the pad, then collapsed back and exploded in a fireball. On 31 January 1958 the Army launched a satellite using essentially the same vehicle configuration as the Stewart Committee had rejected. On being asked for permission to inform Washington of the success, Medaris reputedly said: "Not yet, let them sweat a little." The satellite, Explorer 1, was integrated into the solid rocket of the final stage and inserted into an orbit which ranged between 360 and 2,550 km. The Geiger-Mueller tube it carried was supplied by James van Allen, a physicist at the University of Iowa, and detected the presence of charged-particle radiation trapped within the Earth's magnetic field, far above the atmosphere.

With the development of nuclear-armed intercontinental-range ballistic missiles threatening to make manned strategic bombers obsolete, the Air Force reacted to the prospect of its strike force becoming 'silo rats' by claiming that it needed to develop a manned space flight capability. Its Ballistic Missile Division, headed by General Bernard Schriever, devised Man In Space Soonest. This envisaged a progression of steps that would result in an Air Force officer landing on the Moon in 1965. When this was submitted to the Pentagon in March 1958 the response was lukewarm – in

---

[1] Ironically, *if* the US had been the first to orbit a satellite, then the nation would not have suffered the trauma of *not* being first and subsequent events might have turned out very differently indeed!

On 6 December 1957 the Vanguard rocket explodes within seconds of ignition.

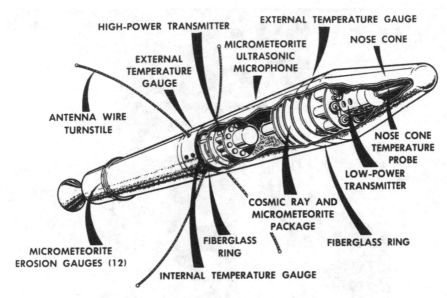

HIGH-POWER TRANSMITTER

EXTERNAL TEMPERATURE GAUGE

NOSE CONE

MICROMETEORITE
ULTRASONIC
MICROPHONE

EXTERNAL
TEMPERATURE
GAUGE

ANTENNA WIRE
TURNSTILE

NOSE CONE
TEMPERATURE
PROBE

LOW-POWER
TRANSMITTER

COSMIC RAY AND
MICROMETEORITE
PACKAGE

MICROMETEORITE
EROSION GAUGES (12)

FIBERGLASS
RING

FIBERGLASS RING

INTERNAL TEMPERATURE GAUGE

Details of the Explorer 1 satellite, with the instrument section integrated with the solid-rocket final stage.

part owing to the estimated cost of $1.5 billion, but also due to the absence of a clear military necessity. In fact, the proposal was an example of what would be referred to in today's parlance as a demonstration of 'the vision thing'.

No sooner had the Army developed its Jupiter intermediate-range ballistic missile than the Pentagon assigned operational control of all land-based missiles with ranges exceeding 320 km to the Air Force, thus limiting the Army to 'battlefield' missiles. In fact, the Air Force had no use for the Jupiter, since it had just developed its own Thor intermediate-range ballistic missile.

The only prospect for the Army Ballistic Missile Agency was therefore to develop powerful launch vehicles for satellites. On 19 December 1957 the Army proposed the *National Integrated Missile and Space Vehicle Development Program*. Like the Air Force, the Army saw itself as the obvious service to explore space. In 1959 it proposed Project Horizon to achieve a manned lunar landing in 1965, but this was received no more enthusiastically than the rival Man In Space Soonest.

AMERICA TRIES FOR THE MOON

Explorer 1 restored national honour, but the Department of Defense was deeply concerned that the Soviet launch vehicle was so much more powerful than its own. On 7 February 1958 President Eisenhower created the Advanced Research Projects Agency headed by Roy W. Johnson, who would report directly to the Secretary of Defense. Its was to develop national goals and coordinate, but not itself conduct, the necessary research.

The Explorer 1 satellite, installed atop the drum-like second stage of the Juno I launch vehicle.

After the successful launch of Explorer 1, W.H. Pickering (farthest away), James van Allen and Wernher von Braun hold aloft a full-scale model of the spacecraft.

Wernher von Braun poses beside the framed *Huntsville Times* announcing the successful launch of America's first satellite.

On 21 October 1957, three weeks after the launch of Sputnik, W.H. Pickering, since 1954 Director of the Jet Propulsion Laboratory (JPL) in Pasadena, California, had proposed that a spin-stabilised probe be launched towards the Moon, possibly as soon as June 1958.[2] The purpose of this Project Red Socks would be to produce "a significant technological advance over the Soviet Union" that would enable America to "regain its stature in the eyes of the world". The Pentagon sent the proposal to Roy Johnson, who was eager "to surpass the Soviet Union in any way possible". The fact that the Soviets had not yet announced a lunar flight prompted him to accept the challenge of America being the first to do so. Neil H. McElroy, who had superseded Charles Wilson as Secretary of Defense just a few days before the Soviets launched Sputnik, announced on 27 March 1958 the US decision to "determine our capability of exploring space in the vicinity of the Moon, to obtain useful data concerning the Moon, and provide a close look at the Moon". It would be undertaken as part of America's contribution to the International Geophysical Year.

The project was named Pioneer. To pre-empt calls for it to be assigned to one or other of the services, the Air Force and Army were to work in parallel on their own contributions. The Air Force would modify its Thor missile to use the upper stages made for Vanguard. Meanwhile, the Army, just as it had used the Jupiter-C variant of the Redstone in a configuration named Juno I to launch Explorer 1, would fit its Jupiter missile with upper stages by clustering solid rockets to create the Juno II. As conceived, there would be five flight opportunities: three for the Air Force and two for the Army.

The Air Force assigned the technical direction of its part of the project, including the provision of the payload, to the Space Technology Laboratories of Redondo Beach, California. This company served as the contract manager for the Air Force's ballistic missile program. The plan was for the launch vehicle to undertake a 'direct ascent' from Earth and release the probe on a trajectory that would enable it to enter orbit around the Moon. The design of the probe was finished in June 1958, just three months after the project was given the go-ahead. It comprised a pair of squat cones with their bases on a short cylindrical section. The body was 74 cm in diameter and 46 cm tall. It was to be spun at 200 rpm for stability. The mass of 38 kg included the solid-fuelled retro-rocket to brake into lunar orbit and 18 kg of scientific payload. The Advanced Research Projects Agency stipulated that the probe have an imaging system, but the scientists considered the primary payload to be their instruments to follow up the discovery by Explorer 1 of charged-particle radiation near Earth, and in this case the 'particles and fields' instruments were a magnetometer to measure magnetic fields in cislunar space and a micrometeoroid impact counter.

At 12:18 GMT on 17 August 1958 the Thor-Able lifted off from Pad 17A at Cape Canaveral, but the seizure of a turbopump bearing 77 seconds later brought the

---

[2] Note that Pickering's proposal was even *before* the US matched the Soviets by launching an Earth satellite.

A model of the Pioneer 1 satellite.

flight to a premature end. Intended to be named Pioneer 1, this inauspicious start entered the history books as Pioneer 0.

Meanwhile, Lyndon Johnson began to argue for a new government agency to run a major space program – it featured in a speech he gave in January 1958 in which he 'signalled' his intention to seek the party's nomination to run for the presidency in 1960.

As Eisenhower's Special Assistant for Science and Technology, James R. Killian chaired the President's Science Advisory Committee. This reported "space" to be "inevitable", citing as reasons: (1) defence implications, (2) national prestige, and (3) opportunities for scientific research. The Committee warned that if the Pentagon was allowed to run a 'national' program, grandiose proposals would jeopardise the scientific work. It would be better to organise the scientific program independently of the military. The Committee recommended assigning it to a body modelled on the National Advisory Council for Aeronautics, which had been established in 1915 to coordinate aeronautical research. On 2 April 1958 Eisenhower signed an executive order calling for the National Advisory Council for Aeronautics to be subsumed into the National Aeronautics and Space Administration (NASA) in order to manage the national civilian space program. He also created the National Aeronautics and Space Board to advise on policy. The National Aeronautics and Space Act was passed by Congress on 16 July, and signed into law on 29 July. The new agency inherited all of its predecessor's facilities: the Langley Aeronautical Laboratory at Langley Field, which had been established in Hampton, Virginia, in 1917, together with its Pilotless Aircraft Research Station at Wallops Island; the Ames Aeronautical Laboratory at Moffett Field, established in 1939 in Mountain View, California; the Lewis Flight Propulsion Laboratory, which was established in 1941 in Cleveland, Ohio; and the High-Speed Flight Station, established in 1949 at Muroc Field in the high desert of California and renamed Edwards Air Force Base in 1950.[3] Although NASA's remit

---

[3]   On transfer to NASA, the Langley Aeronautical Laboratory became the Langley Research Center, the Ames Aeronautical Laboratory became the Ames Research Center, the Lewis Flight Propulsion Laboratory became the Lewis Research Center and the High-Speed Flight Station became the Flight Research Center.

was much broader than that of its predecessor, it did *not* immediately gain control of the rocketry expertise at either JPL or the Army Ballistic Missile Agency.[4]

On 8 August Thomas Keith Glennan, for the last decade President of the Case Institute of Technology in Cleveland, Ohio, was nominated as NASA Administrator. Hugh Latimer Dryden, Director of the National Advisory Council for Aeronautics since 1947, was to provide continuity by serving as his deputy. Congress confirmed the appointments within days. When NASA became operational on 1 October 1958, it inherited the Pioneer project from the Advanced Research Projects Agency.

The Air Force's second probe rose from Pad 17A at 08:32 GMT on 11 October. The Thor performed flawlessly, but a guidance error caused the second stage to shut down prematurely. The third stage took over, but was incapable of making up the 250-m/s shortfall in velocity. Pioneer 1 was successfully released, but upon attaining an altitude of 115,350 km, about one-third of the way to the Moon, it fell back and burned up in the atmosphere on 13 October. The trajectory precluded the electronic TV imager from viewing the Moon. The scientists welcomed the data provided by the magnetometer and micrometeoroid detector. This flight also had an ion chamber supplied by James van Allen, but it developed a leak and the data was difficult to interpret.

The scientists augmented the third probe with a proportional counter supplied by the University of Chicago. The Thor-Able lifted off at 07:30 GMT on 8 November 1958. The first two stages worked, but the engine of the third stage failed to ignite. The trajectory of Pioneer 2 peaked at an altitude of only 1,550 km and it fell back into the atmosphere 6.8 hours after launch, having returned no significant data.

The Army's probe was developed by JPL, which had supplied Explorer 1. It was a cone mated at its base to a cylindrical section, stood 51 cm tall and had a maximum diameter of 24 cm. Whereas the Air Force had used an optical-electronic sensor that scanned as the probe rotated, JPL designed a camera whose film would be wet-developed, scanned optically and transmitted to Earth for reproduction by facsimile methods. The image was to be taken from a lunar altitude of 24,000 km, with the shutter being triggered when a photocell noted the presence of the Moon in its field of view. The first flight was to test this sensor. Its successor would carry the entire camera and loop around the back of the Moon to take a picture of the mysterious far-side at a resolution of 32 km. However, when studies by Explorers 3 and 4 revealed that the charged-particle radiation in the Earth's vicinity would 'fog' the film of the Moon-bound probe, the Army cancelled the film camera in August 1958

---

[4]   On 3 December 1958 Eisenhower ordered that JPL be transferred to NASA. This took effect on 1 January 1959, although only under contract, since the facility was owned by Caltech, which NASA paid. In September 1959 the Pentagon voluntarily yielded the Army Ballistic Missile Agency since the military had decided it did not require the Saturn launch vehicle; it would develop the Titan III instead. On 21 October 1959 NASA announced that it was to gain von Braun's rocket team. On 1 July 1960 the Army Ballistic Missile Agency became the Marshall Space Flight Center.

The Thor-Able launch vehicle with Pioneer 1 being prepared for launch on 11 October 1958.

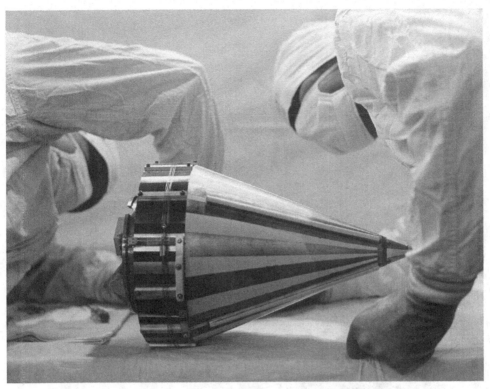

Technicians prepare the Pioneer 3 satellite.

in favour of the development of a lightweight slow-scan TV camera and a magnetic tape recorder to store the image for transmission.

A Juno II launched Pioneer 3 from Pad 5 at 05:45 GMT on 6 December 1958. The probe was intended to make a direct ascent, fly close to the Moon and pass into solar orbit. But the Jupiter first stage shut down prematurely, and the upper stages were unable to make up the 286-m/s shortfall in velocity. The 6-kg probe peaked at an altitude of 102,300 km, fell back and burned up 38 hours 6 minutes after launch. Nevertheless, it produced useful data. In place of the camera, it had a pair of Geiger-Mueller tubes supplied by James van Allen to measure radiation in cislunar space, and these revealed the existence of a second zone of radiation some distance above the one already identified: the intensity peaked at 5,000 km and again at 16,000 km, then diminished to the probe's peak altitude. At van Allen's suggestion, the imaging system was deleted from the second probe to enable his instrument to fly again. In addition, lead shielding was installed on one of the Geiger-Mueller tubes to screen out the low-energy charged particles. With the imaging cancelled, the trajectory was revised from a loop around the back of the Moon to a flyby into solar orbit – as had been intended for the first probe. After lifting off at 05:45 GMT on 3 March 1959, Pioneer 4 successfully flew by the Moon at 22:25 on 4 March. Unfortunately, the range of 60,500 km was twice that planned, with the result that the photocell test

The Juno II launch vehicle with Pioneer 4 is prepared for launch on 3 March 1959.

failed – but as there was no follow-up probe available to carry the camera this was of little consequence. The Geiger-Mueller results provided further support for the hypothesis that the Earth's magnetic field traps charged particles that originate from the Sun.[5]

The idea of streams of particles flowing outward from the Sun was first suggested by British astronomer Richard C. Carrington. In 1859 he made the first observation of what would later be named a solar flare. The occurrence of a geomagnetic storm the following day prompted him to suspect a connection. In the 1950s the German scientist Ludwig Biermann cited the fact that the tail of a comet always points away from the Sun irrespective of the comet's direction of travel, as evidence that the Sun emits particles. In 1958 Eugene Parker in America postulated a supersonic flow of high-energy charged particles, primarily protons and electrons, streaming from the corona in the form of a 'solar wind'. The presence of charged particles circulating in the Earth's magnetic field strongly supported this hypothesis.

SOVIET LUNAR FLYBY

Although the Advanced Research Projects Agency had hoped to beat the Soviets to the vicinity of the Moon, by the time Pioneer 4 became the first American probe to do so this particular race had been won by the Soviets. Luna 1 lifted off at 16:41 GMT on 2 January 1959 on a direct ascent trajectory, and on 4 January flew by the Moon at a range of 5,500 km and passed into solar orbit. In fact, the objective was to hit the Moon, but the Soviets gave the impression that the plan had been to make a flyby. The 1.2-metre-diameter 361-kg spherical probe was not stabilised in flight. Its particles and fields instruments included a magnetometer on a 1-metre-long boom. The transmissions continued for 62 hours, by which time it was 600,000 km from Earth. It detected plasma in interplanetary space, further supporting the existence of the solar wind, but no evidence that the Moon possessed a magnetic field.

NASA EMBRACES LUNAR SCIENCE

The term 'sky science' was coined by JPL historian Cargill Hall to encompass the study of the Earth's upper atmosphere and ionosphere, particles and fields in space and micrometeoroid particles. It included solar and cosmic rays, plasma dynamics and the interaction of the solar and terrestrial magnetic fields. By the late 1950s, sky scientists had a range of instruments with which to pursue their interests. These had been developed and refined initially by balloon-borne packages and, more recently, by sounding rockets. Its members were a cohesive group with impeccable academic pedigrees. By way of the National Academy of Sciences they had played key roles in

---

[5]   The charged particles trapped in the Earth's magnetic field became known as the van Allen radiation belts.

selecting the experiments for the sounding rockets fired by the United States for the International Geophysical Year, and had dominated planning for the Vanguard and Explorer satellites and Pioneer space probes.[6] As a group, they were not interested in physical bodies such as the Moon other than as sources of magnetic fields, and they were certainly not interested in the geological history of the lunar surface.

In early 1958 'planetary scientists' began a series of informal Lunar and Planetary Exploration Colloquia. The first meeting on 13 May 1958 was jointly sponsored by the RAND Corporation, the California Research Corporation and North American Aviation, and it was hosted by the latter in Downey, California. The three principal objectives were (1) to bring together people of common interest for the exchange of scientific and engineering information; (2) to define the scientific and engineering aspects of lunar and planetary exploration and to provide a means for their long-term appraisal; and (3) to make available, nationally, the collective opinions of a qualified group on this subject.[7]

In June 1958 the National Academy of Sciences, a private organisation chartered in 1863 to promote the advancement of science and, when requested, to advise the government on scientific matters, established the Space Science Board, chaired by Lloyd Berkner, to advise the not-yet-active NASA on space research priorities. By the end of 1958, both the President's Science Advisory Committee and the Space Science Board had cited 'lunar exploration' as a worthwhile scientific objective for the new agency.

The only interest in the Moon as a body in its own right expressed by the Pioneer project started by the Advanced Research Projects Agency was to photograph its far-side. But this was not achieved by the Air Force probes, and the discoveries made by the first Army probe led to the deletion of the imaging system of the second probe in order to obtain further particles and fields observations. To be fair, this data was an important contribution to a rapidly developing field of the International Geophysical Year.

The first lunar project authorised by NASA was the Atlas-Able, which was an Air Force launch of a probe supplied by the Space Technology Laboratories. The initial idea had been to use the Atlas, which was much more powerful than the Thor, for a program that would make two launches to send probes towards Venus and then two to insert probes into lunar orbit, but Luna 1's flyby of the Moon prompted NASA to order the planetary payloads to be replaced by lunar orbiters. The 170-kg probes were to have a spin-stabilised 1-metre-diameter spherical structure with four 'paddle wheel' solar arrays around the equator. They would use liquid-propellant engines to make a midcourse correction on the way to the Moon and later to enter orbit around that body. In addition to a suite of particles and fields instruments, they would carry the TV system made for the Thor-launched probes. The first probe was destroyed on 24 September 1959, when the Atlas exploded during a static test. The second lifted

---

[6]   Physicists James van Allen, Homer Newell, Charles Sonett and Lloyd Berkner were notable early members of the 'sky science' community.

[7]   Colloquia were held quarterly at different venues on the West Coast through to May 1963.

off at 07:26 GMT on 26 November from Pad 14 on a direct ascent trajectory, but the aerodynamic shroud protecting the probe failed 45 seconds into the flight. The third lifted off at 15:13 GMT on 25 September 1960 from Pad 12, but the first of the two Able stages malfunctioned and fell into the atmosphere 17 minutes after launch. The final probe was launched at 08:40 GMT on 14 December 1960 from Pad 12 but the vehicle exploded 68 seconds later. Although in each case the plan was to enter lunar orbit, most of the 55-kg scientific payload was for particles and fields investigations; indeed, the TV system was deleted from the second pair of probes to accommodate additional radiation detectors. In fact, the Moon was to serve merely as an 'anchor' in space away from Earth. In no way could this series of probes be said to constitute 'lunar exploration'. However, even as the Atlas-Able probes were being developed the situation was changing.

As Assistant Director of the Lewis Laboratory of the National Advisory Council for Aeronautics, in the summer of 1958 Abraham Silverstein played a leading role in the establishment of NASA. When the new agency set up the Office of Space Flight Development at its headquarters, Silverstein became its Director. He promptly hired Homer E. Newell as his assistant for Space Sciences. Newell had joined the Naval Research Laboratory in 1944, and the next year started to conduct research into the upper atmosphere using sounding rockets – in particular investigating the interaction between the magnetic fields of the Sun and Earth. He served as the Science Program Coordinator for the Vanguard project of the International Geophysical Year.

At NASA, Newell created a division staffed by members of the upper atmosphere research group of the Naval Research Laboratory, to organise the agency's activities in that field. In November, he appointed Robert Jastrow to chair another division to address astronomy, cosmology and planetary sciences. As a sky scientist, Jastrow set out to learn about these topics by visiting the leading proponents.

On reaching the age of 65 in 1958, Harold Urey had retired from the University of Chicago and taken a research position at the University of California at San Diego. He was a member of the Space Science Board of the National Academy of Sciences, and in a paper entitled *The Chemistry of the Moon* given on 29 October 1958 at the third Lunar and Planetary Exploration Colloquia he had explained the significance of sending a probe to photograph the far-side of the Moon.

When Jastrow visited Urey in early January 1959, Urey emphasised the "unique importance" of the Moon for achieving an understanding of the origin of the planets. As Jastrow recalled of this meeting in his 1967 book *Red Giants and White Dwarfs*: "I was fascinated by [his] story, which had never been told to me before in 14 years of study and research in physics." The following week, at Jastrow's invitation, Urey gave a 2-day presentation at NASA headquarters in which he urged that probes be sent to the Moon. After deliberating, Newell concluded that NASA should initiate a program to study the Moon as an object in its own right. He formed an ad hoc Working Group on Lunar Exploration, with Jastrow in the chair and Urey as a member, to evaluate and recommend experiments which should be placed into orbit around the Moon or landed on its surface. This gave the lunar (and later planetary) scientists a presence at NASA headquarters to match that of the sky scientists.

At that time, JPL was drawing up a proposal for a program of a dozen deep-space missions, of which five would study the Moon. On 5 February 1959 Jastrow sent a contingent to JPL to pass the word that NASA was keen to undertake lunar exploration. The options were for probes to report results as they plunged to their destruction by smashing into the Moon, to enter lunar orbit, and to land instruments on the surface. JPL was authorised to initiate preliminary work. It formed a study group chaired by Albert R. Hibbs.

In fact, in 1959 JPL engineers and scientists were more interested in the challenge of sending probes to Venus and Mars than they were in studying the Moon. There were 'windows' for efficiently sending probes to Mars at 25-month intervals and to Venus at 18-month intervals. As the next windows were October 1960 for Mars and January 1961 for Venus, the feeling was that the immediate effort should be devoted to these opportunities, rather than the Moon, which could be reached at almost any time. In 1958 JPL had proposed to NASA the development of a new upper stage for the Atlas. This Vega stage would be powered by the engine of the Viking 'sounding rocket', modified for ignition in the upper atmosphere. The Atlas-Vega was to be used to launch satellites. A third stage would dispatch deep-space probes. The Soviet lunar flyby in January 1959 spurred the US Congress to authorise the development of the Vega stage. On 30 April JPL sent NASA a 5-year plan of deep-space missions using the Atlas-Vega. The aim was to devote the early effort to flybys of Venus and Mars, and postpone lunar science until 1961. A rough lander would be followed up by an orbiter equipped to investigate the space environment near the Moon and to obtain high-resolution pictures to enable a site for a soft lander to be chosen. Jastrow recommended to Newell the development of a seismometer, communication system and power supply for the package that would be delivered to the lunar surface by the rough landing method. It was later decided to operate a magnetometer, gamma-ray spectrometer and X-ray fluorescence spectrometer during the approach phase. On 25 May 1959, Silverstein and Newell decided to follow the rough lander launched by the Atlas-Vega with two soft landers launched by the Atlas-Centaur, the latter using a powerful stage that was expected to enter service in 1962. In June 1959 Silverstein told JPL to cancel the Mars mission and reassign its launcher to a lunar orbiter.

On 23 July Keith Glennan, Hugh Dryden and Associate Administrator Richard E. Horner met George B. Kistiakowsky, who had succeeded James Killian as Eisenhower's Special Assistant for Science of Technology, to discuss the objectives of the space program. Glennan warned that slippage in the development of the Vega stage made planetary flights in 1960–1961 impractical. Windows for the Moon were not only more frequent, the flight time was days rather than months. Glennan recommended that the agency focus on lunar missions in order to address the short-term objectives recently specified by the National Security Council's policy paper, *Preliminary US Policy on Outer Space*.[8] This was agreed. Several days later, Silverstein cancelled the Venus mission and directed JPL to prepare a new schedule which focused on the Moon. Meanwhile, Newell had established the Lunar and

[8]   Document NSC5814/1.

Planetary Programs Office and transformed Jastrow's ad hoc Working Group on Lunar Exploration into a standing committee as the Lunar Science Group.

MORE SOVIET SUCCESSES

Luna 2 was launched at 06:40 GMT on 12 September 1959, and 33.5 hours later it smashed into the Moon in the triangle defined by the craters Archimedes, Aristillus and Autolycus. On striking the ground at 3 km/sec, the probe would have vaporised. It carried similar instruments to its predecessor. In extending the particles and fields survey down to the surface, it found no appreciable magnetic field and no evidence of a radiation belt. The Earth's field is generated by electric currents in molten iron in the core. The absence of a dipole field suggested that either the Moon had never developed a molten core, or it had and either the rate of the Moon's rotation was too slow to generate electric currents or the core had since solidified.[9]

When Luna 3 was launched at 00:44 GMT on 4 October 1959, it was to undertake the eagerly anticipated photographic mission. It was inserted into a highly eccentric orbit of Earth which would provide a view of the far-side of the Moon. The 279-kg probe was 1.3 metres long. The main body was a 1.2-metre-diameter cylinder, and it had solar transducer cells in fixed positions on its exterior. The earlier probes in this series were battery powered, but a battery would not provide the duration required to undertake this particular mission. In addition to particles and fields instruments, the probe had a camera. Its trajectory crossed the radius of the Moon's orbit 6,200 km to the south of the Moon. The point of closest approach was at 14:16 on 6 October, and lunar gravity deflected the trajectory northward. While coasting, the probe was spin stabilised. On approaching the Earth–Moon plane, beyond the Moon, gas jets halted the spin, a Sun sensor locked on, and the vehicle rolled until another sensor enabled the optics to view the Moon.

The camera had a 200-mm f/5.6 lens and a 500-mm f/9.6 narrow-angle lens, with the optical axes coaligned. Starting at 03:30 on 7 October, at a range of 65,200 km, pairs of pictures were captured simultaneously on 35-mm film which had a 'slow' rating in order to resist being 'fogged' by the radiation in the space environment. After 40 minutes of photography, the vehicle resumed its spin for stability. The film was wet-developed, fixed and dried. After the probe had achieved the 480,000-km apogee on 10 October, it headed back for the 47,500-km perigee on 18 October. The film was scanned using a constant-brightness light beam which was detected by a photoelectric multiplier whose output took the form of an analogue signal. The scanner had two transmission rates: a slow rate for when far away from Earth and a higher rate for perigee. Contact with the probe was lost on 22 October. On Earth, the signal was recorded on magnetic tape for further processing. The raw images were marred by bands of 'noise', which had to be 'removed', and the contrast was drawn

---

[9] The belief at that time was that a body had to rotate rapidly to generate a magnetic field.

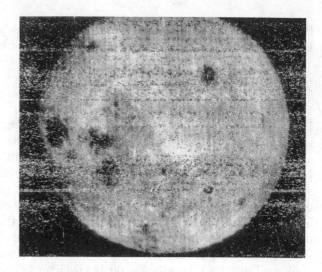

A picture taken by Luna 3 on 7 October 1959 showing Mare Crisium (on the left) and a large portion of the hitherto unobserved hemisphere of the Moon.

out to emphasise detail. In the wide lens the lunar disk spanned 10 mm, and in the narrow lens it was 25 mm.

The phase of the Moon was 'new' on 2 October and 'first quarter' on 9 October. Therefore, when the pictures were taken on 7 October a portion of the near-side was illuminated. The prominent presence of Mare Crisium gave a sense of perspective. The remainder of the illuminated zone was 70 per cent of the hitherto unobserved region. As the Sun was directly behind the camera, the lack of shadow detail made it impossible to discern the topography. However, the results did show there to be few dark features – revealing the hemispheres which faced towards and away from Earth to be different. The maria cover 30 per cent of the near-side, but only 2 per cent of the far-side; in total about 16 per cent. The likelihood of there being fewer maria had been inferred from the paucity of maria on the limb, with those present being patchy rather than major plains, but their virtual absence was a surprise. The fact that the Moon's rotation is 'tidally locked' with Earth was evidently an important factor in creating the dichotomy between the two hemispheres.

The results were published in 1960 as *Atlas Obratnoy Storony Luny* by N.P. Barabashov, A.A. Mikhaylov and Yu.N. Lipskiy. They unilaterally assigned names to the far-side features, with the two most prominent dark patches becoming Mare Moscoviense and Tsiolkovsky.[10] The atlas listed some 500 features of various types on a scale which represented the Moon as a disk 35 cm in diameter.

The Luna 3 mission was a remarkable success for the first attempt at a difficult task.

---

[10] As would later be realised, Mare Moscoviense fills the floor of a 300-km-diameter crater and Tsiolkovsky covers a portion of the floor of a crater which has a prominent central peak.

4

Preparing Ranger

FIRMING UP THE PLAN

The spectacular Soviet achievement of photographing the far-side of the Moon in October 1959 prompted NASA to revise its planning. The Air Force was developing the Agena B as a variant of an upper stage which could restart its engine in space. The payload projection for low Earth orbit of an Atlas-Agena B exceeded that projected for the two-stage Atlas-Vega, and almost matched that of the three-stage variant. On 7 November 1959 NASA decided that all satellites scheduled to use the Atlas-Vega should be transferred to the Atlas-Agena B; the Vega stage would be used only for lunar probes. But on 11 December it was decided that the lunar probes would switch too, and the Vega was cancelled. The Atlas-Agena B was seen as the interim launch vehicle for deep-space probes, pending the introduction of the Atlas-Centaur.

On 21 December 1959 Abe Silverstein of the Office of Space Flight Development at NASA headquarters told JPL to prepare five spacecraft to reconnoitre the Moon in 1961–1962. The objectives now included obtaining high-resolution pictures of the lunar surface during the terminal approach, which would require to be transmitted in real-time since the vehicle would be destroyed on impact. This imaging was to make up for the loss of the orbiter – which Silverstein had ordered in June 1959 and now cancelled because a review had judged it to be too complex for this early point in the program. In addition, Silverstein told JPL to consider adding instruments to perform particles and fields investigations during the cruise to the Moon. This project was to be completed within 36 months, in order to pass on to the next project of the lunar exploration program. It was acknowledged to be a high-risk venture on a short-term schedule, but was intended, in addition to studying the Moon, "to seize the initiative in space exploration from the Soviets".

In January 1960 Silverstein created the Lunar and Planetary Programs Division, with Edgar M. Cortright as its Director. In March Oran W. Nicks was made Chief of Lunar Flight Systems, and given the task of monitoring the project's progress. After NASA formally approved the project in February 1960, Clifford I. Cummings at

JPL proposed that it be named 'Ranger'.[1] Silverstein was not keen, but the moniker was made official and on 4 May Cummings announced it to the *Los Angeles Herald–Examiner*.

At the end of 1959 Keith Glennan reorganised NASA headquarters. The dominant office in terms of budget and activities remained Silverstein's, now called the Office of Space Flight Programs.

W.H. Pickering reorganised JPL at the end of 1959 by creating the Lunar Program Office under Cummings, with James D. Burke as his deputy. Cummings duly made Burke Ranger Spacecraft Project Manager. JPL managed the new NASA projects by superimposing small 'project offices' on the existing functionally arrayed technical ('line') divisions that comprised the laboratory's core expertise. Indeed, the Ranger Project Office initially comprised only two men and a secretary. Burke's task was to allocate funds, plan, schedule, assign tasks to the divisions and review progress, but he had no direct supervisory authority over the work since each division did its own design and development and divisional chiefs set their own priorities and assigned engineers within their bailiwicks. Initially, Ranger was the focus of activity, but as other projects claimed attention, most notably the Mariner interplanetary missions, engineers were reassigned and Ranger suffered.

As another element of his restructuring, Pickering added a Systems Division to develop, build and test spacecraft, and a Space Sciences Division to install scientific experiments. For Ranger, the Systems Division would be responsible for systems analysis, including launch to orbit, departure to the Moon, and the requirements of midcourse and terminal manoeuvres; the design and integration of the spacecraft's systems, including qualification and performance testing, and quality assurance; and assembly and checkout for launch. It would call upon other divisions as subcontractors. Harris M. Schurmeier was the Chief of the Systems Division, and as such he became Burke's main point of contact with the technical side of the laboratory. In February 1960 Schurmeier appointed Gordon P. Kautz as the Project Engineer for Ranger in the Systems Division, but in October Kautz was reassigned as Burke's deputy and Allen E. Wolfe took the vacated post. The Space Sciences Division, led by Albert R. Hibbs, consolidated JPL's experimenters into a single group. The Guidance Division was headed by Eugene Giberson. The Engineering Mechanics Division was under Charles Cole. The Telecommunications Division was under Eberhardt Rechtin. The Propulsion Division was under Geoffrey Robillard.

In May 1960 NASA directed JPL to start work on the Surveyor project. As it was doing with Ranger and Mariner, JPL sought to maximise commonality of systems between the two forms of the Surveyor – one for orbital reconnaissance of the Moon and the other to soft land and investigate the physical and chemical properties of the surface. It was expected that because a rough landing from a direct approach would be simpler than entering orbit or soft landing, Ranger would be able to be completed while Surveyor was in development.

---

[1]   The name Ranger set a trend for lunar projects with the names Surveyor and Prospector; in contrast to Mariner for planetary missions – that is 'land' names as against 'sea' names.

MANAGEMENT ISSUES

As Silverstein at NASA headquarters had arranged things, JPL reported to him for the Ranger spacecraft, deep-space tracking and control, in-flight operations and data processing. The procurement of the launch vehicle would be managed by the Office of Launch Vehicle Programs. This was directed by Donald R. Ostrander, who, as a Major General in the Air Force assigned to NASA, was well qualified to liaise with the military and its contractors. Ostrander delegated the task of procuring Agena and Centaur stages to Wernher von Braun in Huntsville. By dividing the spacecraft (JPL and the Office of Space Flight Programs) from the launchers (Huntsville and the Office of Launch Vehicle Programs), this arrangement provided considerable scope for confusion and conflict.

On 29 December 1959 Associate Administrator Richard Horner created the Space Exploration Program Council with himself in the chair. It was to seek to improve the management of space flight projects, and to reconcile the inevitable differences that would arise between headquarters and the centres managing individual projects. Its members were Abe Silverstein, Donald Ostrander, and Wernher von Braun (launch vehicles), Harry Goett of the Goddard Space Flight Center (satellites in Earth orbit) and W.H. Pickering (deep-space missions). At its inaugural meeting on 10 February 1960 it discussed an internal review sent to Ostrander on 15 January that warned of potential difficulties in the procurement of the Agena B, and how this might affect Ranger. It was decided that Silverstein's technical assistant, William A. Fleming, should chair a steering committee. This Agena B Coordination Board was formed on 19 February, and drew its membership from von Braun's team in Huntsville (which was in the process of transferring to NASA, and in July would become the Marshall Space Flight Center), the Goddard Space Flight Center and JPL. The Council also decided that a NASA project engineer should be assigned to the plant in Sunnyvale, California, where the Missile and Space Division of Lockheed manufactured the Agena. In Huntsville, Hans Heuter was made head the Light and Medium Vehicle Office, which was to manage procurement of the Agena B and Centaur stages, and Friedrich Duerr became its Agena Systems Manager. However, whilst von Braun's team would plan and *supervise* procurement, the fact that the Air Force did not want an independent line of authority leading to its supplier meant the Ballistic Missile Division would *implement* procurement from the contractor. The Air Force Space Systems Division in Inglewood, California, of which the Ballistic Missile Division was a part, was commanded by Major General Osmond J. Ritland. In April 1960 Major John E. Albert was assigned to assist NASA in procuring the Agena B, which meant he had responsibility for all Air Force technical matters relating to Ranger. He would work with Duerr in Huntsville for the launch vehicle and Burke at JPL for the interface between the Agena and the Ranger spacecraft. Duerr sent Robert Pace to Sunnyvale as the resident project engineer. Lockheed appointed Harold T. Luskin to work with Albert and Pace. The final contract, which was agreed only on 6 February 1961, was for nine Agena B vehicles. In terms of a production line that was making Agenas for several Air Force programs, this was a small order – and it was treated as such by the company until NASA complained. As Huntsville was responsible for NASA's launch operations, in addition to procuring the Atlas-Agena B it had to obtain ground support equipment and the systems required to track the vehicle in its ascent to orbit.[2]

In March 1960, as the Army Ballistic Missile Agency was being incorporated into NASA, Donald Ostrander's Office of Launch Vehicle Programs created the Launch Operations Directorate to manage NASA launches in Florida. In essence it was an expansion of the Army's Missile Firing Laboratory, and being based at Huntsville it answered to von Braun. Kurt H. Debus, Director of Launch Operations, was keenly aware that he was responsible for activities he could not actually control, because in reality NASA was merely a tenant at the Cape and as such was limited to monitoring the preparation and launch of the vehicles by the 6555th Aerospace Test Wing. His counterpart on the Air Force side was Major General Leighton I. Davis. That same month, the Office of Space Flight Programs set up its own office at the Cape to coordinate the on-site activities of the flight project teams.

On 1 September 1960 Richard Horner resigned from NASA. He was succeeded as Associate Administrator by Robert C. Seamans. Noting criticism that the Agena B Coordination Board had proved ineffective at resolving disputes, Seamans ordered a review. On 19 October, Albert Siepert of the Office of Business Administration submitted *A NASA Structure for Project Management*. On 19 January 1961, the day before he left office with the other Eisenhower political appointees, Keith Glennan endorsed the recommendations. In this new scheme, Silverstein's office would set budgets for flight projects, establish objectives and review progress. The Marshall Space Flight Center, reporting to Ostrander's office, would provide launch vehicles and launch operations in support of a *project manager* at a field centre. In the event of disputes, Seamans would personally decide the issue. The Agena B Coordination Board was dissolved. This revision gave JPL direct authority and responsibility for Ranger. NASA named Burke as its Ranger Project Manager, thereby giving him greater authority than he had when he was simply JPL's Ranger Spacecraft Project Manager.

When James E. Webb became NASA Administrator in February 1961, he argued that although the Air Force might procure the rockets for NASA, the agency should be wholly responsible for preparing and launching them. On 17 July the Air Force conceded that in due course NASA could install its own launch groups to supersede the 6555th Aerospace Test Wing.[3]

THE SPACECRAFT

In March 1958 Major General John B. Medaris of the Army Ballistic Missile Agency requested JPL to compute the payload that could be dispatched into deep space by a configuration of the Jupiter launch vehicle that would later be named the Juno IV.

---

[2]   Later, launch operations would be made a separate field centre.
[3]   In early 1962 the entire NASA launch organisation was restructured.

The result prompted W.H. Pickering to ask Daniel Schneiderman, head of a payload design group that included James Burke, to study preliminary concepts for a spacecraft capable of a flyby of Mars. They outlined a 3-axis stabilised vehicle that would face flat arrays of transducer cells to the Sun for power and maintain a high-gain antenna pointing at Earth. It was decided the craft would require a small rocket engine to correct a modest trajectory error inherited from the launcher. These points were accepted by a review in June. When the Juno IV was cancelled in October, JPL proposed that the Atlas-Vega be developed with the capability to dispatch 265 kg to the Moon or 200 kg to either Venus or Mars. To reduce the development costs and improve reliability, JPL decided that all deep-space missions should use a common 'bus' to provide not only the main structure but also electrical power, command and control functions, communications, 3-axis attitude control and midcourse correction. Peripheral structures would be held against the bus for launch, and be deployed in space.

When in July 1959 NASA told JPL to focus on the Moon in the short-term, the laboratory decided to stick with the bus concept so as to exploit this development to prepare for planetary missions in 1962. This made Ranger much more sophisticated than strictly required for a flight to impact on the Moon. Although recognised to be a high-risk venture in the short term, this strategy was expected to pay off in the long term. The preliminary outline was submitted to NASA on 1 August. The scientific payload for Ranger was to be specified by the Office of Space Flight Programs, and Silverstein delegated this task to Homer Newell. By the autumn, Newell's working groups and JPL had agreed on a rough priority for experiments to be carried on six Vega-launched lunar flights. The first two spacecraft, designated Block I, would not be sent towards the Moon, but would test the basic spacecraft systems in the deep-space environment by using Earth orbits with apogees significantly beyond the orbit of the Moon. The spacecraft engineers had argued for not carrying any experiments, in order to increase the mass available to provide redundancy in key systems until their reliability could be determined, but Newell insisted that such orbits provided an excellent opportunity to study particles and fields in space. The Block II spacecraft would be devoted exclusively to studying the Moon. However, then the Vega was cancelled and Silverstein directed JPL to prepare a five-flight series for the Atlas-Agena B. On 28 December 1959 Pickering proposed that the bus it had planned for the Vega-launched spacecraft be revised to suit the Agena B, and NASA concurred. Once again, the first two missions would be test flights carrying particles and fields payloads.

Daniel Schneiderman issued the design concept for Ranger on 1 February 1960. Reflecting the fact that the frame of the third stage of the Atlas-Vega was to have been hexagonal, the bus was a hexagonal disk 1.5 metres in diameter. The systems would be contained in rectangular boxes on its sides. For the test flights, a 'tower' was to be mounted on top of the hexagon with a platform to support the experiments and a fixed low-gain antenna on the tip. It would not, however, have the midcourse engine. On the underside would be a wider hexagonal frame scaled to mate with the Agena, and this would have a pair of solar panels that would be held against the bus for launch and hinged out in space, and a 1.2-metre-diameter high-gain antenna dish

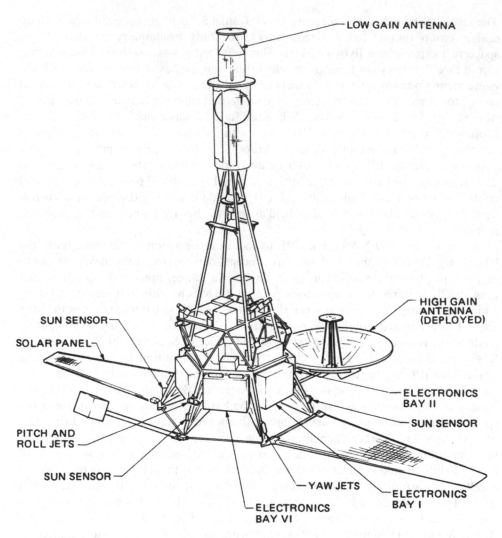

LOW GAIN ANTENNA

HIGH GAIN
ANTENNA
(DEPLOYED)

SUN SENSOR

SOLAR PANEL

ELECTRONICS
BAY II

SUN SENSOR

PITCH AND
ROLL JETS

SUN SENSOR

YAW JETS

ELECTRONICS
BAY I

ELECTRONICS
BAY VI

Details of the Block I Ranger spacecraft.

that would be stowed directly beneath the bus for launch and gimballed out in space. The Systems Division at JPL completed the design of the Block I in May 1960, and Burke froze it. The design was then split up in terms of functions and the individual tasks assigned to the various divisions. These missions were to provide data on flight performance in space that could not be gained by testing on the ground – at least not with the facilities then in existence.

The Block II spacecraft would have the same hexagonal frame, solar panels and high-gain antenna, but include the midcourse engine and a structure to accommodate a rough landing package. The operational flights were to activate a TV camera and a gamma-ray spectrometer during the terminal approach and,

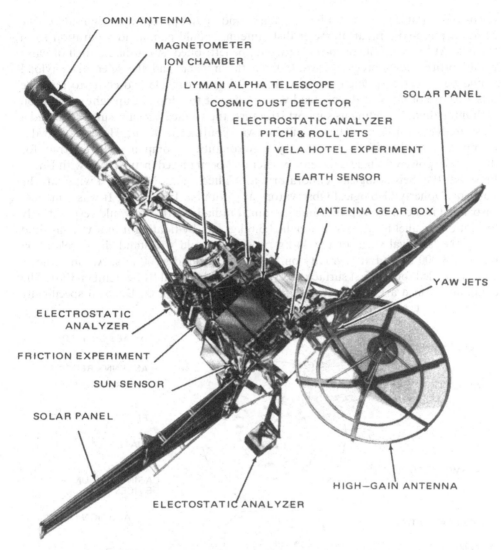

OMNI ANTENNA
MAGNETOMETER
ION CHAMBER
LYMAN ALPHA TELESCOPE
COSMIC DUST DETECTOR
SOLAR PANEL
ELECTROSTATIC ANALYZER
PITCH & ROLL JETS
VELA HOTEL EXPERIMENT
EARTH SENSOR
ANTENNA GEAR BOX
YAW JETS
ELECTROSTATIC ANALYZER
FRICTION EXPERIMENT
SUN SENSOR
SOLAR PANEL
HIGH—GAIN ANTENNA
ELECTOSTATIC ANALYZER

The configuration of the Block I Ranger spacecraft.

shortly prior to impact, eject the landing package that contained a single-axis seismometer. Conceived at a time when America's only experience in deep space was the Pioneer probes, this was an extremely ambitious design.

JPL contracted out the development of the surface package. It was an *experiment*, and as such represented a distraction to the spacecraft engineers since it would not directly contribute to the exploitation of the bus for planetary missions. In February 1960 JPL issued three competitive contracts. It received the proposals on 15 April, and on 25 April selected the one submitted by the Aeronutronic Division of the Ford Motor Company. The design mounted the spherical capsule above a solid-propellant retro-rocket (later revised to a liquid engine) and at an appropriate height above the Moon a pulse-type radar altimeter dish antenna would command separation from the bus. At burnout, the retro-rocket would be jettisoned. A crushable shell of balsa would protect the fibreglass capsule from the impact, and the scientific payload would be immersed in high-viscosity fluid. Once the capsule had come to rest, the offset centre of mass of the payload would allow it to adopt an upright orientation within the fluid. Including its support structure, the surface package subsystem had a mass budget of 136 kg. The propulsion was contracted to the Hercules Powder Company, and the altimeter to the Ryan Aeronautical Company. The contract for the battery powered single-axis seismometer had been placed in July 1959 with Frank Press at the Seismological Laboratory at Caltech and Maurice Ewing at the Lamont–Doherty Geological Observatory at Columbia University. It was a magnet suspended in a coil by a spring, and restrained radially so that it would respond only to motion parallel to its axis. It weighed 3.6 kg, and by floating it in a viscous fluid inside the 'survival shell' of the 44-kg capsule it could withstand the deceleration force of 3,000 times Earth gravity on hitting the ground at 60 m/s. Aeronutronic's contract called for the first surface subsystem to be delivered by September 1961. The company formed a Lunar Systems group headed by Frank G. Denison specifically for the project, and he was to report to Burke at JPL.

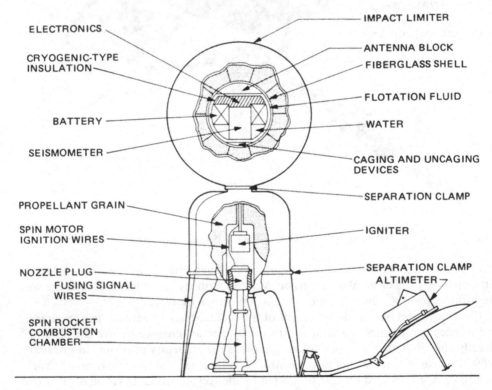

Details of the surface package subsystem of the Block II Ranger spacecraft.

The Ranger spacecraft would require 100 to 150 watts of power when fully active, which was to be generated by a pair of solar panels. Attitude determination would be by photocells and gyroscopes. Attitude control would be by thrusters squirting cold nitrogen gas. After being released by the Agena, the spacecraft would deploy its solar panels and orient itself to face its longitudinal axis towards the Sun for power generation, then aim its high-gain antenna at Earth. Later, it would have to adopt the attitude required for the midcourse manoeuvre, then re-establish its cruise attitude. This burn would be made by a hydrazine monopropellant rocket engine. Because the engine was installed beneath the hexagonal framework, it could not be fired until the high-gain antenna had swung clear. The autopilot would hold the vehicle's attitude during the manoeuvre using vanes in the rocket exhaust. For the Block I, a simple 'alarm clock' would prompt a sequencer to perform the various actions of deploying the solar panels, locking onto the Sun and deploying the high-gain antenna. For the more demanding Block II, a 'computer' would drive the sequencer – but the attitude for the midcourse manoeuvre and the magnitude of the burn would be specified by Earth. The low-gain antenna was for use after the craft was released by the Agena and before it adopted cruise attitude, and later for the midcourse manoeuvre during which the high-gain antenna would not attempt to hold its 'lock' on Earth. The high-gain antenna would send telemetry while cruising, and scientific data (including TV) during the terminal approach. On the Block II, a boom would swing the low-gain antenna away from its initial position above the surface package shortly prior to the terminal approach – by which time the high-gain antenna would have locked on – in order to clear the way for the separation of the surface package, which would occur when the radar altimeter indicated that the slant range had reduced to 24 km.

The Vega-launched Ranger had been allocated a launch mass of 364 kg in order to match the predicted performance of the three-stage Atlas-Vega. NASA's decision to switch to the Agena B for deep-space missions was based on the expectation that the performance of the Atlas-Agena B would almost match this. But whereas the Air Force's Agena payloads operated in Earth orbit, NASA wanted the Agena to remain in 'parking orbit' only briefly before reigniting its engine to head for deep space. On 11 July 1960 Lockheed announced a cut in its estimate of the Agena B's capability in this role by 34 kg to 330 kg. This did not affect the 307-kg Block I, but the design of the Block II had started out as 364 kg – including the 136-kg surface package subsystem. The issue of the Block II mass remained in doubt through the remainder of the year, with the Agena B Coordinating Board failing to achieve a resolution. Meanwhile, James Burke told Harris Schurmeier's Space Division to do all it could to lighten the Block II, short of deleting equipment. But by the end of 1960 the mass of the Block II design still exceeded 330 kg. Worse, Aeronutronic warned that their subsystem may well exceed its mass limit. On 14 December the Space Technology Laboratories were asked to re-evaluate the capacity of the Agena B by combining Lockheed's data for launch to translunar injection and JPL's data for the remainder of the mission. The calculation would take several months to perform.

In early 1961, awaiting this report, Burke faced a dilemma. To match the overall project timescale of 36 months specified by Silverstein, the schedule called for the test

A model of the Block II Ranger spacecraft, with the boom swinging the low-gain antenna off its axial position above the surface package subsystem.

flights in July and October 1961, and the lunar flights in January, April and July 1962. To achieve this, he would have to commit the Block II to a mass heavier than the currently projected capability of the Atlas-Agena B. Alternatively, he could let the schedule slip until he had a definitive mass figure. Part of NASA's rationale for Ranger was to steal a march on the Soviets. However, on 12 February 1961 they dispatched a 454-kg spacecraft on an interplanetary trajectory towards Venus. It was 3-axis stabilised, had solar panels and a high-gain antenna, and utilised the parking orbit technique rather than direct ascent – all of which JPL had hoped Ranger would pioneer.[4] Under pressure, on 16 February Burke told Schurmeier that the Systems Division "must begin removing items" from the Block II in order to lighten it. He also told Aeronutronic that if their subsystem exceeded its allotted mass it would not be carried.[5] The deletions that Burke specified would cut the mass of the Block II to 332 kg, which was still marginally greater than Lockheed's revised estimate. The design of the Block II was frozen in April 1961 at this mass. Ironically, in late May 1961 the Space Technology Laboratories reported that the Agena B would be able to place 382 kg on the Block II trajectory, which was even more than the Vega figure! Rather than introduce a further delay by trying to exploit this situation, Burke let the development proceed. If the actual capability of the Atlas-Agena B in this role had been known when JPL drew up the specifications for the Block II in early 1960, a significant degree of redundancy would have been built into the spacecraft. In fact, the mass-saving exercise had deleted those redundant systems that existed, making it even more technologically risky than initially envisaged. In effect, the design of the Block II was predicated on the assumption that all of its systems would work, which in turn put it at risk of loss if a single system were to fail.

In January 1958 Joshua Lederberg, a geneticist at the University of Wisconsin, warned the National Academy of Sciences that if a spacecraft transported terrestrial microbes to another body of the solar system, this would preclude a later experiment to determine whether life independently originated there. The International Council of Scientific Unions established an ad hoc committee to consider the issue, and then urged all nations to sterilise their spacecraft. On 15 October 1959, with the approval of Keith Glennan, Abe Silverstein directed that any NASA "payloads which might impact a celestial body must be sterilised before launching".

The case for such a precaution on planetary missions was self-evident, because at that time the seasonal variation of the dark areas on Mars was widely believed to be due to vegetation, and although the atmosphere of Venus was permanently cloudy, one idea was that conditions on the surface might resemble the carboniferous period of Earth's past. As regards the Moon, one theory (not widely held, but impossible to

---

[4] The Soviet spacecraft fell silent on 27 February 1961, at a distance of 23 million km from Earth. A launch on 4 February had stranded a similar spacecraft in parking orbit, but its role was disguised by naming it Sputnik 7.

[5] Surface science was only one of the objectives; there were the investigations to be made during the terminal approach, and achieving these would mark an acceptable compromise on the first mission.

falsify with the data available) posited there might be water ice at a shallow depth beneath the surface, and this environment might be conducive to microbial life that would not be able to survive on the surface. It was therefore decided that spacecraft destined for the Moon must be sterilised.

JPL soon decided that NASA's recommendation of immersing the spacecraft in lethal gaseous ethylene oxide would have to be augmented by 'dry heat' treatment. In April 1960 George L. Hobby, a research biologist in the Space Sciences Division, was assigned to work with James Burke to draw up sterilisation procedures for the Block II spacecraft. The first task was to define the term 'sterile' – which basically was an issue of deciding how efficient the process had to be. Ideally all components and subassemblies would be subjected to a temperature of 125°C for a period of 24 hours, but this proved impractical owing to its deleterious effects on the electronics, and one by one waivers were issued to protect particularly sensitive items. Once the spacecraft had been fully assembled (using the sterilised subassemblies) it would be thoroughly cleaned with alcohol. It would be transported to the Cape in a controlled environment. After the spacecraft had been installed on its launcher and passed its final checks, the aerodynamic shroud would be sealed and gaseous ethylene oxide pumped in for a time to complete the sterilisation process. On 26 June 1961 Robert Seamans approved these procedures.

As a matter of policy, the choice of experiments for space missions was made by Homer Newell's Space Sciences Steering Committee. Abe Silverstein retained the final approval. Experimenters therefore submitted proposals to NASA headquarters, and, if successful, worked with a field centre to implement the experiment. In the case of JPL, this was the Space Sciences Division. While James Burke concentrated on the spacecraft, the launch vehicle and its support systems, Albert Hibbs prepared the scientific experiments to be carried. The particles and fields experiments did not require much development because they were by now being flown widely, but a lot of care was required to integrate them into the spacecraft. The final list of scientific experiments for the Block I was a solar plasma detector, a magnetometer, a trapped-radiation detector, an ion chamber, a cosmic-ray telescope, a Lyman-alpha detector, a micrometeoroid detector and the Vela Hotel package. The Vela Hotel provided by the Atomic Energy Commission was a late addition which Silverstein approved on 29 June 1960. A network of satellites were to carry X-ray and gamma-ray sensors to detect above-ground nuclear tests. But if the Sun issued microsecond-duration bursts of X-rays, then these might cause the satellites to report false detections. The highly elliptical orbits of the Ranger test flights would have their apogees above the van Allen belts, and the 3-axis-stabilised spacecraft would enable the Vela Hotel sensor to 'stare' at the Sun to determine whether it produced such emissions.

The in-flight experiments for Block II consisted of a TV camera and a gamma-ray spectrometer. The TV camera was to be activated by command from Earth when the spacecraft was within 4,000 km of the Moon. The Astro-Electronic Division of the Radio Corporation of America in Hightstown, New Jersey, which had supplied the TV system for the Tiros meteorological satellites, was hired to provide a slow-scan vidicon imaging tube and related electronics. The Space Sciences Division at JPL built the optical element, which was essentially a telescope with a focal length of 1 metre and an aperture ratio of f/6. The image scanned off the vidicon tube would have 200 'lines'. The plan was for the camera to provide about 100 pictures during the terminal approach, with the transmission terminating when the separation of the surface package perturbed the attitude of the bus and the high-gain antenna lost its lock. In excellent 'seeing', the best telescopes had a lunar surface resolution of about 300 metres. An image taken by this camera at an altitude of 50 km was expected to provide a resolution 100 times better. On 16 October 1961 Newell's Space Sciences Steering Committee named Gerard Kuiper, Gene Shoemaker and Harold Urey as the experimenters who would receive and interpret the pictures transmitted by this camera.[6]

The gamma-ray spectrometer experiment was led by James R. Arnold, a chemist at the University of California at San Diego. Its scintillation counter was to detect the natural radioactivity originating from the uppermost layer of the lunar surface. In particular, it would detect gamma rays issued by the decay of uranium, thorium and potassium. If these large-ion lithophile elements were widespread, this would imply that the interior of the Moon had undergone significant thermal differentiation. The instrument was mounted on an 18-metre-long boom that would be deployed after the midcourse manoeuvre in order to determine the 'background' from the spacecraft itself and from celestial sources. It would operate until the high-gain antenna lost its lock. In fact though, such a study would be better done from a polar orbit in order to obtain global coverage.

On 27 March 1961 Walter E. Brown, head of the Data Automation Systems Group of JPL's Space Sciences Division, pointed out that if the signal from the radar altimeter were to be telemetered to Earth, the radar echo could be correlated with the imagery to gain insight into the density, conductivity and thickness of the material at the surface – in particular its dustiness. This information would be of use to the team planning the soft lander. On 28 April, after it was established that this modification would not adversely affect operations, James Burke authorised the modification. The Space Sciences Steering Committee in Washington duly designated Brown as the investigator for the radar reflectivity experiment.

DEEP-SPACE TRACKING

When the Advanced Research Projects Agency decided in 1958 that a series of probes should be launched towards the Moon, Eberhardt Rechtin led a team at JPL in the development of radio tracking, telemetry and command facilities. When W.H. Pickering created the Telecommunications Division, he made Rechtin its chief. JPL recognised early on that it would need a world-wide network of antennas to maintain contact with deep-space missions. In late 1958 NASA approved the proposal by JPL to develop the Deep Space Instrumentation Facility. To oversee this activity, Abe

---

[6] Note that Kuiper, Shoemaker and Urey did not actually *originate* the experiment.

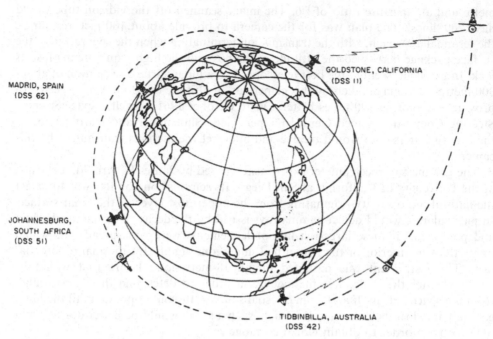

MADRID, SPAIN
(DSS 62)

GOLDSTONE, CALIFORNIA
(DSS II)

JOHANNESBURG,
SOUTH AFRICA
(DSS 51)

TIDBINBILLA, AUSTRALIA
(DSS 42)

The Deep Space Instrumentation Facility had large antennas in California, Spain, South Africa and Australia to provide continuous communication with lunar and interplanetary missions.

Silverstein appointed his assistant for Space Flight Operations, Edmond C. Buckley, who had experience of tracking and instrumentation at the rocket range on Wallops Island.[7]

As the Ranger project geared up in 1960, Rechtin was given the additional role of Program Director for the Deep Space Instrumentation Facility. It was decided to build three stations located approximately 120 degrees apart in longitude to provide a continuous line of sight to a spacecraft in deep space. The main station was built near the Goldstone Dry Lake in the Mojave Desert of California, 160 km east of JPL and on the far side of a range of mountains which would shield the antenna from the 'noise' of the coastal cities. The other stations were at the Woomera Test Range in Australia and near Johannesburg in South Africa. Later, a fourth station was added near Madrid in Spain.

Because the antennas had to be both large and fully steerable, it was decided to adapt a radio-telescope design. Although the dish was mounted like a telescope, the steering system was designed to hold the antenna pointing precisely at a spacecraft travelling against the background of stars, rather than to maintain sidereal rate. This

[7] In November 1961, James Webb introduced the Office of Tracking and Data Acquisition at NASA headquarters and made Buckley its Director.

was to be done by having the antenna lock onto the spacecraft's radio transmission and maximise the received signal strength. A 26-metre-diameter dish was required to track the 3-watt transmitter of the Ranger spacecraft. In addition, a system was installed to enable the antenna to simultaneously send 'uplink' at one frequency and to receive 'downlink' at another. This allowed not only the position of the spacecraft in the sky to be determined, but also both its range and radial velocity along the line of sight. This data would enable the vehicle's location and motion in space, together known as its state vector, to be monitored continuously in real-time.

Planning for the Space Flight Operations Centre at JPL began in May 1960 and was finished in November 1960. As it fell within the remit of the Systems Division, Harris Schurmeier appointed Marshall S. Johnson to supervise its construction in Building 125 of the campus. The Space Flight Operations Centre, together with the terrestrial communications network (initially by voice lines and teletype) to link it to the Deep Space Instrumentation Facility, were declared operational on 4 July 1961.

In February 1961 W.H. Pickering ordered a study of future requirements for flight operations, and the recommendation was to construct a new building specifically for this role. In July, NASA gave the go-ahead. This Space Flight Operations Facility entered service in the summer of 1964. Meanwhile, on 24 December 1963 the Deep Space Instrumentation Facility, the terrestrial communications network and the JPL control centre were integrated under the umbrella of the Deep Space Network.[8]

---

[8]   Note that the Deep Space Network *comprised* the Deep Space Instrumentation Facility and Space Flight Operations Facility, it did not supersede them.

5

The Apollo commitment

THE ROLE OF MAN IN SPACE

In 1958 NASA was assigned the task of exploring space for scientific purposes, but no immediate objectives were specified. The National Aeronautics and Space Act left the agency to set its own goals.

Accordingly, within days Keith Glennan established the Space Task Group at the Langley Research Center to manage Project Mercury, which was to launch a man into orbit as soon as possible. This was not exactly what President Eisenhower had had in mind, but he saw it as a one-off venture. At a packed press conference on 9 April 1959, Glennan introduced the seven military test pilots who had been chosen to be astronauts.[1]

On 25–26 May 1959 Harry J. Goett of the Ames Research Center chaired the first meeting of the Research Steering Committee on Manned Space Flight, which was to consider possible man-in-space objectives for the coming decade. These included:

- launching and operating a small orbital laboratory
- assembling a large permanent space station
- flying circumlunar and lunar orbital missions
- making a lunar landing.

George M. Low, who represented headquarters, was firmly of the view that *only* a manned lunar landing provided a reasonable ultimate objective – it was an 'end', not just an intermediate step along a path. Whilst less demanding, the alternatives were not as definitive – either a man landed on the Moon or he did not; there was no way

---

[1] They were Lieutenant Commander Alan Bartlett Shepard Jr, Lieutenant Malcolm Scott Carpenter and Lieutenant Commander Walter Marty Schirra Jr from the Navy; Lieutenant Colonel John Herschel Glenn Jr from the Marines; and Captain Virgil Ivan 'Gus' Grissom, Captain Donald Kent 'Deke' Slayton and Captain Leroy Gordon Cooper Jr from the Air Force.

to redefine it as something simpler and assert this to be equivalent.[2] It was therefore decided to set the *long-range* goal of achieving a manned lunar landing in the 1970s. This would gave a framework in which to define a series of intermediate objectives for the 1960s.

On 5 July 1960 the House Committee on Science and Astronautics said, "NASA's 10-year program is a good program as far as it goes, but it does not go far enough." In other words, it urged NASA to accelerate its long-range plan. In the committee's view, "A high priority program should be undertaken to place a manned expedition on the Moon in this decade. A firm plan with this goal in view should be drawn up and submitted to the Congress by NASA." But it warned that this plan, "should be completely integrated with other goals, to minimise total costs. The modular concept deserves close study. Particular attention should be paid immediately to long lead-time phases of such a program."

Eisenhower responded by asking his science advisor, James R. Killian, whether a manned lunar landing represented a scientific venture which could be justified in the same manner as launching a satellite for the International Geophysical Year. Killian convened a meeting of scientists, and their report, written by Donald F. Hornig of Princeton University, was dismissive: "At the present time, man-in-space cannot be justified on purely scientific grounds." The rationale for sending men to the Moon seemed to be "emotional compulsion and national aspirations". Hence Eisenhower refused funding for manned space flight beyond Project Mercury, and in particular the proposal for a three-man Apollo spacecraft. He had no problem with the agency using the new Saturn booster to launch heavy satellites, but he withdrew funding for the upper stages intended to enable this to launch a manned spacecraft. Nevertheless, on 12–13 September the Space Task Group held a briefing for potential bidders to develop the Apollo spacecraft with the Moon as the ultimate objective, and released the formal request for proposals. On 17 October, Low told Abe Silverstein, Director of the Office of Space Flight Programs, that he was going to set up a committee to study the circumlunar objective in greater detail, to ensure that the Apollo spacecraft would be capable of supporting a landing mission.[3] On 25 October NASA issued contracts to three companies to provide feasibility studies for the Apollo spacecraft.

The national election of November 1960 was won by John F. Kennedy. He was inaugurated on 20 January 1961. In giving his final budget speech prior to leaving office, Eisenhower said on 18 January 1961 that Congress would have to determine "whether there are any vital scientific reasons for extending manned space flight beyond Mercury". In a campaign statement, Kennedy had said: "We're in a strategic space race with the Russians, and we are losing. If a man orbits Earth this year, his name will be Ivan. If the Soviets control space they can control the Earth, as in past centuries the nation that controlled the seas has dominated the continents. We

---

[2]   This reasoning would resurface when John F. Kennedy asked for a worthy challenge.
[3]   In a reorganisation on 8 December 1959, the Office of Space Flight Development had become the Office of Space Flight Programs.

cannot afford to run second in this vital race. To insure peace and freedom we must be first. Space is our great New Frontier." The contrast with Eisenhower's view was stark. Kennedy also had an appreciation of national prestige, which in the Cold War meant a comparison with the achievements of the Soviet Union. The issue of prestige had been dismissed by Eisenhower.

During the transitional period, Kennedy assigned a number of task forces to draw up policy recommendations. The Committee on Space was chaired by Jerome B. Wiesner, who served on the President's Science Advisory Committee during James Killian's chairmanship and was to become Kennedy's Special Assistant for Science and Technology. The Committee on Space in turn set up the Panel on Man-in-Space, composed largely of scientists, and its report on 18 November 1960 criticised the program envisaged by NASA.[4] Although it agreed the need for large launch vehicles and urged an emphasis on space science and applications, it criticised "the popular belief that man in space is the most important aim of our non-military space effort". Wiesner recommended that Project Mercury be ended as soon as it had achieved its objective of placing a man in orbit, and that there should be no follow-on. However, Kennedy had made Vice President Lyndon B. Johnson chairman of the National Aeronautics and Space Council, and Johnson was in favour of expanding the space program.

And when the Space Science Board of the National Academy of Sciences issued a position paper on *Man's Role in the National Space Program* on 27 March 1961 it said, "scientific exploration of the Moon and planets should be clearly stated as the ultimate objective of the US space program for the foreseeable future. This objective should be promptly adopted as the official goal of the United States space program and clearly announced, discussed and supported." It also advised that whilst it was "not now possible to decide whether man will be able to accompany expeditions to the Moon and planets", NASA should proceed with its planning "on the premise that man will be included". Taking the broader view, it said that such exploration would be "potentially the greatest inspirational venture of the century and one in which the entire world can share; inherent here are great and fundamental philosophical and spiritual values which find a response in man's questing spirit". Clearly this national scientific body, established to advise NASA on policy, was taking a much broader view than the sky scientists involved in space research at that time.

KENNEDY'S CHALLENGE

NASA was well placed to exploit the new administration's willingness to expand the space program. Its long-term planning was impressive for its detail, in particular

---

[4]   This report was excerpted in the *New York Times* on 12 January 1961, and is sometimes wrongly dated as such.

because George Low's committee had costed the accelerated plan – concluding that it would require $7 billion to land a man on the Moon by the end of the decade. In January 1961 Low briefed Keith Glennan on the forthcoming hearings for NASA's budget, but Glennan expected to be replaced by the new administration and so was in a weak position.

On Johnson becoming Kennedy's Vice President, Robert S. Kerr took over from him the chairmanship of the Senate Committee on Aeronautical and Space Sciences. After consulting Kerr, Johnson recommended James E. Webb to succeed Glennan as NASA administrator, and Webb took over on 14 February. Whereas Glennan was a scientific administrator with a conservative outlook, Webb was a political operator. He had served as Director of the Bureau of the Budget between 1946 and 1949 and Undersecretary of State from then until 1952 in the Truman administration. He had been a director of Kerr's oil and uranium conglomerate, Kerr–McGee Oil Industries, and simultaneously a director of the McDonnell Aircraft Company.

Webb immediately set out to obtain the funding that was earlier denied for Apollo and the Saturn launch vehicle. When the Bureau of Budget refused, Webb wrote to Kennedy in early March that Eisenhower had "emasculated the 10-Year Plan before it was one year old", and if the funding were not made available it would "guarantee that the Russians will, for the next five to ten years, beat us to every exploratory space flight". To ram home the message in terms that Kennedy would appreciate, Webb said, "We have already felt the effects of the fact that they were the first to place a satellite into orbit, have intercepted the Moon, photographed the back side of the Moon, and have sent a large spacecraft to Venus. They can now orbit seven and a half ton vehicles about the Earth, compared to our two and a half tons, and they have successfully recovered animals from flights of as much as 24 hours. Their present position is one from which further substantial accomplishments can be expected, and our best information points to a steadily increasing pace of successful effort on a realistic timetable."

On 23 March Kennedy met with Lyndon Johnson, Jerome Wiesner, David Bell of the Bureau of Budget and Edward C. Welsh, a former aide to Johnson who was now serving as Executive Director of the National Aeronautics and Space Council, of which Johnson was chairman. Kennedy agreed to increase funding for the Saturn launch vehicle, but said he would need to deliberate further on the Apollo spacecraft – he would decide in the autumn, he said.

Just when NASA began to think that it might beat the Soviets to a manned space flight, on 12 April 1961 Yuri Alexseyevich Gagarin made a single orbit and landed safely. Webb told Congress, in budget hearings then underway, that NASA could certainly work faster if its funding was increased.

The next evening Kennedy met at the White House with Jerome Wiesner, David Bell, James Webb, Hugh Dryden, Theodore Sorensen, who was a friend and advisor, and Hugh Sidey, a journalist for *Life* magazine who was one of Kennedy's friends, and put to them the question, "at what point we can overtake the Russians". NASA opened with a space station to be assembled in Earth orbit to serve as a jumping off point for a future mission to the Moon. But, it pointed out, if the Soviets were on the same plan they would likely remain in the lead for some considerable time. Kennedy

Feature Index

The Huntsville Times

Where Progress... Covers The Valley!

HUNTSVILLE, ALABAMA, WEDNESDAY, APR. 12, 1961

Man Enters Space

'So Close, Yet So Far,' Sighs Cape

U. S. Had Hoped For Own Launch

Hobbs Admits 1944 Slaying

Soviet Officer Orbits Globe In 5-Ton Ship

Maximum Height Reached Reported As 188 Miles

VON BRAUN'S REACTION:

'To Keep Up, U. S. A. Must Run Like Hell'

Praise Is Heaped On Major Gagarin

First Man To Enter Space Is 27, Married, Father Of Two

'Worker' Stands By Story

Reds Deny Spacemen Have Died

No Astronaut Signal Received At Ft. Monmouth

Reds Win Running Lead In Race To Control Space

Today's Chuckle

The *Huntsville Times* reports the first man in space.

wanted to minimise this period, either by accelerating or by short circuiting the plan. Dryden said a 'crash' program might land a man on the Moon ahead of the Soviets, but it might cost as much as $40 billion. "The cost! That's what gets me," Kennedy mused. "When we know more, I can decide if it's worth it or not. If somebody can just tell me how to catch up." As the meeting broke up, Sorensen remained behind to discuss what had been said, and upon emerging told the others, "We're going to the Moon!"

On 19 April Kennedy summoned Johnson and told him he had decided to issue a momentous challenge. The next day, Kennedy sent a memo to Johnson seeking "an overall survey of where we stand in space". Specifically:

1.   Do we have a chance of beating the Soviets by putting a laboratory in space, or by a trip around the moon, or by a rocket to land on the moon, or by a rocket to go to the moon and back with a man. Is there any other space program which promises dramatic results in which we could win?
2.   How much additional would it cost?
3.   Are we working 24 hours a day on existing programs. If not, why not? If not, will you make recommendations to me as to how work can be speeded up.
4.   In building large boosters should we put [our] emphasis on nuclear, chemical or liquid fuel, or a combination of these three?
5.   Are we making maximum effort? Are we achieving necessary results?

On 21 April Kennedy told reporters that his administration was considering the options and cost of space, and said, "If we can get to the Moon before the Russians, we should."

Johnson consulted NASA first, which said there was little chance of beating the Russians to a space station; it might be possible to beat them to lunar orbit; the best bet was a lunar landing. This matched Johnson's thinking. NASA suggested 1967 as a target date because it was expected that the Soviets would attempt to make a lunar landing then in order to mark the 50th anniversary of the Bolshevik Revolution. As a result of the additional analysis by Low, the costing had been increased from the $7 billion estimate for a landing in 1969 to $22 billion; but a landing in 1967 would be $34 billion. Next, Johnson consulted the Pentagon, and the Air Force agreed that a manned lunar landing would be appropriate – even although the Air Force would not be allowed to perform it. Finally, Johnson consulted three businessmen whose judgement he trusted: Frank Stanton of the Columbia Broadcasting System; Donald Cook of the American Electric Power Service Corporation; and George Brown of Brown and Root, which was a construction company in Texas. The fact that none of them was involved in the aerospace industry that would be called upon to build the hardware for the program was a point in their favour, since it meant they were unbiased. At the National Aeronautics and Space Council on 24 April, Johnson, as Wiesner later described it, "went around the room saying, 'We've got a terribly important decision to make. Shall we put a man on the Moon?' And everybody said 'yes'. And he said 'Thank you'."

The scientific community was represented in the White House by Wiesner. The majority of space scientists were interested in particles and fields, and because this

C
O
P
Y

April 20, 1961

MEMORANDUM FOR THE VICE PRESIDENT

In accordance with our conversation I would like for you as Chairman of the Space Council to be in charge of making an overall survey of where we stand in space.

1. Do we have a chance of beating the Soviets by putting a laboratory in space, or by a trip around the moon, or by a rocket to land on the moon, or by a rocket to go to the moon and back with a man. Is there any other space program which promises dramatic results in which we could win?

2. How much additional would it cost?

3. Are we working 24 hours a day on existing programs. If not, why not? If not, will you make recommendations to me as to how work can be speeded up.

4. In building large boosters should we put out emphasis on nuclear, chemical or liquid fuel, or a combination of these three?

5. Are we making maximum effort? Are we achieving necessary results?

I have asked Jim Webb, Dr. Wiesner, Secretary McNamara and other responsible officials to cooperate with you fully. I would appreciate a report on this at the earliest possible moment.

/s/ John F. Kennedy

The historic memo to Lyndon B. Johnson which led John F. Kennedy to challenge his nation to land a man on the Moon before the decade was out.

research did not require a human presence, money spent on sending men into space was by definition wasted. But Kennedy wanted "dramatic results" and the scientists were unable to offer this. To be fair, Kennedy invited Wiesner to suggest a terrestrial challenge that would serve the purpose, "... something with an overseas impact, like desalination or feeding the hungry". However, Wiesner could see that the Moon was shaping up to be the challenge, and advised the President "never to refer publicly to the Moon landing as a scientific enterprise".

On 28 April Johnson submitted the National Aeronautics and Space Council's recommendation:

> Largely due to their concerted efforts and their earlier emphasis upon the development of large rocket engines, the Soviets are ahead of the United States in world prestige attained through impressive technological accomplishments in space. The US has greater resources than the USSR, etc. The country should be realistic and recognize that other nations, regardless of their appreciation of our idealistic values, will tend to align themselves with a country which they believe will be the world leader. The US can, if it will firm up its objectives and employ its resources, have a reasonable chance of attaining world leadership in space. If we don't make a strong effort now, the time will soon be reached when the margin of control over space and other men's minds through space accomplishment will have swung so far on the Russian side that we will not be able to catch up. Even in those areas in which the Soviets already have the capability to be first and are likely to improve upon such capability, the United States should make aggressive efforts, as the technological gains as well as the international rewards are essential steps in gaining leadership. Manned exploration of the Moon, for example, is not only an achievement with great national propaganda value, but is essential as an objective, whether or not we are first in its accomplishment – and we may be able to be first.

Kennedy was receptive to Johnson's recommendation, but he postponed a formal decision until after the first manned Mercury mission, which came on Friday, 5 May 1961 when Al Shepard rode a Redstone missile on a suborbital arc.[5]

Over the weekend, Johnson met James Webb and Secretary of Defense Robert S. McNamara to draw up a formal recommendation to Kennedy's memo of 20 April. *Recommendations for our National Space Program: Changes, Policies and Goals*, jointly authored by Webb and McNamara, said, "It is man, not merely machines, in space that captures the imagination of the world. All large-scale projects require the mobilization of resources on a national scale. They require the development and successful application of the most advanced technologies. Dramatic achievements in space, therefore, symbolize the technological power and organizing capacity of a nation. It is for reasons such as these that major achievements in space contribute to

[5]   In fact, NASA could have launched Shepard several weeks ahead of Gagarin's flight. If this had been done, Kennedy may well not have issued the challenge of landing a man on the Moon before the decade was out. The fact that Shepard's flight had been only suborbital whereas Gagarin's was orbital, would probably not have mattered, since the world's first 'spaceman' would have been an American. The fact that America 'lost' *both* the first satellite and the first man into space could be said to be directly responsible for the race to the Moon. It serves to illustrate that history is not an irresistible tide, it can be extremely sensitive to the outcome of singular events.

On 5 May 1961 a Redstone rocket lifts off with a Mercury capsule containing Alan B. Shepard for a suborbital mission.

After his successful Mercury flight, Alan B. Shepard shakes hands with John F. Kennedy at the White House.

In a speech to Congress on 25 May 1961 John F. Kennedy challenged his nation to land a man on the Moon before the decade was out.

national prestige." They wrote, "even though the scientific, commercial or military value of [such an] undertaking may by ordinary standards be marginal or economically unjustified", it nevertheless generated "national prestige", which had value in its own right. Furthermore, "The non-military, non-commercial, non-scientific but 'civilian' projects such as lunar and planetary exploration are, in this sense, part of the battle along the fluid front of the Cold War." This echoed Kennedy's criticism of Eisenhower: whereas Eisenhower had been conscious of the cost and dismissive of national prestige, to Kennedy national prestige was the issue and the cost was secondary.

On 25 May Kennedy gave a speech to a joint session of Congress on the theme of *Urgent National Needs*. In view of recent space achievements by the Soviets, he proclaimed, "Now it is time to take longer strides, time for a great new American enterprise, time for this nation to take a clearly leading role in space achievement, which in many ways may hold the key to our future on Earth." Having outlined the political background, he laid down the gauntlet. "I believe that this nation should commit itself to achieving the goal, before this decade is out, of landing a man on the Moon, and returning him, safely, to the Earth." He had opted for a lunar landing precisely because it posed a great technical challenge. By literally 'shooting for the Moon', he was betting that America would not only catch up with the Soviet Union in space, but forge ahead. Having concluded that space was the arena of superpower politics, he was challenging his rival, Nikita Khrushchev, for world leadership. He had imposed the deadline to ensure that reaching the Moon was perceived as a race. He was also well aware of the magnitude of the task. "No single space project in this period will be more impressive to mankind, or more important for the long-range exploration of space; and none will be so difficult or expensive to accomplish." The sending of a man to the Moon was to be the modern form of the ancient practice of 'single combat', whereby opposing armies lined up and each dispatched a single warrior to decide the issue. To indicate that it was a matter of national honour, he added, "In a very real sense, it will not be one man going to the Moon; if we make this judgment affirmatively it will be an entire nation, for all of us must work to put him there." And in order to emphasise what was at stake, he warned, "If we are to go only halfway, or reduce our sights in the face of difficulty, in my judgment it would be better not to go at all."

For Kennedy the Moon was a symbol and, in terms of what he wished to achieve it was an excellent symbol. He had the impression that the applause in Congress was "something less than enthusiastic", as he told Sorensen immediately after giving the speech. But Johnson had read the mood well: there was only minor opposition in the House of Representatives, and the debate in the Senate lasted less than an hour – only five of the 96 senators spoke, and the floor was dominated by Robert Kerr, who was Johnson's man. NASA's budget was doubled without a formal vote being taken.

RANGERS FOR APOLLO

When on 25 May 1961 President Kennedy challenged his nation to land a man on the Moon before the decade was out, the sky scientists were unimpressed but the geologists were delighted.

On 8 June Hugh Dryden advised the Senate Committee on Aeronautics and Space Sciences that NASA intended to make use of automated spacecraft to strengthen the manned lunar program. In particular, it was essential to find out whether the surface would support the weight of the Apollo lander. As Dryden put it, "We want to know something about the character of the surface on which the landing is to be made, and obtain as much information as we can before man actually gets there." Following up, Abe Silverstein provided some details. For a start, Ranger would be extended by four Block III missions. Congress authorised the funding for these missions several weeks later.

Clifford Cummings, JPL's Lunar Program Director, visited NASA on 21 June and told Edgar Cortright and Oran Nicks, the two managers in Silverstein's office who were responsible for Ranger, that the greatest single contribution this project could make to Apollo would be to provide high-resolution imagery to enable the nature of the lunar surface to be characterised to provide the information needed to design the landing gear of the Apollo lander. For this, the Block III would replace the surface package subassembly with a TV system that was more sophisticated than that made for the Block II. In the interim, some insight would be provided by the Block II radar altimeter and the accelerometers of the surface capsule as this impacted and rolled to a halt.

JPL recommended that the contract to develop the high-resolution TV system go to the same company that supplied the camera for the Block II, and this was agreed.

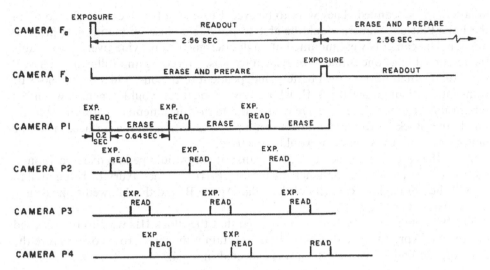

The shuttering sequence of the six cameras of the Block III Ranger spacecraft's high-resolution TV system.

On 5 July 1961 JPL discussed the design of the system with the Radio Corporation of America, and it was decided to use a shutter (which was not a standard feature on a continuous-scan TV system) to define a 'frame' on a vidicon tube. The contract was signed on 25 August. Responsibility for the design, fabrication and testing of the system was delegated to the company. Harris Schurmeier's Systems Division would monitor the work. On 31 August, Cummings appointed Allen E. Wolfe as the Ranger Spacecraft Systems Manager to assist James Burke with the increased work. Wolfe had replaced Gordon Kautz as Project Engineer in the Systems Division when Kautz was made Burke's deputy. Wolfe's first responsibility would be to steer the remaining Block II spacecraft through all phases of assembly and testing, and then supervise the development of the Block III.

The design of the high-resolution TV subsystem was finished in September 1961. It had three major assemblies: a tower superstructure incorporating a thermal shield to stand on the top of the hexagonal bus; a central box to house the main electronics; and, above, a battery of six cameras and their individual electronic systems. It used two types of camera. The 'A' type had a lens with an aperture ratio of f/1 and a focal length of 25 mm. The 'B' type had an f/2 lens with a focal length of 75 mm. There were two 'A' cameras and four 'B' cameras. The vidicons were all the same, but the entire 11-mm square image would be used for the full (F) frame and only the central 3-mm square for the partial (P) frame. One 'A' and one 'B' camera would operate a 5.12-second cycle in which the shutter fired to expose its vidicon and this was read out over an interval of 2.56 seconds, then erased over the next 2.56 seconds. They were to operate out of phase so that a frame was taken every 2.56 seconds. The other cameras would require 0.2 second to fire the shutter and perform the readout, and 0.6 second to erase. The faster cycle time for these cameras was because a smaller area was to be scanned. They were to be cycled to take a frame every 0.2 second, in the hope that one camera would be able to provide a close-up picture just prior to impact. The cameras were mounted at angles designed to provide overlap to enable the relationship of one frame to be related to those preceding and following. The TV subsystem would have its own battery, independent of the bus, and a pair of 60-watt transmitters. Unlike the Block II, whose flow of pictures would conclude when the separation of the surface package caused the high-gain antenna to lose its lock on Earth, the Block III would continue to send pictures until it hit the surface. In all, the high-resolution TV subsystem would be 160 kg.[6]

As in the case of the Block I, the low-gain antenna would be in a fixed position at the top of the tower. The designers of the Block III had the luxury of being able to exploit the full payload capacity of the Atlas-Agena B, and this allowed some degree of redundancy in the basic systems.

On 19 September 1961 NASA announced that the Block IIIs were to be launched in January, April, May and August 1963 – certainly they were to be over before the first soft-landing Surveyor, which was expected in 1964.

APOLLO UNDERWAY

On 24 September 1961 NASA announced that the Manned Spacecraft Center to be built near Houston, Texas, would supersede the Space Task Group. It would not only design, develop, evaluate and test manned spacecraft, but also train astronauts and manage mission operations. Robert R. Gilruth, head of the Space Task Group, was made Director of this new centre.

On 1 November, NASA restructured its headquarters. As part of this review, the offices of Space Flight Programs and Launch Vehicle Programs were wrapped up, and new program offices were created for Manned Space Flight, Space Sciences, and Applications. This raised Manned Space Flight to office status, as opposed to a subdivision of Space Flight Programs. The effect was to put the administration of all the agency's activities (some of which were aeronautical) on a par with the Office of Manned Space Flight, although that office had fully three-quarters of the budget. In effect, James Webb had gathered the power of decision-making into headquarters, since the directors of all the 'offices' and 'centres' would report to Robert Seamans, the Associate Administrator who, as the agency's 'general manager', would have budgetary control.

The obvious candidates to be Director of the Office of Manned Space Flight were Abe Silverstein and Wernher von Braun, but because their relationship was stormy Webb had sought an outsider, and on 21 September hired Dyer Brainerd Holmes. As general manager of the Major Defense Systems Division of the Radio Corporation

---

[6]   Specifically, the cameras were designated F_a (25-mm), F_b (76-mm), P1/P2 (76-mm) and P3/P4 (25-mm).

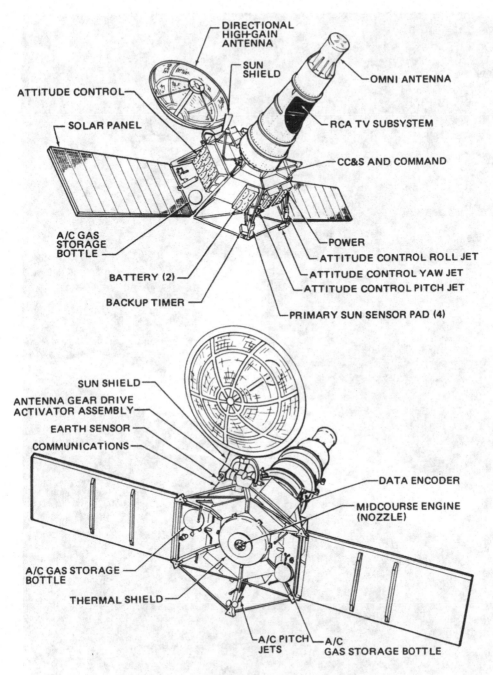

Detail of the Block III Ranger spacecraft.

A model of the Block III Ranger spacecraft.

On 20 February 1962 an Atlas rocket lifts off with a Mercury capsule containing John H. Glenn for an orbital mission.

of America, Holmes had built the Ballistic Missile Early Warning System on time and on budget, which was no mean feat.[7] Silverstein returned to the Lewis Research Center, this time as its Director.

Homer Newell was promoted from Silverstein's deputy to become Director of the Office of Space Sciences. Edgar Cortright became Newell's deputy, and Oran Nicks superseded Cortright as Director of the Lunar and Planetary Programs Division. As one of his first acts, Nicks established individual offices in the Lunar and Planetary Programs Division for Ranger and Surveyor, and also for the Mariner interplanetary program. For Ranger, William Cunningham was Program Chief, Walter Jakobowski was Program Engineer and Charles Sonett served in an interim capacity as Program Scientist. James Burke at JPL was delighted with this structure, because it integrated engineering and science in a single program office and greatly simplified his relationship with NASA headquarters.

Holmes promptly assigned Joseph F. Shea, a systems engineer who had run the development of the inertial guidance system for the Titan intercontinental-range ballistic missile, to resolve the protracted debate about how Apollo would fly to the Moon – the 'mission mode' issue.

On 28 November, NASA announced that North American Aviation of Downey, California, had been awarded the contract to develop the Apollo spacecraft. On 21 December, Holmes set up the Manned Space Flight Management Council. Drawing on senior managers at headquarters and the field centres, this would set policy for manned space planning. At its first meeting, the Council decided on a launch vehicle which would become known as the Saturn V. A single launch would be capable of dispatching an Apollo circumlunar mission. It might even be possible to undertake a lunar landing with a single launch. A landing mission involving Earth orbit rendezvous could certainly be done using just two launches.

On 20 February 1962, America finally inserted a man into orbit, with John Glenn riding an Atlas missile to circle the globe three times. On 7 June NASA decided on lunar orbit rendezvous as the mode for Apollo. On 7 November, it announced that the Grumman Aircraft Engineering Corporation of Bethpage, New York, had been awarded the contract to develop the Apollo lunar module.

By the end of 1962, therefore, NASA had taken all the key decisions that defined how it would address Kennedy's challenge.

[7]   BMEWS used large radar stations in Alaska, Greenland and England to provide the US with the famous "fifteen minute" warning of a Soviet ICBM strike over the north pole.

6

Ranger struggles

STRANDED

After Ranger 1 passed its qualification tests at JPL in May 1961, Oran Nicks, Chief of Lunar Flight Systems at NASA headquarters, authorised its transportation to Cape Canaveral, where the Air Force had assigned Hangar AE to the project. The launch window ran from 26 July to 2 August. In late June the Atlas was erected on Pad 12, the Agena added, and the spacecraft in its aerodynamic shroud installed to complete the stack. The combined systems tests of the fully assembled space vehicle were concluded on 13 July.

The countdown was delayed three days by a variety of problems, and was unable to start until the evening of 28 July with the intention of launching at dawn the next day, but a problem with the Cape's electrical power supply meant that the clock had to be halted with 28 minutes remaining. After two other counts were frustrated, the attempt to launch on 2 August was abandoned when, as high voltage was applied to the spacecraft's scientific instruments for calibration purposes, an electrical failure caused the explosive bolts to fire to deploy the solar panels inside the shroud. The spacecraft had to be retrieved and returned to the hangar. It was concluded that there had been an electrical arc to the spacecraft's frame, but the precise source was not evident. The damaged parts were replaced. The launch was rescheduled for the start of the window for the next lunation.

The countdown began on the evening of 22 August and ran smoothly to liftoff at 10:04:10 GMT the next morning. With Ranger 1 on its way, James Burke became Mission Director at the Hangar AE command post.

The Atlas ignited its sustainer, the two side-mounted boosters and the two vernier control engines, and was held on the pad until verified to be running satisfactorily. For the first 2 seconds the vehicle rose vertically, and then it rolled for 13 seconds to swing its guidance system onto the flight azimuth. After 15 seconds the autopilot pitched the vehicle in that direction so as to arc out over the Atlantic. When a sensor detected that the acceleration had reached 5.7 times that of

Earth gravity,[1] about 142 seconds into the flight, the Atlas shut off its boosters, and 3 seconds later jettisoned its tail to shed 6,000 pounds of 'dead weight'. The sustainer engine continued to fire. In the boost phase, the vehicle had been tracked by a radar at the Cape to enable the Air Force to calculate its initial trajectory, and as the sustainer flew on it acted upon steering commands radioed by the ground. When the sustainer shut down, the two verniers on the side of the Atlas fired as appropriate to refine the final velocity. As it did not have the power to insert the Agena directly into orbit, the upper stage was to be released on a high ballistic arc. Once free, the Agena, now above the dense lower atmosphere, jettisoned the aerodynamic shroud to shed dead weight, and ignited its engine. It then achieved the desired circular parking orbit at an altitude of 160 km. Meanwhile, the Air Force's computer processed the tracking provided by the radars of the downrange stations of the Eastern Test Range in order to calculate the length of time the Agena should spend in parking orbit and the parameters required for its second manoeuvre. This information was transmitted to the vehicle.

The plan for this test flight was for the Agena B to use its second burn to enter an elliptical orbit with an apogee of 1 million km, far beyond the orbit of the Moon, and for simplicity the orbit would be oriented not to venture near the Moon. The primary objective was to evaluate the spacecraft's systems in the deep-space environment, in particular its 3-axis stabilisation using Earth, Sun and star sensors, the pointing of its high-gain antenna, and the performance of the solar panels. Each Block I Ranger was expected to have an operating life of several months, and to provide worthwhile data for the sky scientists.

After its second burn, the Agena was to fire explosive bolts in order to release the spacecraft, which would be pushed away by springs. Then the spent stage was to use its thrusters to make its trajectory diverge. Radio interference prevented the tracking site at Ascension Island in the South Atlantic from monitoring the reignition. When Johannesburg reported detecting the spacecraft several minutes ahead of schedule, it became evident that the second burn had failed and the spacecraft was still in a low orbit. When Goldstone picked it up, the orbit was calculated to have a perigee of 168 km and an apogee of 500 km. Although the Agena had reignited, it had shut down prematurely and then released the spacecraft. It was encouraging that the spacecraft had deployed its solar panels, locked onto the Sun, rolled to acquire Earth and then deployed its antenna, but because it was 'stranded' in a low orbit it soon entered the Earth's shadow and lost both power and attitude lock. On re-emerging into sunlight it fired its thrusters to restabilise itself. This occurred on every shadow passage, with the result that after only one day the nitrogen was exhausted and, unable to stabilise itself to face its solar panels to the Sun, the battery, intended only for launch and the brief midcourse manoeuvre, expired. The inert spacecraft re-entered the atmosphere on 30 August.

A study of the telemetry tapes confirmed that the Agena reignition sequence had started at the proper time, but almost immediately the flow of oxidiser had ceased. The small amount of oxidiser which had entered the engine gave the 70-m/s velocity

---

[1]   Earth imparts a gravitational acceleration of 32.2 ft/sec^2.

Preparing the Ranger 1 spacecraft.

increment that slightly raised the apogee. The premature cutoff was classified as a one-off failure.

Although Ranger 1 flew in an environment different to that intended, its designers were encouraged that it had correctly deployed its appendages and (repeatedly) been able to adopt cruise attitude. But the sky scientists received nothing of value from the mission.

On 5 October, as a result of lessons learned from Ranger 1 when various lines of authority had penetrated the Space Flight Operations Center, Marshall Johnson was appointed Chief of the Space Flight Operations Section and, with it, sole authority to direct the control team while a mission was underway.

The launch window for Ranger 2 was 20–28 October 1961. The tests on the fully assembled space vehicle on Pad 12 were completed on 11 October. The countdown began on time in the evening of 19 October, but was scrubbed with 40 minutes on the clock owing to a fault with the Atlas. Although this was readily repaired, the fact that another Atlas was due to leave from another pad the next day meant Ranger 2 had to wait. The countdown on 23 October was abandoned because of another issue with the Atlas. At this point, a Thor-Agena B launched from Vandenberg Air Force Base in California was lost as a result of the failure of the hydraulics of the Agena's engine, and NASA decided to await the outcome of that investigation. The problem was diagnosed and fixed in time for the next window, and Ranger 2 lifted off on the first attempt at 08:12 GMT on 18 November. As before, the spacecraft rose above the horizon at Johannesburg early, indicating that the second burn had failed – this time without even producing a modest apogee. Ranger 2 performed perfectly, but it was doomed and re-entered the atmosphere on 19 November.

An Air Force analysis of the telemetry indicated that the roll gyroscope of the Agena B's guidance system had been inoperative at liftoff, most probably due to a faulty relay in its power supply. The attitude control system had compensated for the roll control failure by using its thrusters, and in so doing had exhausted the supply of gas. As a result, the Agena had tumbled in parking orbit. This caused the propellants to slosh in their tanks, which in turn prevented them from flowing into the engine when it tried to reignite. On 4 December 1961 the Air Force informed NASA of its findings, and Lockheed promised to report within a month on how it would fix the fault. When NASA decided in December 1959 to use the Agena B, it had presumed the Air Force would have worked the bugs out of the vehicle by the time it was needed, but only one had been launched prior to Ranger 1 and, in effect, NASA was testing it for the Air Force!

Although some aspects of the Block I tests had not been achieved, the engineers at JPL were encouraged that on both occasions the spacecraft had worked as well as could be expected in the circumstances. If the Agena was fixed as soon as Lockheed hoped, then it should be possible to proceed with Ranger 3 as planned.

LUNAR FLYBY

In May 1961 Aeronutronic began to drop balsa encapsulated 'survival capsules' containing sterilised systems immersed in viscous fluid from aircraft flying over the Mojave Desert – with disappointing results: those that fell on rocky ground failed to operate. Further tests in October showed that even seismometers which survived the impact often suffered electronic issues. With the launch of Ranger 3 only months away, this was disconcerting. On 6 November 1961 Don B. Duncan replaced Frank Denison in charge of developing the capsule. NASA issued heat-treatment waivers for the most sensitive components of the radar altimeter, retro-rocket and capsule. In 1961 Albert Hibbs, Chief of the Space Sciences Division at JPL, appointed Harold W. Washburn as Ranger Project Scientist to liaise between the spacecraft engineers and the experimenters and coordinate their activities during a mission.[2]

Meanwhile, after the Ranger 3 bus was assembled at JPL in July 1961 it suffered much greater component failure rates than the case of the proof-test model, with the only difference between them being heat-sterilisation. NASA issued further waivers for the most sensitive components. It was clear that the sterilisation process efficacy specified in 1960 was unattainable. On 15 November NASA formally accepted the repaired bus. It was driven by truck in an environmentally regulated container, and arrived at the Cape on 20 November. On 6 December Oran Nicks opined to Edgar Cortright that one of the three Block II flights would be successful, and one, perhaps two, of the four Block III flights. This underscored the perceived technological risk of the venture. James Burke expected one of each to be a fully successful flight. On 2 January 1962 Clifford Cummings told Robert Seamans that it was "likely that the sterilization procedures have compromised spacecraft reliability". However, when Lockheed announced that it had fixed the problem with the Agena B, NASA decided to try to launch Ranger 3 on schedule.

The major objectives for the mission were to perform the midcourse and terminal manoeuvres. Given the performance of its predecessors, it was fully expected that on being set free by the Agena, Ranger 3 would deploy its appendages and adopt cruise attitude. If it flew to the Moon as planned, then Burke's engineers would be content. If the surface package functioned properly, this would constitute a bonus – if not, the package would have two more opportunities to achieve its scientific objectives.

Although it would be a considerable technical feat to reach the Moon at all, and in a sense anywhere would satisfy the mission, the trajectory was very limited. The fact that the retro-rocket of the surface package could not deal with a lateral velocity component meant the bus needed to make a vertical descent over the target. This, in turn, meant a site near the equator on the leading hemisphere. Prior to the space age, astronomers had defined lunar longitude in terms of how the Moon appeared in the terrestrial sky, with the leading limb (i.e. the one that faces the Moon's direction of travel as it pursues its monthly orbit of Earth) being east. However, in August 1961 the International Astronomical Union had redefined the system to match the point of view of an observer on the lunar surface, with east in the direction of sunrise; thus reversing the old scheme. For a Ranger Block II the

---

[2] In 1963 Thomas Vrebalovich would succeed Washburn as Ranger Project Scientist at JPL.

target would therefore have to be in the western hemisphere. In fact, a vehicle launched from Florida would expend less energy in approaching Oceanus Procellarum between 10 and 50 degrees west of the meridian and within 16 degrees of the equator than it would for any other region. Because the Moon maintains one hemisphere facing towards Earth, at any given site Earth remains in a more or less fixed position in the sky. The target could not be so far towards the limb that at unfavourable librations the signal from the surface would be too weak to be read. Another constraint was that the timing had to be such that the Moon was visible to Goldstone when the bus transmitted its scientific data. The phase of the Moon would be 'full' on 20 January 1962 and 'last quarter' on 28 January. The window was set for 22–26 January. If Ranger 3 managed to lift off on the first day, it would make its approach on 25 January.

After Ranger 3 completed its final systems checks in Hangar AE, it was driven to Pad 12 on 18 January and installed on its launch vehicle. When kerosene was loaded into the Atlas on 19 January, a leak was discovered in the bulkhead between the fuel tank and the liquid oxygen tank. Over the next few days the Air Force removed the centre engine and built a wooden frame up through the exposed aperture at the base of the fuel tank in order to allow technicians wearing masks and oxygen cylinders to replace the ruptured bulkhead. This round-the-clock effort made the vehicle ready in time to attempt to launch on the last day of the window. Meanwhile, as this was the first mission of the project intended to reach the Moon, the spacecraft was sealed in the aerodynamic shroud atop the Agena and bathed in gaseous ethylene oxide for 11 hours as the final stage of the sterilisation process, then the shroud was purged with dry nitrogen passed in through a sterile filter.

The countdown on 26 January proceeded smoothly, and Ranger 3 was launched at 20:30 GMT. The Air Force tracked its ascent and calculated the steering commands, but when these were transmitted to the Atlas it failed to act on them. The autopilot flew on, ignorant of deviations from the planned trajectory. In particular, it was not possible to command the moment of shutdown to optimise the final velocity, and when the autopilot ordered this using its programmed parameters it was both higher and faster than required. As a result of this discrepancy, the parking orbit attained by the Agena was slightly different to that planned. The limited ability of the spacecraft to correct its trajectory meant that in making the translunar injection the Agena had to pass through a 'key hole' in the sky that was only 16 km wide, and attain a speed which differed from the desired 40,000 km/hour by no more than 25 km/hour. In the event, Woomera's tracking indicated that the spacecraft would cross the orbit of the Moon at a point 32,000 km ahead of that body – a discrepancy which exceeded the spacecraft's 44-m/s midcourse manoeuvre capability. Nevertheless, engineers were encouraged that Ranger 3 had deployed its solar panels, locked onto the Sun, rolled to acquire Earth and then deployed its high-gain antenna. The flight would provide an opportunity to evaluate the spacecraft's performance in deep space, essentially as had been intended for the Block I.

James Burke flew from the Cape to JPL to operate the spacecraft from the Space Flight Operations Center. It was decided to exercise all of the functions, including the midcourse and terminal manoeuvres. It was not possible to test the retro-rocket

Preparing the surface package subsystem of the Block II Ranger spacecraft.

of the surface package, since its separation from the bus could be triggered only by its own radar altimeter. On 27 January the sequence of commands was uplinked for a midcourse manoeuvre designed to reduce the flyby range. Ranger 3 executed the preliminary roll and pitch changes as specified, fired its engine, then resumed cruise attitude. However, tracking revealed that the burn was the opposite of that intended and had increased the miss distance to 36,750 km. The error was an inverted sign in the computer program used by JPL to calculate the burn. Regardless of the outcome, the engineers were delighted that the spacecraft had made a manoeuvre and resumed its cruise attitude. Shortly after this, the boom holding the gamma-ray experiment completed its deployment. It had hinged down after separation from the Agena, and now, as planned, a gas generator extended it in a telescopic manner. The instrument was able to calibrate the emissions by the spacecraft and then make the first direct measurement of the flux of gamma rays in space.

A plan had been devised to perform a terminal manoeuvre which would orient the spacecraft to enable it to photograph the Moon during the flyby – in much the same manner, in fact, as had been intended for some of the Pioneer probes. In this case, it would view the illuminated leading hemisphere, and reveal that portion of the far-side which was in darkness for Luna 3 in 1959. The unplanned trajectory meant that Ranger 3 drew close to the Moon on 28 January. The cover for the optics was commanded to open, and power was applied to warm up the TV system. Goldstone uplinked the commands for the terminal manoeuvre to turn the spacecraft in order to point the camera at the Moon. The inverted sign had been corrected prior to making the calculation. An hour later, the spacecraft was told to make the manoeuvre. It initiated the pitch change in the correct direction, but soon thereafter the downlink began to intermittently drop out – a computer/sequencer fault had denied the vehicle the use of its Earth and Sun sensors, the gyroscopes directing the turn did so in an uncontrolled manner and the spacecraft was left spinning. At the appointed time, the TV system took pictures. Some frames were received at a very weak signal strength, and the fact that it was possible to see the black reference marks on a pane of glass in front of the focal plane silhouetted against a soft glow of sunlight glinting off the structure of the spacecraft provided welcome confirmation to the engineers from the Radio Corporation of America that their system had worked. At 23:23 GMT, some 6 hours after the attempted manoeuvre, Ranger 3 crossed the orbit of the Moon and passed on into solar orbit.

On 8 February 1962 JPL informed Oran Nicks of the preliminary findings of its investigation into the loss of Ranger 3. It had been concluded that the malfunction of the computer/sequencer was a result of the heat treatment required for sterilisation. It would therefore be necessary to issue waivers for the components believed to have failed. Although the mission had not reached the Moon, it had nevertheless provided an opportunity for the flight control team to compute a deep-space trajectory and then uplink the commands for a midcourse manoeuvre, which the spacecraft – the most sophisticated American deep-space vehicle to date – had executed as directed. The engineers were therefore confident that the next mission would reach the Moon.

Meanwhile

Homer Newell's Space Sciences Steering Committee decided on 1 December 1961 to consolidate the TV experiments of the Block II and Block III versions of Ranger. Previously, Gerard Kuiper, Gene Shoemaker and Harold Urey had been named to receive and interpret such pictures as were returned by the Block II flights – having played no part in the development of the camera. Now it was decided that they should form a team, together with Ray Heacock of the Space Sciences Division at JPL, and work with the Radio Corporation of America in the development of the high-resolution TV system for the Block III. In effect, Newell wished to integrate the engineers and scientists at the project level at JPL to match the recent integration at the program level in Washington, in the expectation that this would enhance the scientific results. Unfortunately for James Burke, in early 1962 Newell also sought to augment the Block III to recover some of the particles and fields research lost on the Block I flights. On 14 March 1962 the Steering Committee decided to add eight experiments to the Block III. Burke learned of this from Oran Nicks on 20 March. The solar panels of the Block I and Block II were triangular with truncated tips. It was necessary to fit larger rectangular panels to provide the power to operate all the additional Block III experiments.

LOBOTOMY

Ranger 4 arrived at the Cape on 26 February 1962. The countdown ran smoothly to liftoff at 20:50 GMT on 23 April. The Atlas performed satisfactorily, the Agena achieved the desired parking orbit and then made the translunar injection manoeuvre as planned. But when the spacecraft appeared above the horizon at Johannesburg it was transmitting a carrier signal without encoded telemetry, which meant that it was not possible to determine the state of the systems. Unable to lock onto the Sun, the initial instabilities imparted by separation from the spent stage caused the spacecraft to tumble. It was concluded that the master clock in the computer/sequencer must have failed. Ranger 4 had transmitted telemetry during the ascent, but the clock had stopped at some point in the gap in coverage between the vehicle passing beyond the final station of the Eastern Test Range and its coming into range of Johannesburg. In effect, the Agena had released a lobotomised robot. Ironically, the radar tracking by Woomera showed that the slight discrepancy in the trajectory would have been well within the capacity of the spacecraft to correct.

The target for Ranger 4 was the same as for its predecessor. The Moon was 'full' on 20 April and would be 'last quarter' on 27 April. James Webb, W.H. Pickering, Oran Nicks, Clifford Cummings and James Burke congregated at Goldstone on 26 April as the spacecraft approached the Moon. Without solar power, the battery had expired, terminating the carrier wave from the main transmitter, but the fact that the independently powered surface package was transmitting enabled radio tracking to continue. At 12:47 GMT, some 64 hours after launch, the spacecraft passed behind the leading limb of the Moon. Calculations showed that it impacted 2 minutes later. For the first time, American hardware had hit the Moon, marking a success for

On 23 April 1962 an Atlas-Agena lifts off with the Ranger 4 spacecraft.

the launch vehicle – but this was little consolation for the spacecraft engineers, and the scientists gained nothing. On the one hand, in the original concept of the Block II it was deemed that three flights would be necessary to have a reasonable probability of achieving a fully successful flight that would deliver the scientific objectives of the project. However, the scientists appeared to think that every flight should succeed! It was difficult to precisely determine why Ranger 4's clock had stopped, because the failure occurred when out of communication and it prevented the transmission of telemetry. But because the telemetry was present when the spacecraft was last seen on the Agena and absent following its release, attention focused on the separation procedure – in particular, when the umbilical plug was withdrawn from the bus and the onboard computer/sequencer issued the power-up command to the systems. It was inferred that there must have been a short circuit at this time, and the obvious root cause was the heat sterilisation process. Additional waivers were issued, but the computer/sequencer for Ranger 5 had already been treated. With mass no longer an issue for the Block II, it was decided to install a backup clock in order to ensure that telemetry would be produced in the event of the computer/sequencer being disabled.

Meanwhile

On 15 June 1962 Brainerd Holmes issued *Requirements for Data in Support of Project Apollo*, in which he called for three types of information about the Moon as a matter of priority, certainly within the next few years. First, environmental data on particles and fields in space near the Moon to assist in the design of manned spacecraft and assure the safety of crews both in flight and on the Moon. Second, information on the physical properties of the lunar surface in order to confirm the design of the Apollo landing gear. Third, photo-reconnaissance and topographical data in order to facilitate early selection of Apollo landing sites. Holmes had not consulted Homer Newell in drawing up this list of requirements, he simply expected that since Apollo was the agency's pre-eminent program Newell would arrange for the information to be provided as soon as possible – and pay the bill out of his own office's budget. But Newell's Space Sciences Steering Committee had its own priorities.

Holmes supported the unmanned lunar projects which would provide information for Apollo, but opposed those intended to undertake tasks which astronauts would soon be able to do. He therefore opposed Prospector, which was to collect and return lunar samples to Earth. Accepting this logic, Newell set out to ensure that astronauts performed useful science while on the Moon.[3] Soon after being appointed Director of the new Office of Space Sciences, Newell arranged for the Space Science Board of the National Academy of Sciences, now chaired by the Princeton geologist Harry H. Hess, to arrange a series of joint workshops to discuss the best way to undertake space science. The first Summer Study was held at the University of Iowa between 17 June and 10 August 1962, with over 100 representatives of NASA, academia and industry. The aim was to evaluate past and current programs, and recommend future

---

[3]   Newell also wished to maximise the amount of science on manned flights in Earth orbit.

programs. Afterwards, the Space Science Board issued a summary report, *A Review of Space Research*, in which it acknowledged that Apollo would start off as "an engineering effort", but expressed the hope that "scientific investigations will later become the primary goals".

In September 1962 Gene Shoemaker began a 12-month secondment to NASA to assist the Office of Space Sciences. His motivation for taking this post was partly to increase his chances of becoming one of the astronauts who would have the good fortune to undertake field geology on the Moon. Don Elston served as Acting Chief of the Branch of Astrogeology in Shoemaker's absence.

On 11 October 1962 Robert Seamans called in Homer Newell, Brainerd Holmes, W.H. Pickering and Oran Nicks, and told the Office of Manned Space Flight and the Office of Space Sciences to coordinate their lunar activities. Newell was told that his priority was to support Apollo's requirement for data about the Moon. Nicks was to coordinate, and report how unmanned missions could best contribute to Apollo. In particular, could further Rangers provide some of this data by delivering a surface package incorporating a penetrometer to measure the strength of the lunar surface. When on 15 October Newell publicly announced five additional Rangers for 1964 equipped with the high-resolution TV system, he emphasised they "would increase the probability of obtaining lunar surface detail information that could be used in the manned landing system". Newell was also considering another series for 1965 which would deliver surface capsules. If these rough landers were funded, they would be primarily for scientific research. For these, Newell asked the Aeronutronic Division of Ford, which had developed the seismometer, to investigate a small TV camera capable of being delivered to the lunar surface in a capsule. On 22 October Holmes and Newell announced that a Joint Working Group would be formed, composed of representatives of their two offices. Chaired by Gene Shoemaker, at least during his period of secondment to NASA, it would be responsible for recommending to the Office of Manned Space Flight "a detailed program of scientific exploration" and for recommending to the Office of Space Sciences "a program of data acquisition to assure a timely flow of environmental information into planning for manned projects". It would also be responsible "for establishing and maintaining close liaison with field centers, government agencies and universities in the development of an integrated scientific program for manned space flights".

Seamans's directive that the Office of Space Sciences fly lunar missions primarily in support of Apollo, rather than for purely scientific purposes, renewed Newell's determination to ensure that Apollo crews conducted proper science whilst orbiting the Moon and on its surface – the Manned Spacecraft Center, being fully occupied with the engineering challenge of sending men to the Moon, was slow to pursue this aspect of the program.

POWER FAILURE

There was eagerness for the final Block II to provide close-up pictures and radar reflectivity of the Moon's surface, as well as (hopefully) seismometry. Ranger 5 had been scheduled for June 1962, but was postponed to allow the first pair of Mariner interplanetary missions to be dispatched in July and August.

On 30 August Rolph Hastrup, in charge of sterilisation, recommended that heat-treatment not be applied to the Block III. The use of 'clean rooms' to assemble the spacecraft, and the infusion of gaseous ethylene oxide to sterilise it within the Agena shroud shortly prior to launch should be continued. Clifford Cummings postponed a decision until after the next mission.

Ranger 5 arrived at the Cape on 27 August. As a result of recent modifications, it was about 10 kg heavier than its predecessors. The countdown on 16 October was scrubbed when a short circuit occurred in the spacecraft's radio system. A launch the next day was ruled out by high winds. The mission got underway at 18:00 GMT on 18 October. Despite suffering a glitch, the Atlas responded to steering commands from the Cape, and the Agena achieved the desired parking orbit. This time, tracking ships were stationed in the Atlantic in order to provide continuous monitoring of the spacecraft's telemetry. It had been decided that if the trajectory from the Agena's second burn were to be beyond the spacecraft's ability to correct, then the scientific priority would be to obtain gamma-ray data, rather than to snap flyby pictures of the Moon. This was because the Block III would provide TV, whereas there would be no gamma-ray spectrometers on any spacecraft that would head into deep space any time soon. The Agena made the translunar injection and released its payload. For the first time, the Deep Space Instrumentation Facility had two missions to keep track of in space. Mariner 2 was cruising to Venus, but the lunar mission would have priority call on resources during its 3-day flight.

When the Woomera tracking station acquired Ranger 5, it had deployed its solar panels and locked onto the Sun. The next task was to roll in order to acquire Earth as the second point of reference. But the temperature in the power switching and logic module of the computer/sequencer rose sharply and power from the solar panels was lost – there had been a short circuit. Patrick Rygh had replaced Marshall Johnson in charge of the Space Flight Operations Center, to free Johnson to manage the design and construction of the new Space Flight Operations Facility. James Burke, at the Cape, directed Rygh to have the spacecraft make a midcourse manoeuvre before its battery expired, to ensure that it would hit the Moon. But because the spacecraft had not acquired Earth its actual orientation in space was indeterminate. It was therefore decided to set up the manoeuvre using only the Sun as a reference. A command was uplinked to gimbal the high-gain antenna away from the nozzle of the engine on the base of the bus. The spacecraft initiated the ad hoc 30-minute manoeuvre sequence, but before it could be completed the transmitter fell silent. It appeared that electrical shorts had drained the battery. The Moon was 'last quarter' on 20 October. The inert vehicle flew by the trailing limb on 21 October at an altitude of 720 km and passed on into solar orbit – its progress once again being tracked by the transmitter in the surface package.

On 22 October W.H. Pickering ordered an investigation staffed by JPL personnel who were not involved in the project. When this issued its report on 13 November, it lamented that the mass limit imposed on the Block II prevented it from having any redundancy – in order to achieve its mission, the spacecraft required every system to

The Space Flight Operations Center at JPL during the Ranger 5 mission, with Patrick Rygh in command.

work. Burke was criticised for (in the opinion of people not involved) having spent too much of his time on launch vehicles, launch operations and space experiments, as opposed to the spacecraft. Burke was also criticised for the importance he gave to meeting schedules. However, in this he had merely been reflecting NASA's desire to get ahead of the Soviets within the 36 months that had been assigned to the project. The structure of JPL was also criticised, in that engineers assigned to work on flight projects by the technical divisions often lacked vital experience, and section chiefs unfamiliar with either the project management or the subsystems that their engineers worked on had inadequately reviewed this work. Remarkably, despite the fact that a lack of commonality in the failures implied a reliability issue in the components, the investigation did not address the issue of heat sterilisation, and Hastrup's memo to Cummings was not discussed. The report concluded that the Block III was unlikely to perform any better. To remedy the situation, it recommended (in part) that Burke be replaced and that his successor review the Block III design, add redundancy, and introduce new project management, inspection and testing procedures.

Neither of the two Block Is had achieved the intended high-apogee orbits (owing to Agena problems) and only one of the three Block IIs had reached the Moon (in an inert state). The project had been acknowledged to be technologically risky when it was commissioned, but no one had expected such poor performance. The spacecraft failures undoubtedly resulted from heat-sterilisation. The only scientific result from the entire exercise was provided by the gamma-ray spectrometer of Ranger 3, which established the existence of 'hard' radiation in space. However, absolutely nothing had been learned about the Moon. Nevertheless, the sense of 'crisis' would not have come about if the final Block II mission had been a complete success.

Responding to the mood, Homer Newell asked Oran Nicks to establish a Board of Inquiry to review the past performance and future prospects of the Ranger project. It was chaired by Albert J. Kelley, Director of the Electronics and Control Division of the Office of Advanced Research and Technology, and drew its membership from headquarters, field centres not involved in the project, and analysts from Bellcomm Incorporated – a systems engineering group established by the American Telephone & Telegraph Company in March 1962 at the request of the Office of Manned Space Flight to conduct independent analyses in support of Apollo. In particular, it was to submit recommendations "necessary to achieve successful Ranger operation". No thought was given to cancelling the project, because the high-resolution TV from the Block III was required for Apollo. On 30 November the Board issued its report. As regards JPL, it said that because the laboratory was attempting to use a common bus for its lunar and planetary projects, Ranger was more complex than strictly required, and as yet the high order of engineering skill and fabrication technology required for this not to represent an issue had yet to be achieved. It also said that the degree of ground testing was inadequate – the laboratory's tradition with military missiles was to iron out problems by test flights; this was impractical with spacecraft. The Board judged heat sterilisation to have been a significant factor in the failure rate. Of course, the lack of redundancy in the spacecraft was criticised. JPL was also criticised for trying to run such a major venture simply by superimposing a small project office on top of its divisional structure. The recommendations therefore included strengthening the project office at JPL and revising the procedures for design review, design change control, testing and quality assurance. Heat sterilisation should cease. The Block III objectives should be restated, and all activities which did not directly contribute put aside. If additional versions of the spacecraft were required, then JPL should hire an industrial contractor.

On 7 December 1962 JPL relieved both Clifford Cummings and James Burke of their posts. On 12 December, Brian Sparks, Deputy Director of the laboratory, led a delegation to Washington to discuss the Kelley report with Homer Newell. At this and a second meeting on 17 December it was decided (in part) to delete the eight particles and fields experiments which Newell had added to the Block III in March; to discontinue heat sterilisation and the use of gaseous ethylene oxide; to discard all heat-treated hardware; and that (as originally intended) the sole goal of the Block III would be to obtain high-resolution TV of the lunar surface in support of Apollo. On 21 January 1963 William Cunningham, the Ranger Program Chief at headquarters, told the scientists that their experiments had been deleted from the Block III and, to ease the blow, pointed out that they would be favourably considered for carriage on possible future missions.

Morale at JPL was boosted on 14 December 1962 when Mariner 2 made a close flyby of Venus and became the first deep-space mission to make in-situ observations of another planet, along the way establishing that the solar wind was 'gusty'.

On 18 December Robert Parks superseded Cummings, and Harris Schurmeier was made Ranger Project Manager – having been Chief of the Systems Division that had handled most of the work, he was the obvious choice. He immediately instituted a Ranger System Design Review Board involving Burke (who remained on the project staff), Gordon Kautz, Allen Wolfe and section chiefs of the supporting engineering divisions. Its primary task was to increase reliability by identifying and eliminating potential weak points in subsystems. The deletion of the experiments from the Block III released 22.5 kg of mass to accommodate redundancy. The enlarged solar panels were retained to provide a healthy power margin. Meanwhile, Bernard P. Miller of the Radio Corporation of America held a thorough review of the high-resolution TV package and recommended that the various wide-angle and narrow-angle cameras, together with their associated electronic assemblies, be split into two independent electrical chains so as to ensure that *some* pictures would be obtained even if an electrical problem were to disable one chain. Furthermore, to guard against the failure of the computer/sequencer, Miller recommended that a backup timer be added to start the TV system. Schurmeier accepted these recommendations. He also duplicated the gas supply of the attitude control system, and increased the capability of the main engine to make the Block III better able to correct a discrepancy in the translunar injection. And as arcing discharges were the single most worrisome cause of in-flight failures, he ordered that plastic covers be placed over all exposed terminals. W.H. Pickering strengthened the project office by revising the lines of authority and responsibility within the technical divisions so as to make the section chiefs personally involved in project activities, accountable for the quality of their engineers' work, and no longer able to reassign personnel without the consent of the project manager. Pickering also made Ranger the laboratory's highest priority flight project – thereby guaranteeing Schurmeier the authority he needed (and Burke had lacked) to drive work through in the manner desired. On 13 February 1963 NASA approved the long list of changes to be made to the Block III. In October the schedule for Block III was set, calling for missions in late January, March, May and July 1964.

Meanwhile
Since the Kelley Board had also been critical of the record of the Atlas-Agena B launch vehicle, this was also reviewed. The procurement process had proved fraught with problems. James Webb moved responsibility for monitoring the procurement of the Agena from the Marshall Space Flight Center to the Lewis Research Center. His rationale was that Wernher von Braun's people were fully occupied with the Saturn vehicles for Apollo, whilst Lewis, which had taken over development of the Centaur in early 1962, was eager for more work and Webb knew that Abe Silverstein (now in charge at Lewis) would get on top of the problems which had afflicted the Atlas-Agena B. As a result, the Light and Medium Vehicle Office at Marshall was closed. Silverstein appointed Seymour C. Himmel as Agena Systems Manager, and Joseph Ziemanski as Agena Project Engineer. The agency's presence was greatly increased at both the Lockheed factory in Sunnyvale and the General Dynamics-Astronautics factory in San Diego. Significantly, although Himmel monitored the procurement and launch of the Atlas-Agena Bs for Ranger, he reported to Harris Schurmeier at JPL. To improve the reliability of the Atlas, it was decided that vehicles intended for

NASA satellites and space probes should incorporate some of the systems developed to 'man rate' it for Project Mercury.

On 17 January 1963 NASA and the Air Force agreed to revise launch operations. The Air Force would continue as the single manager of Cape Canaveral, but NASA would be responsible for Merritt Island to the north, where the agency was to build its Saturn V launch facilities. Furthermore, all pre-launch activities, launch and post-launch evaluations for NASA missions would be directed by NASA, irrespective of whether they used Cape Canaveral or Merritt Island. In the case of Atlas-Agena launches, the General Dynamics-Astronautics and Lockheed personnel who actually prepared the vehicles and performed launch operations would no longer report to the 6555th Aerospace Test Wing but to NASA's on-site team, which would report to Himmel at Lewis. The Air Force would still provide services such as range tracking, however. On 9 August 1963 the Air Force assigned to NASA full responsibility for procuring Atlas and Agena vehicles – Himmel would no longer 'monitor' via the Air Force, he would directly supervise procurement. And because Himmel reported to Schurmeier, this gave the latter direct control over all elements of Ranger: the spacecraft, procuring the launch vehicle and launch operations.

On 28 January 1963 the high-resolution TV team met at JPL to discuss the formal requirement for the Block III system. The best telescopic resolution was 300 metres. A resolution of 30 metres would provide scientific insight, but not satisfy Apollo. It was therefore decided that the final frames in the sequence would have to provide a resolution of better than 3 metres. On 1 March 1963 William Cunningham notified JPL that the formal objective would be "at least an order of magnitude" better than Earth-based photography. The package was to provide continuous overlapping fields of view, starting at a resolution comparable to that of a telescope and running right down to the view from an altitude of about 100 metres, a fraction of a second prior to impact. The spacecraft designers and astronaut trainers at the Manned Spacecraft Center wished to assess the roughness of the surface (to evaluate the empirical rule of thumb that an area that looked smooth at the limiting telescopic resolution would be smooth in detail), to measure slopes, to estimate crater and boulder coverage, and to estimate the bearing strength of the surface from the size of the rocks lying around in order to determine whether it would support the weight of a spacecraft.

In March 1963 the number of additional TV imaging spacecraft was cut from five to three. They were to be launched in the second half of 1964, immediately after the Block IIIs. However, the number of rough landings was increased from four to six, all of which were to be flown in 1965. In addition to the high-resolution TV system, the additional imaging spacecraft would be augmented by instruments which would use "non-visual techniques" to provide information on the topography, environment and composition of the Moon. After the Block II failures, James Arnold had argued for his gamma-ray spectrometer, which had proved itself in space, to be carried over to the Block III, but had been refused. When the Space Sciences Steering Committee met on 15 April to select experiments for the three additional spacecraft, it chose the gamma-ray spectrometer and a repackaged form of the Block II altimeter to obtain radar reflectance data. This particularly pleased Harold Urey, who wished to determine the chemical composition of the surface.

Because W.H. Pickering had assigned the restructured Ranger project JPL's top priority, the decision was taken to abandon the intention to create commonality between the Surveyor intended to soft land on the Moon and that which would undertake orbital photography. On 25 April Edgar Cortright proposed to Homer Newell that since (as he put it) one successful orbiter would be worth "dozens of successful Ranger TV impactors", the three additional imaging missions should be cancelled, and work on the rough-landing capsules should proceed only as a contingency against a serious delay in the development of the Surveyor lander. The two non-visual experiments should be reassigned to the rough-landing Rangers. Newell accepted Cortright's reasoning, and on 12 July Robert Seamans concurred. To enable Harris Schurmeier to focus his efforts on the Block III, on 9 August JPL assigned project management of the follow-on Rangers to Geoffrey Robillard, then Chief of the Propulsion Division. Cortright had also recommended that the development of the lunar orbiter be transferred to the Langley Research Center, and this was approved by Seamans on 30 August 1963.

Meanwhile, on 19 July Newell restructured the high-resolution TV 'experiment' of Ranger Block III to match the agency's new *modus operandi* – when the TV was selected, NASA had not yet adopted the Principal Investigator role. Gerard Kuiper became PI. The science team included Gene Shoemaker, Harold Urey and Ewen A. Whitaker – one of Kuiper's colleagues at the Lunar and Planetary Laboratory of the University of Arizona. At the same time, it was agreed that Gene Shoemaker would be PI for the TV on the Surveyor lander. On 30 July 1963 Newell reorganised the Joint Working Group of the Offices of Manned Space Flight and Space Sciences as the Manned Space Science Division.

Brainerd Holmes, upon realising in November 1962 that the funding for manned space flight put at risk Apollo's ability to reach the Moon within the decade, urged James Webb to seek a $400 million supplemental appropriation from Congress, but Webb refused, saying that they should wait for the Fiscal Year 1964 budget and gain the necessary funds in a manner calculated not to irritate Congress – as an 'old hand' in Washington, Webb had a keen sense of the congressional mood. When Holmes suggested that the requisite funding be transferred from other parts of the agency's budget, including science, Webb insisted on a "balanced" program. In March 1963 NASA submitted its budget request – which represented a 54 per cent increase over the previous year's request, with 80 per cent of the increase intended for the manned space program. Congress was sceptical. It was the first time that NASA funding had been seriously questioned. On 9 April President Kennedy asked Lyndon Johnson to review NASA and report whether, and to what extent, its budget could be trimmed without detriment to Apollo. When Kennedy met Webb a few weeks later to discuss the budget request, Webb said that he, Hugh Dryden and Robert Seamans were in agreement that the agency must pursue a balanced program, not focus exclusively on Apollo. The Senate Committee on Aeronautical and Space Sciences took testimony from a number of scientists who disapproved of Apollo. The debate ran through the summer. But Kennedy had warned in his speech that if the nation were to accept his challenge, it must not falter, and the truth was that the US could afford whatever it had the *will* to pursue. In the end, the budget was passed with a 40 per cent increase –

which was not bad. But on 12 June Holmes had submitted his resignation, to come into effect on 1 September. On 23 July NASA announced that George E. Mueller, Vice President of the Space Technology Laboratories (now part of the Thompson Ramo Woodridge conglomerate), would take over from Holmes as Director of the Office of Manned Space Flight.

On 24 October 1963, Newell, facing difficulty in funding the follow-on Rangers in the constrained budget for Fiscal Year 1964, began to consider cancelling them in order to transfer the funding to Langley for the reassigned orbiter. On 13 December he told Pickering that this had been decided – it was not a criticism of the scientific investigations of the rough landers, simply a financial constraint.

As part of yet another major reorganisation of top management, on 1 November 1963 Newell's role was expanded to Associate Administrator for the merged Office of Space Sciences and Applications. That same month, Gene Shoemaker concluded his 12-month secondment to NASA, and Willis B. Foster took over as chairman of the Manned Space Science Division.

TV FAILURE

Assembly of the first Block III began on 1 July 1963. The Radio Corporation of America delivered the high-resolution TV subsystem on 15 August. At 366 kg, the spacecraft was about 25 kg heavier than its immediate predecessor. On 6 December W.H. Pickering suggested to Homer Newell that NASA appoint a small group for an independent assessment of Ranger 6, which had just completed its pre-acceptance testing. Newell sent some members of the Kelley Board, with William Cunningham (Program Chief) and Walter Jakobowski (Program Engineer) representing the Office of Space Sciences and Applications. After being accepted, the spacecraft left JPL by truck on 19 December and arrived at the Cape on 23 December.

As the Block III did not have a surface capsule, it could tolerate a lateral velocity component in its terminal dive, but at the expense of smearing in the final images – those of greatest interest to Apollo. The launch window for Ranger 6 was 30 January to 6 February 1964. The Moon was 'full' on 28 January and would be 'last quarter' on 5 February. The target longitude would vary with the date of launch, migrating westward with the evening terminator. The constraints on latitude were less strict, but the Apollo planners were primarily interested in the equatorial maria. The target for a launch at the start of the window was in the equatorial zone 15 degrees east of the lunar meridian, in Mare Tranquillitatis.

The countdown started in the morning darkness of 30 January, and ran smoothly to liftoff at 15:49 GMT. The Atlas delivered a flawless performance. The Agena made translunar injection as planned. The only anomaly was about 2 minutes after launch, when the spacecraft's telemetry showed that the TV subsystem had switched on for a period of 67 seconds. When Johannesburg picked up Ranger 6, it was on its way to the Moon and gave every appearance of being healthy. After locking onto the Sun and Earth, it deployed its high-gain antenna. A small midcourse manoeuvre was

The auditorium at JPL awaits news of Ranger 6's fate.

made on 31 January. "I'm cautiously optimistic," Pickering told reporters at a press conference shortly after the manoeuvre.

As Ranger 6 neared the Moon on 2 February, it was accelerated by that body's gravity. Radio tracking indicated that it would hit within a few kilometres of the aim point. Homer Newell and Edgar Cortright were observers in the VIP gallery of the Space Flight Operations Center. Walter Downhower, Chief of the Systems Design Section, gave a running commentary for the journalists in the auditorium. Since the spacecraft's cruise attitude was compatible with imaging, Harris Schurmeier decided not to attempt the terminal manoeuvre lest this fail and ruin the mission. With 18 minutes to the predicted impact, the wide-angle cameras began their 5-minute warm-up, followed a few minutes later by the narrow-angle cameras. They were to switch over to full power at T–13 minutes and T–10 minutes respectively, and start to take pictures.

"Thirteen minutes to impact," noted Downhower. "There is no indication of full power." In due course, he followed up with, "Ten minutes to impact. We're still awaiting transmission from the spacecraft of full-power video."

At this point Schurmeier told Goldstone to issue an emergency command to the spacecraft to switch on its TV system. This was done. Ranger 6 accepted the uplink and executed the command, but to no effect. When an audio representation of the downlink telemetry suddenly ceased at 09:24:32 GMT, Downhower observed, "We have our first report of impact. Still no indication of full-power video." On striking the surface at a speed of 9,500 km/hour, the spacecraft vaporised. A movie camera had been mounted on a telescope in an effort to record any sign of the impact, but no flash or cloud of dust was evident.

A few hours later, Pickering set up an investigation headed by Donald Kindt, the JPL project engineer for the TV subsystem, and the next day Pickering appointed a group of section chiefs, chaired by Downhower, to monitor the investigation and to study its conclusions and recommendations. It was found that the failure occurred when the TV subsystem had briefly switched on during the ascent to orbit. Electrical arcing had destroyed the high-voltage power supply of the cameras and transmitters. The likely cause was shorting across the exposed pins of the umbilical connector of the Agena fairing which gave electrical access to the TV subsystem prior to launch. In the absence of a positive identification of the cause of the arcing, the investigation recommended (in part) that the subsystem be 'locked out' during the ascent to orbit, and enabled only after the spacecraft had separated from the Agena. On 11 February 1964 Pickering told Newell that Ranger 7 would have to be postponed, pending a definitive resolution of the issue.

Meanwhile, on 3 February Robert Seamans had established a NASA Board of Inquiry chaired by one of his deputies, Earl D. Hilburn. Concerned that JPL had not been able to positively identify the reason for the TV subsystem's failure to transmit pictures, the Board reviewed the situation and on 14 February Hilburn alerted Hugh Dryden to the fact that his investigation had uncovered a number of deficiencies in the design and testing of the TV subsystem, pointing out in particular that the 'split' architecture was not entirely redundant. Hilburn judged JPL's proposal to 'lock out' the TV subsystem during the ascent to be inadequate, and instead recommended that the system be completely redesigned – which would mean delaying the next mission by a year or more. Dryden was appalled at the prospect of such a long delay. Homer Newell feared that it would be decided simply to abandon the Ranger project. After considering the matter further, on 17 March Hilburn submitted his final report. This concluded that there must have been "two or more failures" in the TV subsystem; that the system was not as redundant as the designers had believed; and that testing had been inadequate – in particular, the report pointed out that the system had not been verified at full power during the pre-launch checks. In fact, JPL had decided early on in the project *not* to apply full power to 'experiments' in pre-launch checks lest a short circuit ignite the midcourse engine with a fuelled launch vehicle below. The recommendation was to redesign the TV subsystem. James Webb received the report, but took no immediate action.

On 23 March Harris Schurmeier, having seen Hilburn's report, directed Maurice Piroumian of the Launch Vehicle Systems Section to further investigate the arcing issue. At liftoff, the plug of the ground equipment had withdrawn from the multi-pin connector and a flap had swung shut and latched to protect the connector. As this was the first flight of the TV subsystem and the connector was a new feature of the vehicle, it was possible that some aspect of its design was flawed. Tests were made over the next several months to try to determine how arcing might have taken place across these pins.

Alexander Bratenahl of the Space Sciences Division drew attention to the fact that the anomaly had coincided with the Atlas jettisoning its booster section. A study of long-range tracking camera footage showed that when this occurred the vehicle was briefly obscured by a large white cloud. On being informed by General Dynamics-

Astronautics that 180 kg of propellant drained out of the feed pipes when the lines were severed, Bratenahl speculated that suddenly dumping so much liquid into the rarefied air had produced a physical shockwave that was able to momentarily buckle the hinged flap inwards and mechanically short the pins; but an analysis showed that this was not feasible. At the end of June, Schurmeier terminated the investigation and classified the anomaly as a one-off.

Meanwhile, despite Hilburn's report, it was decided to accept the Kindt team's recommendation to 'lock out' the TV subsystem during the ascent; and on 11 May Schurmeier scheduled Ranger 7 for the window that would open on 27 July – as late as possible before priority would have to be assigned to the two Mariner missions to Mars scheduled for later in the year.

Bratenahl, however, continued to ponder the manner in which the Atlas staged. Intrigued when a more detailed analysis of the film showed flashes within the white cloud, he realised that the fluid dump had comprised both kerosene and oxygen, and that what he had naively presumed to be a simple physical shockwave was actually a detonation flash as the plume of the still-firing sustainer engine ignited the dumped propellants. The rapidly expanding spherical flashwave had washed over the vehicle, allowing *plasma* to penetrate the umbilical compartment to induce short circuiting. The timing was compelling: the Atlas shed its booster section at T + 140.008 seconds and the TV subsystem switched on at 140.498, coinciding with the progress of the flashwave up the length of the vehicle. On 30 July Bratenahl wrote a memo pointing out that arcing could be precluded if the cover flap were revised to form a hermetic seal. But by then Ranger 7 was in-flight to the Moon and the memo remained buried in an 'in tray' until after that mission.

In effect, NASA was learning by experience the many ways in which a spacecraft could be disabled. Although the chances of success increased as the failure modes were eliminated, the issue was whether Ranger would run out of spacecraft before it could deliver useful data!

7

Ranger triumphs

SUCCESS AT LAST

Ranger 7 was mated with its launch vehicle on 6 July 1964. The countdown began early in the morning of 27 July, but was scrubbed owing to a problem with the Atlas. It lifted off at 16:50 GMT on 28 July, the Atlas performed flawlessly, and the Agena achieved a circular parking orbit at 185 km then performed the translunar injection.

As to the target, Maxime Faget of the Manned Spacecraft Center had suggested investigating the crater made by its predecessor in order to use that calibrated impact to calculate the strength of the surface material, but the Moon was 'full' on 24 July and 'last quarter' on 1 August, with the result that the Sun would have set for Mare Tranquillitatis. After an analysis found that the best region would be in the vicinity of Guericke in Mare Nubium, in early July Homer Newell approved aiming for a point where the mare was crossed by bright rays from both Copernicus and Tycho. The translunar injection of 39,461 km/hour was within 6.5 km/hour of that planned. It would have caused the spacecraft to skim the leading limb of the Moon and crash on the far-side. The 50-second midcourse manoeuvre at 10:27 on 29 July established the desired trajectory. Afterwards, the terminal approach was analysed in terms of the angle of illumination of the lunar surface, the direction of the velocity vector of the spacecraft and the optical axis of the camera system, and it was decided that no terminal manoeuvre would be required for the photographic operation on 31 July.

The 3-storey Space Flight Operations Facility had been completed several months earlier. It had a main room for engineers and controllers, and nearby science support rooms for scientists to receive and analyse the incoming data. All these rooms were windowless for 24-hour use. Homer Newell, Edgar Cortright and Oran Nicks joined W.H. Pickering in the VIP gallery. Harris Schurmeier and Patrick Rygh supervised the flight control team. Gerard Kuiper's experimenters congregated in their room. With 20 minutes remaining, George Nichols informed the auditorium that the wide-angle cameras had started to warm up. The duration of this process had been halved, and 90 seconds later he reported they were on full power, prompting a round of applause. These cameras began to take pictures at 13:08:36. The narrow-angle

The Space Flight Operations Facility at JPL became operational in May 1964.

cameras followed suit at 13:12:09. It was early morning at JPL, but there was a large crowd of technical staff on hand, and they applauded and cheered when Goldstone announced that both video streams were coming in. Impact was at 13:25:49, within 12 km of the aim point.[1] This was the first American lunar spacecraft of any type to fully achieve its mission. Schurmeier produced several cases of champagne for the flight controllers, then led the VIPs to join the experimenters.

When Pickering, Newell and Schurmeier entered the auditorium an hour later for a press conference, they received a standing ovation. A total of 4,316 pictures were received. The first was taken at an altitude of 2,100 km and the last at about 500 metres. Goldstone converted the signal to TV format, and simultaneously stored this on magnetic tape and displayed it on a high-speed monitor. A camera whose action was synchronised with the incoming frame rate recorded each TV frame on 35-mm film – this film was then transferred to a vault. As this was going on, a technician 'sampled' another screen using a Polaroid camera to provide an initial evaluation of the quality of the results. The tape was replayed to make another film that was flown to the Hollywood–Burbank Airport for processing by Consolidated Film Industries. In the late afternoon local time, prints and slides were driven to JPL. While the experimenters examined the masters, the VIPs viewed copies. The press conference

---

[1]   The crater that Ranger 7 made was identified in Apollo 16 photography in 1972.

by the experimenters that evening was broadcast 'live' by the national TV networks. After Pickering introduced the team, Kuiper, the principal investigator, began the presentation: "This is a great day for science, and this is a great day for the United States. We have made progress in resolution of lunar detail not by a factor of 10, as hoped would be possible with this flight, nor by a factor of 100, which would have been already very remarkable, but by a factor of 1,000." In fact, he was being a little optimistic, as the resolution of the final frame was about half a metre.

The experimenters were obliged to provide 'instant science', to explain what the pictures showed.

Harold Urey pointed out that he was "pleasantly surprised" at the amount of information that could be inferred from the pictures. The surface was cratered, right down to the limiting resolution of the final frame. The bright rays appeared to have been formed by 'secondary' impacts as ejecta fell back on low-energy ballistic trajectories from an energetic 'primary' impact. Science fiction authors had reasoned that since the Moon had no atmosphere there could be no erosion, and this had led artists to depict an extremely rugged landscape. The fact that the surface was gently undulating was clear evidence that the incessant rain of meteoritic material was a potent form of erosion. Urey introduced the term 'gardening' to describe the process by which impacts 'turn over' the material.

Kuiper said the pictures supported his belief that the mare plains were lava flows, the surface of which, having been exposed to the vacuum of space, must be 'frothy'. This was based on laboratory experiments in which fluids of various viscosities had been exposed to vacuum. He ventured that when an astronaut walked on the surface, the experience would be similar to walking on crunchy snow. He accepted that there would be a layer of impact-generated fragmental debris, at least in some places, but thought it would be very thin, perhaps no more than a few centimetres.

Given the presence of large and obviously heavy blocks of rock, Gene Shoemaker was confident that the surface would bear the weight of a lander.

Nevertheless, Thomas Gold of Cornell University, who was not on the experiment team and not at the conference, had a theory that the maria were deep accumulations of fine dust. On seeing the Ranger 7 pictures Gold told reporters that whilst the dust would flow and tend to smooth out small-scale deformations, the rate at which it did so might be as little as a millimetre per year. As evidence of such flow, he pointed out that the rims of the older-looking craters were softer than the rims of the newer-looking ones.[2]

Although the Ranger 7 results were consistent with the 'hot Moon' hypothesis, in which the maria were volcanic lava flows, Urey did not give up on his 'cold Moon' hypothesis in which they were splashes of impact melt.

Several days later, Robert Gilruth, the Director of the Manned Spacecraft Center,

---

[2] Despite Gold's assertion that the dust would react only slowly upon being loaded, reporters would remain fascinated by the possibility that a lander would rapidly become submerged by it!

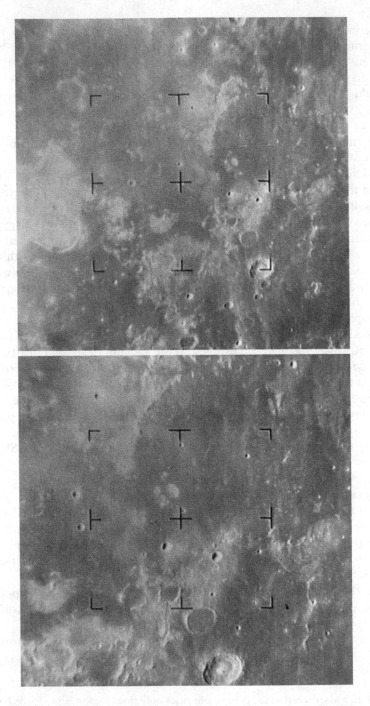

Images taken from altitudes of 2,095 km (top) and 1,223 km by the F_a-camera of Ranger 7 as the spacecraft plunged towards Mare Nubium on 31 July 1964.

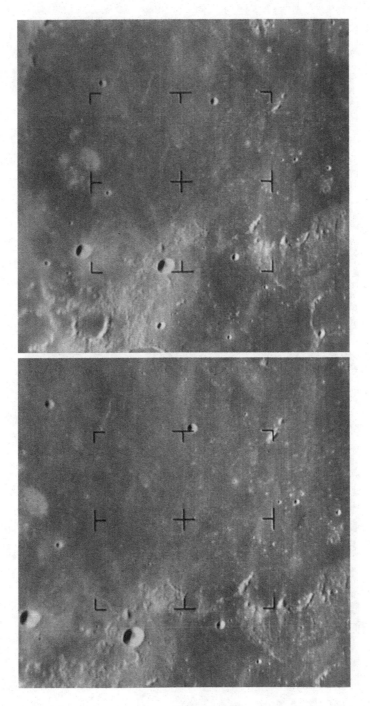

Images taken from altitudes of 747 km (top) and 582 km by the F_a-camera of Ranger 7 as the spacecraft plunged towards Mare Nubium on 31 July 1964.

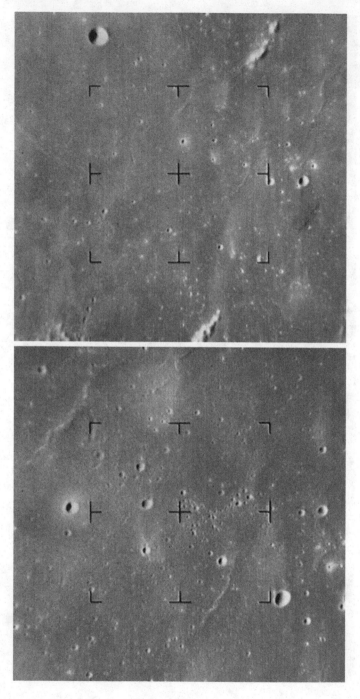

Images taken from altitudes of 300 km (top) and 125 km by the F_a-camera of Ranger 7 as the spacecraft plunged towards Mare Nubium on 31 July 1964.

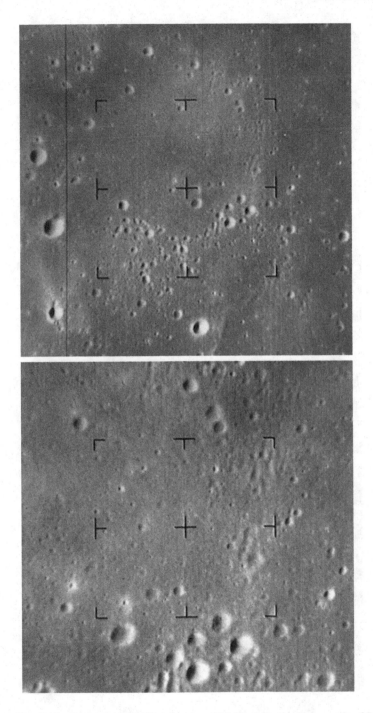

Images taken from altitudes of 65.8 km (top) and 29.9 km by the F_a-camera of Ranger 7 as the spacecraft plunged towards Mare Nubium on 31 July 1964.

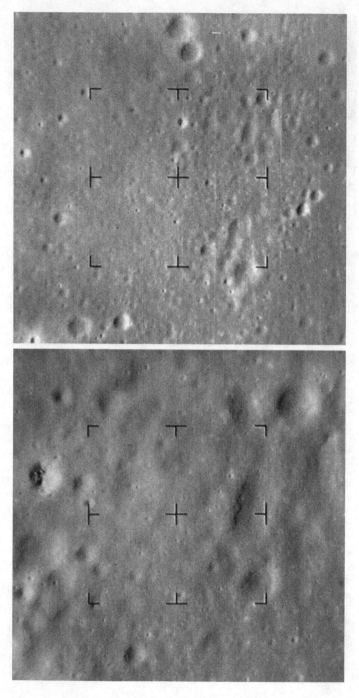

Images taken from altitudes of 17.9 km (top) and 5.9 km by the F_a-camera of Ranger 7 as the spacecraft plunged towards Mare Nubium on 31 July 1964.

An oblique view of the plaster model of the lunar surface made by the US Geological Survey's Branch of Astrogeology on the basis of the final high-resolution images from the Ranger 7 spacecraft.

and Joseph Shea, the Apollo Spacecraft Program Manager in Houston, visited JPL to be briefed by the experimenters. Shea later told reporters that in the pictures much of the mare plain appeared "relatively benign". Overall, Gilruth said, an Apollo landing should be "easier than we thought".

On 28 August NASA hosted the Interim Scientific Results Conference, chaired by Oran Nicks. The visual highlight was a 5-minute movie of successive frames of the picture sequence showing Ranger 7's dive. Immediately afterwards, the experiment team went to the International Astronomical Union meeting in Hamburg, Germany, and showed the movie again. The Union designated the Ranger 7 target area Mare Cognitum – the Known Sea. Shortly afterwards, as the pictures were being used to make an atlas of this region, Gerard Kuiper marvelled, "To have looked at the Moon for so many years, and then to see this ... it's a tremendous experience."

Ranger 7 gave a flood of data to a community which had been starving for years. The population statistics for primary projectiles is such that there is a small number of really large ones and increasing numbers of ever smaller ones. An airless surface exposed to such a population will gain many small craters, and progressively fewer larger ones. Given *actual numbers* for the populations, the cratering density (derived from counting) provides an age estimate – at least in terms of a presumption of how the population of projectiles changed over time. The simplest hypothesis is that this has remained constant, so that although the sheer number of potential projectiles has declined, their *proportions* have remained the same. It was evident that large caters would stand apart, but there would be a size of crater at which a given surface would be *saturated* – i.e. such craters would be so numerous that they would be rim to rim, and each new crater would mask an old one. This saturation would produce a sharp transition in the 'crater curve'. Dating by this method can be done only using craters which exceed the saturation size. It was not possible to determine this telescopically, but Ranger 7's pictures enabled the saturation size for this patch of Mare Nubium to be estimated at 300 metres. At less than this size, the surface was in a 'steady state'. The areal coverage of Ranger 7 did not include many craters larger than 300 metres, but William Hartmann at the Lunar and Planetary Laboratory of the University of Arizona attempted the measurement and derived an age of 3.6 billion years. During the 1950s C.C. Patterson at Caltech had radiologically dated a number of meteorites, and reached the conclusion that the solar system formed 4.55 ($\pm$0.07) billion years ago. If the Moon formed at that time, Hartmann's crater counting indicated that the surface of Mare Nubium was about a billion years younger.

While the scientists digested the Ranger 7 results, there was a hiatus in the project to accommodate the launches of Mariner 3 and Mariner 4 in November 1964, which also used the Atlas-Agena and hence the same pad facilities at the Cape.

REPEAT PERFORMANCE

Now that Ranger Block III had proved itself, the Office of Manned Space Flight asserted its right to specify the requirements for future targets. On 16 October 1964 Sam Phillips, the Apollo Program Director, told Oran Nicks that Ranger 8 should investigate a mare plain in the Apollo zone at a position which was *not* crossed by rays. At a meeting at JPL on 19 November, the scientists argued for comparing the 'reddish' Mare Nubium with a 'bluish' one in the eastern hemisphere. Nicks made the formal recommendation to Homer Newell on 9 February 1965, who concurred. The target for the first day of the launch window was to be Mare Tranquillitatis, but if the launch were delayed then it would move westward along the Apollo zone to keep pace with the migrating terminator. The launch window for Ranger 8 opened on 17 February. The Moon was 'full' on 16 February and would be 'last quarter' on 23 February.

Launch was at 17:05 GMT on 17 February. The translunar injection produced a flyby at a range of 1,828 km. The trajectory was refined by a 59-second midcourse manoeuvre at 10:27 on 18 February. When Ranger 8 made its approach to the Moon, it was decided to start the cameras several minutes early so that the initial pictures would be comparable to the best attainable by a terrestrial telescope. A total of 7,137 pictures were received – almost twice as many as from Ranger 7 owing to the extended sequence. Whereas Ranger 7 had made a near-vertical descent west of the meridian, the target for Ranger 8 was 24 degrees east of the meridian and to reach it the spacecraft had to make a slanting approach. Whilst this significantly increased the areal coverage, in particular depicting the central highlands at an unprecedented resolution, it meant there was no overlap between the frames later on and the lateral velocity smeared the final frames. The impact occurred at 09:57:38 on 20 February. Don Wilhelms was watching through the 36-inch refractor at the Lick Observatory near San Jose in California and listening to a radio countdown from JPL, but saw no flash. Alika Herring of the Lunar and Planetary Laboratory was using the 84-inch reflector at the Kitt Peak National Observatory in Arizona, but did not see anything either. As a result of the lack of overlap, the impact point was not actually within the final frame – not that it really mattered, it was calculated from the trajectory as being 24 km from the aim point.[3] In this case, owing to the smearing of the final images, the best resolution was 1.5 metres.

At a press conference 30 minutes later given by W.H. Pickering, Edgar Cortright, William Cunningham and Harris Schurmeier, the latter delightedly summed up the mission as "another textbook flight".

The experimenters presented some of the pictures later in the day. There were more rocks than at the Mare Nubium site, once again indicating a substantial bearing strength. In fact, this area was one of the 'hot spots' in near-infrared measurements made during the lunar eclipse of 19 December 1964, and the exposed rock supported the interpretation of such thermal anomalies as being due to rocks slowly radiating their heat when the Moon entered into the Earth's shadow.

Despite the intention to investigate a mare site free of rays, it was discovered that Ranger 8 came down over a faint ray from Theophilus. One striking observation was that the surface of Mare Tranquillitatis appeared remarkably similar to that of Mare Nubium. In fact, Gerard Kuiper, showing one of the final frames, remarked, "If you didn't know that this was taken by Ranger 8, you'd think it was one of the Ranger 7 pictures." It was ventured that "probably all lunar maria are pretty much this way".

Harold Urey introduced the term 'dimple crater' for irregular rimless pits which, it was speculated, might be where loose surficial material had drained into a cavity in much the same way as sand drains in an hourglass. Noting that tubes and cavities occur in terrestrial lava fields, and believing the lunar maria to be lava flows, Kuiper speculated that such pits might pose a "treacherous" threat to an Apollo lander. To Gene Shoemaker, however, they appeared merely to be degraded secondary impact craters.

On the larger scale, the scientists were pleased that as Ranger 8 made its slanting approach it provided views of Ritter and Sabine in unprecedented detail. These two 30-km-diameter craters in the southwestern Mare Tranquillitatis lacked radial ejecta and secondary craters, and their depth-to-diameter ratios made them anomalously shallow in terms of the curve plotted for impacts by Ralph Baldwin. The fact that they were aligned along the Hypatia rilles, were located on the fringe of a mare and had 'raised floors' had led some people to interpret them as volcanic calderas. Even people who favoured the impact origin of craters allowed that Ritter and Sabine

---

[3] The crater made by Ranger 8 was photographed by Lunar Orbiter 2, and found to be about 13.5 metres in diameter with a mound at its centre.

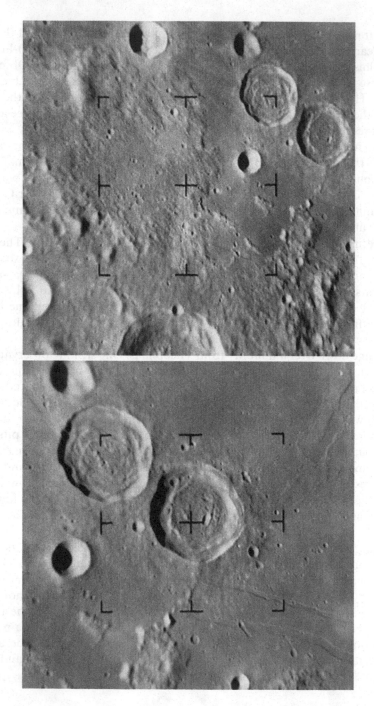

Images taken from altitudes of 385 km (top) and 262 km by the F_a-camera of Ranger 8 on a slanting trajectory to Mare Tranquillitatis on 20 February 1965.

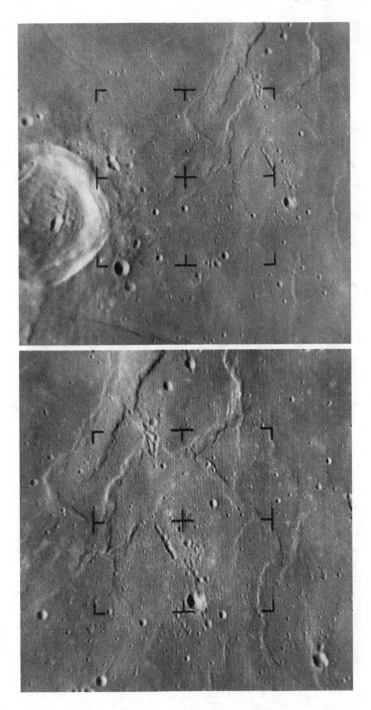

Images taken from altitudes of 197 km (top) and 150 km by the F_a-camera of Ranger 8 on a slanting trajectory to Mare Tranquillitatis on 20 February 1965.

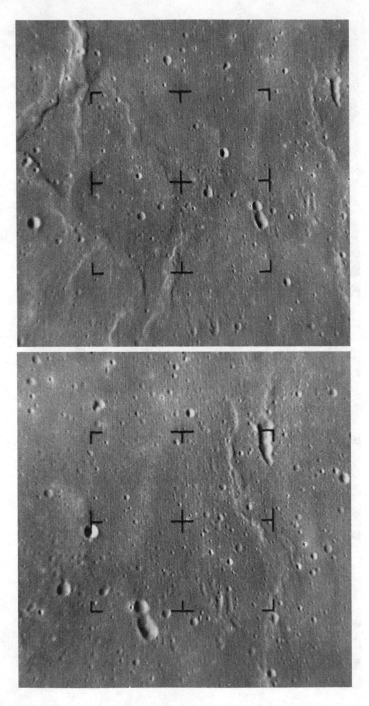

Images taken from altitudes of 105 km (top) and 77 km by the F_a-camera of Ranger 8 on a slanting trajectory to Mare Tranquillitatis on 20 February 1965.

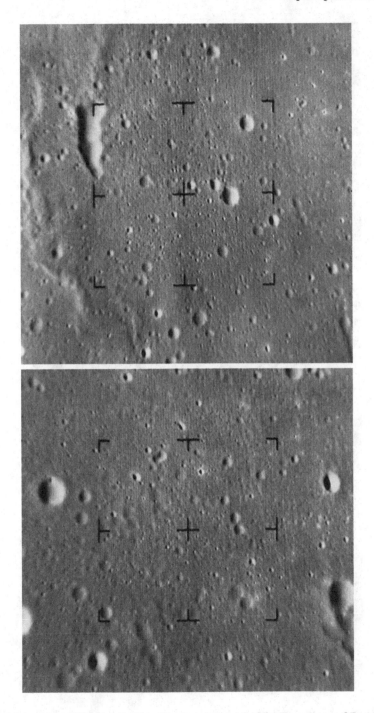

Images taken from altitudes of 49 km (top) and 31 km by the F_a-camera of Ranger 8 on a slanting trajectory to Mare Tranquillitatis on 20 February 1965.

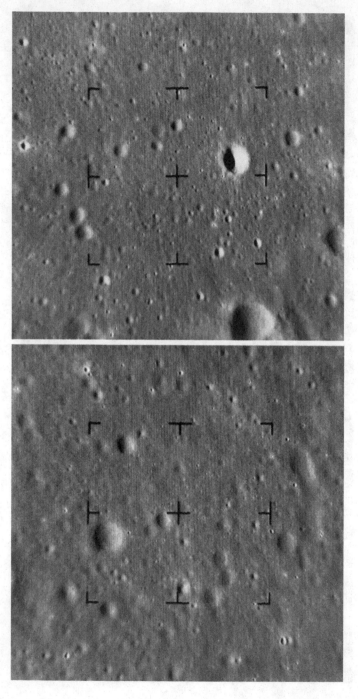

Images taken from altitudes of 21.8 km (top) and 12.7 km by the F_a-camera of Ranger 8 on a slanting trajectory to Mare Tranquillitatis on 20 February 1965.

might be 'hybrid craters' that were excavated by impacts and later modified by volcanism stimulated by their formation – indeed, they were the exemplars of this hypothesis.

TV SPECTACULAR

On 27 February 1965 the Ranger experimenters met at JPL with representatives of the Surveyor and Apollo projects to consider the target for the final Block III with a window that would open on 19 March. The Moon would be 'full' on 17 March and 'last quarter' on 25 March. The Surveyor people argued for Oceanus Procellarum at the western end of the Apollo zone, to identify a safe target for their first soft-lander, but this was rejected. Accepting that the maria were probably all much the same, the Apollo people suggested inspecting a blanket of ejecta, to gain an impression of the roughness of such terrain. The Ranger experimenters themselves argued for a target of particular scientific interest, and this was accepted.

The Ranger team met again on 2 March to consider specific features. To obtain unique data, they considered a variety of locations that were unlikely to be visited by either Surveyor or Apollo – three being Copernicus, Kepler, and Schröter's Valley near Aristarchus. However, Harold Urey and Gerard Kuiper were both in favour of the crater Alphonsus. Its floor was generally flat, but contained irregular rilles and a number of small 'dark-halo' craters which some people thought might be of volcanic origin. Following reports by Dinsmore Alter in America in 1956 of a slight "veiling" of the floor of Alphonsus, Nikolai Kozyrev had monitored the crater for any further such 'transient events', and on 3 November 1958 obtained a spectrogram of a "glow" obscuring the 1,100-metre-tall central peak using the 48-inch reflector of the Crimean Observatory. The spectrogram was disputed, but Kozyrev interpreted it as a release of gas. Alphonsus was therefore selected as the primary target for Ranger 9.

The experimenters could not agree a target east of the meridian for early in the window, but a launch on 21 March was compatible with Alphonsus, and thereafter Copernicus, Kepler and Aristarchus on successive days. This list was sent to the Office of Space Sciences and Applications on 10 March. Oran Nicks endorsed it, and passed it to Homer Newell, who concurred. However, NASA had scheduled the Gemini 3 manned mission for 22 March, and the Air Force required a clear 24 hours to reconfigure the Eastern Test Range for a different type of launch vehicle. On 15 March Robert Seamans ordered Gemini 3 postponed to 23 March to give Ranger 9 a chance at its primary target. When the countdown began, the Cape was cloudy and the low-altitude winds were gusty. The clock was held to await an improvement in conditions. Although it remained cloudy, when the winds declined it was decided to proceed.

Soon after lifting off at 21:37 GMT on 21 March (just before that day's window closed) the vehicle penetrated the cloud deck and was lost from sight. But everything went to plan. The midcourse manoeuvre was deferred to enable radio tracking by the Deep Space Network to precisely define the initial trajectory. It was calculated that

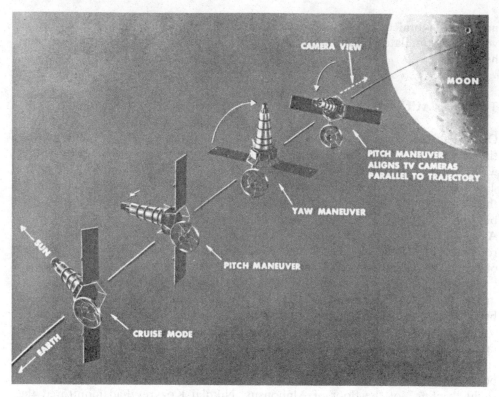

CAMERA VIEW

MOON

PITCH MANEUVER
ALIGNS TV CAMERAS
PARALLEL TO TRAJECTORY

YAW MANEUVER

PITCH MANEUVER

CRUISE MODE

SUN

EARTH

As Ranger 9 approached the Moon it performed a terminal manoeuvre to point its cameras along the velocity vector.

the spacecraft would impact some 640 km north of Alphonsus. The 31-second burn at 12:30 on 23 March ensured that it would fall into the 112-km-diameter crater. The aim point was midway between the central peak, the rilles and the dark-halo craters so that they would all be visible during the approach at a resolution better than was possible telescopically; but they would not appear in the final images, which would give a view of the floor of the crater.

As Ranger 9 approached the Moon on 24 March it became the first spacecraft in the project to make a terminal manoeuvre. At 13:31 the vehicle departed its cruise attitude by pitching, yawing, and pitching again whilst holding its high-gain antenna pointing at Earth. By aiming the cameras along the velocity vector, this manoeuvre would optimise the resolution of the final frames. Ray Heacock, the JPL member of the experiment team, provided the commentary in the auditorium. The electronic scan converter made for the Surveyor project had been hastily modified to process the Ranger video for 'live' broadcast by the TV networks. In essence, this comprised two sets of vidicon tubes (one for the wide-angle stream and the other one for the narrow-angle stream) and in each case one vidicon viewed the image displayed on its counterpart, in the process converting the 1,132 lines per frame received from the spacecraft into the 500 lines of the commercial

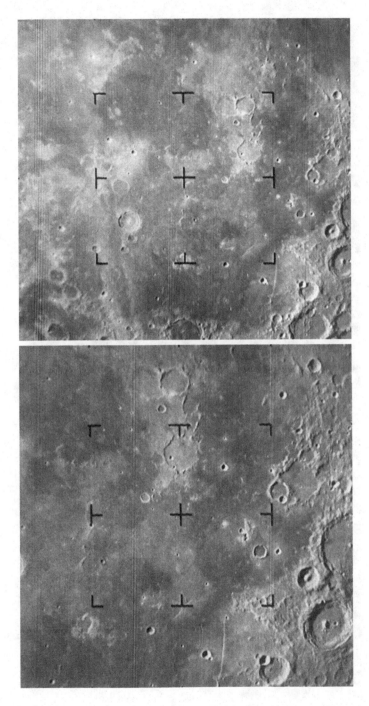

Images taken from altitudes of 2,377 km (top) and 1,600 km by the F_a-camera of Ranger 9 as it dived into the crater Alphonsus on 24 March 1965.

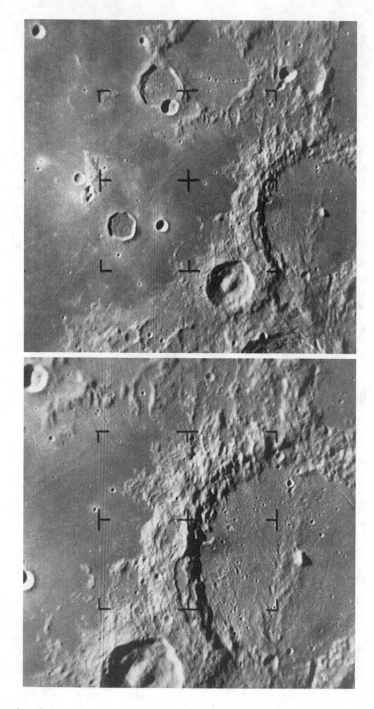

Images taken from altitudes of 703 km (top) and 416 km by the F_a-camera of Ranger 9 as it dived into the crater Alphonsus on 24 March 1965.

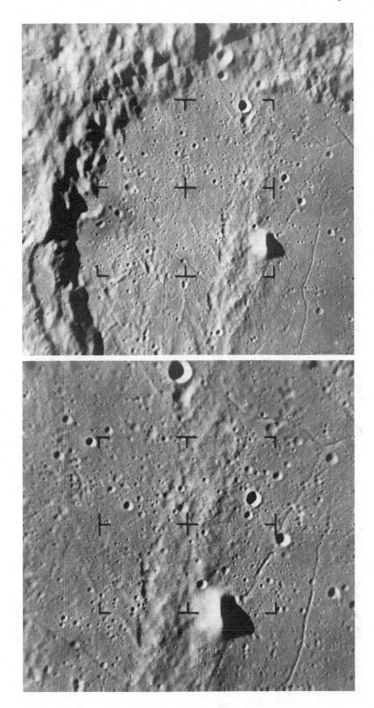

Images taken from altitudes of 227 km (top) and 130 km by the F_a-camera of Ranger 9 as it dived into the crater Alphonsus on 24 March 1965.

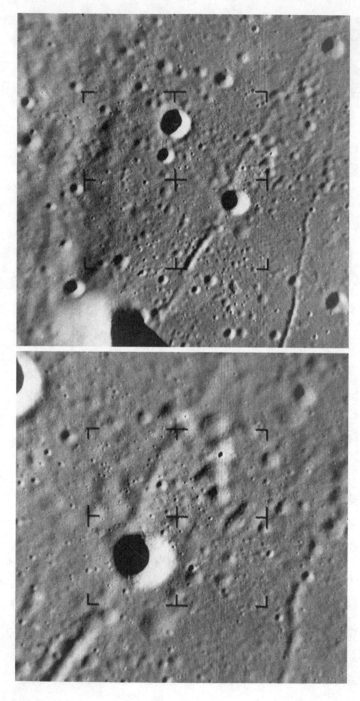

Images taken from altitudes of 69 km (top) and 32 km by the F_a-camera of Ranger 9 as it dived into the crater Alphonsus on 24 March 1965.

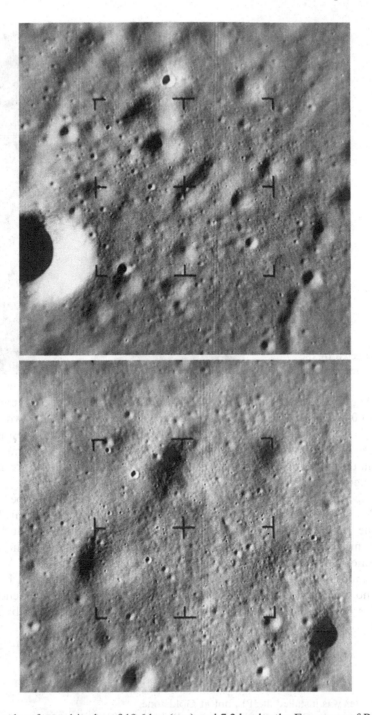

Images taken from altitudes of 19.6 km (top) and 7.2 km by the F_a-camera of Ranger 9 as it dived into the crater Alphonsus, on 24 March 1965.

The sites inspected by the
successful Ranger missions.

system.[4] Imaging began at an altitude of 2,400 km, lasted 18 minutes, and ended with
impact at 14:08:20 when the spacecraft hit the ground at 9,617 km/hour within 6 km
of the aim point.[5] The final picture of the 5,814-frame sequence had a resolution of
25 cm. It was a spectacular way to conclude the Ranger project! The transmission
impressed not only the public, but also those scientists who had not appreciated the
value of imagery.

At the experimenters' press conference later in the day, Gerard Kuiper said the
'dark halo' craters on the floor of Alphonsus appeared to be volcanic. Harold Urey,
no fan of the 'hot' Moon hypothesis, allowed that they were probably "due to some
sort of plutonic activity". The rilles were revealed not to be clefts but chains of small
irregular craters, which some people argued were volcanic. The walls of Alphonsus
were very smooth. The central peak was not the harsh edifice depicted by artists.
There was no indication that it was a volcano, but neither was there evidence that it
was not. There was no clue as to the cause of the 'transient events' seen by Alter or
Kozyrev.

---

[4] The converter was installed at JPL, not at Goldstone.
[5] The crater made by Ranger 9 was photographed by Apollo 16 in 1972. At 14 metres in
diameter, it was similar to that of its predecessor. .

MISSION ACCOMPLISHED

The three successful Rangers satisfied the objective set for the Block III series in terms of supporting Apollo. The maria proved to be cratered on all scales, but with a smoothly undulating surface of generally shallow slopes. And the presence of large blocks of rock lying on the surface suggested sufficient bearing strength to support a lander. In addition, radio tracking had enabled the estimate of the mass of the Moon to be much improved. It also established the axis that is aligned towards Earth to be about 1 km longer, with the centre of mass being offset several kilometres from the geometric centre in a direction away from Earth.

When some 200 scientists gathered at the Goddard Space Flight Center in April 1965 to discuss the combined results of the Ranger project, Harold Urey and Gerard Kuiper still disagreed about whether the Moon was thermally differentiated. Thomas Gold insightfully noted that the pictures represented a mirror in which each person saw evidence to support his own hypothesis.

One lesson of Ranger was that lunar geological units were so severely blurred by impact 'tilling' at the fine scale that in undertaking photogeological mapping it was better to use a medium scale of 1:1,000,000, as was used by the Air Force Chart and Information Center in St Louis for its Lunar Astronautical Chart series.

Ranger had provided a close look at several sites, but what was required next was for an orbiter to provide a broader view at better-than-telescopic resolution and for a soft-lander to provide 'ground truth'.

8

Soviet activity

THE SECOND GENERATION

After the success of Luna 3, the Soviets developed a new spacecraft designed to deliver a capsule to the lunar surface using the rough landing technique. Luna 4 was launched at 08:16 GMT on 2 April 1962, and after cruising in parking orbit it set off for the Moon. After an ineffective midcourse manoeuvre, the 1,422-kg vehicle made a flyby at 13:25 on 5 April at a range of 8,500 km and passed into solar orbit. After several further failures, the Soviet Union conducted a deliberate flyby mission.

FILLING THE GAP

Zond 3 lifted off at 14:38 GMT on 18 July 1965. After parking orbit, it was sent on a trajectory to pass by the illuminated leading limb of the Moon. Imaging began at 01:24 on 20 July at an altitude of 11,570 km and ended at 02:32 at 9,960 km, with the closest point of approach at 9,220 km. It had been intended to launch this probe in 1964 as a companion to Zond 2 on a mission to Mars, but it was held back. The pictures were not transmitted until the narrow-beam of the high-gain antenna was able to lock onto Earth, which occurred on 29 July at a range of 2.2 million km. The objective of this flight was to test deep-space communications for an interplanetary mission, and the Moon was merely a convenient photographic target. It transmitted two dozen pictures of Oceanus Procellarum and around onto the far-side to view the area which had not been visible to Luna 3.

The results indicated that although there were few maria on the far-side, and those were small, there were multiple-ring structures which for some reason had not been flooded by lava. The Orientale basin was seen in its entirety for the first time, since even at the most favourable libration barely half of it was observable in reprojected telescopic pictures. In addition to the concentric rings, there were radial patterns in evidence. There was a small patch of mare material inside the central ring, and small patches between the rings, but otherwise the entire structure was 'on display' in its magnificence. It boggled the mind that Earth must once have been disfigured by such structures!

ON THE SURFACE

The Soviet effort to deliver a capsule to the lunar surface using the rough landing technique finally succeeded with Luna 9. This was launched at 11:42 GMT on 31 January 1966. Its mass was 1,538 kg, including the surface capsule. The midcourse manoeuvre was made at 19:29 on 1 February. As with the Block II Ranger, the inability to deal with a lateral velocity component in the descent limited targets to longitudes of about 64°W and fairly near the equator. In this case, the target was in Oceanus Procellarum, near Hevelius. At an altitude of 8,300 km, with about half an hour to go, the spacecraft aligned its main axis to local vertical. The radar altimeter initiated the retro manoeuvre at 18:44:42 on 3 February, at an altitude of 75 km. At 18:45:30, after slowing by 2.6 km/sec, the engine was cut off when a 5-metre-long probe made contact with the surface, and simultaneously the payload was ejected upward and to the side. The bus hit the ground at 6 m/s and its transmission ceased.

The 250-foot-diameter radio dish at Jodrell Bank in England was the largest fully steerable antenna in the world, and it was monitoring the transmission. When the signal ceased, Bernard Lovell, the head of the facility, wrote it off as another failed landing. But the shock-proof 58-cm-diameter spheroidal capsule rolled to a halt and, some 250 seconds after being released, initiated its own transmission. Four petals opened to right and stabilise the capsule and to expose its contents, which comprised a radiation detector and a line-scan TV camera that pointed upward and viewed the landscape using a nodding mirror that could rotate in azimuth.

Between 01:50 and 03:37 on 4 February a panoramic picture was built up line by line and the data transmitted in real-time. Jodrell Bank recorded the transmission. On a hunch, Bernard Lovell asked the local office of the *Daily Express* to provide a commercial wire-facsimile machine, and the signal was fed into it. Even before the Soviets announced their probe had transmitted a picture, the 'scoop' was published in Britain with the headline: *From Luna 9 to Manchester – The Express Catches the Moon*. Unfortunately, not knowing how to extract the aspect ratio of the image from the telemetry, they had guessed, and caused the horizontal scale to be compressed by a factor of 2.5, and since it was consistent with the popular expectation of the lunar surface, the resulting jagged landscape seemed 'right'. The ruggedness was further emphasised by the fact that the Sun was just 7 degrees above the horizon and cast very long, very dark shadows. The surface *looked* like glassy scoriaceous lava that would be very treacherous for an astronaut to walk on – much like the 'aa' lava in Hawaii, so named because a person walking on it tends to cry that sound!

When the official version was issued later using the true aspect ratio, the jagged 'spikes' were seen to be just rocks resting on the surface, and the scene was rather less dramatic. The capsule had come to rest oriented 16.5 degrees off vertical. The field of view spanned 11 degrees above and 18 degrees below the perpendicular to the capsule's axis, with a series of 6,000 vertical lines spanning a full 360 degrees of

A model of the Luna 9 spacecraft showing the spheroidal surface capsule attached to the bus, and (right) the capsule in its deployed configuration. The camera is the cylindrical unit on the axis.

Two sections of a panoramic image transmitted by Luna 9. (Courtesy of Philip J. Stooke, adapted from *International Atlas of Lunar Exploration*, 2007)

azimuth. As the mirror was only 60 cm off the ground, the perspective was very low, with objects in the foreground appearing larger than they were, and the horizon was very close as a result of the capsule having landed in a shallow 25-metre-diameter crater. There was no sign of the bus.

Gerard Kuiper claimed that there were vesicles in the rocks, which supported his idea that the maria were volcanic. As meteors were particles of dust that penetrated the Earth's atmosphere, it seemed only reasonable that the airless Moon would have accumulated a blanket of such material, but this did not seem to be the case. Thomas Gold responded to the evident absence of dust (on this patch of mare, where it could reasonably have been expected to be very thick) by suggesting that the 'rocks' were not fragments of lava but fine powder which had adhered to form clods. The surface clearly had sufficient bearing strength to support the capsule's 100-kg mass – but in the weak lunar gravity its *weight* was one sixth of this value. To the Apollo planners, this was the most significant result of the mission. Gold argued that the capsule was spreading this load across the four deployed panels, and *in time* it would sink from sight. The geologists of the Branch of Astrogeology inferred that the surface was (to use Harold Urey's term) gardened by impacts. The site was on a dark geological unit that Jack McCauley, in making the Lunar Astronautical Chart for this area, had interpreted as a pyroclastic blanket with lava flows. Although there was nothing in the image to suggest pyroclastic, it did indeed look like a lava flow, and judging by the sharpness of the rocks and the absence of dust it was relatively young in terms of lunar history.

A second panorama was taken between 14:00 and 16:54 on 4 February, and this showed that the capsule had increased its tilt to an angle of 22.5 degrees, altering the angle of the horizon. Gold claimed this was evidence of the capsule sinking into the dust. The offset had the benefit of facilitating limited stereoscopic analysis. Before the battery expired on 6 February, further partial pans were made to observe how the illumination changed as the elevation of the Sun in the lunar sky increased, thereby demonstrating the value of repeatedly imaging a scene from a fixed vantage point.

IN LUNAR ORBIT

Having achieved the first delivery of a scientific capsule to the lunar surface, the Soviets moved on to attempt to be first to put a satellite into orbit around the Moon. Luna 10 was launched at 10:47 GMT on 31 March 1966. After cruising in parking orbit, it set off for the Moon. It was the same type of bus as Luna 9, but was ferrying an instrument capsule instead of the landing capsule. The midcourse manoeuvre on 1 April refined the trajectory to aim for the point in space at which the 850-m/s orbit insertion burn would be made. When 8,000 km out, it oriented itself for braking. The burn was initiated at 18:44 on 3 April, and slowed the spacecraft sufficiently for it to enter a 350 × 1,017-km orbit with a period of 178 minutes, in a plane inclined at 72 degrees to the lunar equator. The fact that the change in velocity to enter orbit was considerably less than that to land meant that propellant could be traded in favour of an increase in the payload to 245 kg. Shortly after entering orbit the bus released the 1.5-metre-long capsule containing a micrometeoroid detector, radiation detectors, an infrared sensor to measure the heat flux from the Moon, a gamma-ray spectrometer to detect radioactive isotopes, and a magnetometer to follow up the measurements by the early flyby probes. The mission ended when the battery expired on 30 May, after 56 days during which the capsule made 460 revolutions.

СХЕМА ПОЛЕТА АВТОМАТИЧЕСКОЙ СТАНЦИИ «ЛУНА-10»

The instrument compartment of the Luna 10 bus was released once the vehicle had entered orbit around the Moon. The orbit was steeply inclined to the lunar equator in order to enable the gamma-ray spectrometer to survey the surface composition.

The gamma-ray spectrometer was similar to that of the Ranger Block II, but more useful by virtue of being placed into orbit to survey a wide area. The instrument was a scintillation spectrometer with a resolution of 32 channels within the energy range 0.3–3.0 MeV. The surface resolution was rudimentary. The data was consistent with the proposition that the maria were basaltic, but was inconclusive. About the only positive conclusion was there were no large surface exposures of acidic rock such as granite. The question for the 'hot Moon' hypothesis advocated by Gerard Kuiper, was why the process which produced 'continental' material on Earth had seemingly not done so on the Moon. The *mystery* was the composition of the highland material. If a global magnetic field existed, then it was weaker than 1/100,000th that of Earth. Radio occultations on crossing the limb showed no hint of the Moon having even a tenuous envelope of gas. Intriguingly, radio tracking revealed the gravitational field to be uneven.

9

Developing Lunar Orbiter

CHANGING HORSES

In January 1961 the Hughes Aircraft Company was selected to build the Surveyor spacecraft. With a planned mass of about 1,125 kg at translunar injection, it would require the Atlas-Centaur. The plan was for the orbital version to provide wide-area mapping and reconnaissance of potential landing sites for the surface Surveyors and, later, for Apollo. The mass at touchdown was expected to be about 340 kg, of which 114 kg would be scientific instruments which would not only transmit pictures but also provide data on the physical, chemical, mineralogical and biological properties of the surface material. The initial schedule called for the first flight in 1964. It was also envisaged that as the project matured, an orbital variant would be equipped to serve as a communications relay for landers investigating sites on the far-side of the Moon.

On 23 March the Lunar Science Subcommittee at the Office of Space Sciences recommended that the orbiter have a TV system which was capable of providing: (1) full coverage of the limb areas (highly foreshortened to terrestrial observers) and of the far-side at a resolution of 1 km, (2) wide-area reconnaissance at a resolution of 100 metres, and (3) stereoscopic pairs of selected areas with sufficient resolution to discern objects 10 metres in size. On 5 December Charles Sonett, Chief of Lunar and Planetary Sciences at the Office of Space Sciences, asked William Cunningham to determine the status of the orbiter. On 12 January 1962 Cunningham reported that JPL would not be able simply to adapt the vidicon system developed for Ranger; a new TV system would be required, the development of which had not yet begun.

NASA called for JPL to define the design requirements for the Surveyor orbiter, maximising its commonality with the lander, by 1 September 1962, but owing to the problems the laboratory was facing with Ranger the orbiter received little attention.

Meanwhile, on 15 June 1962 the Office of Manned Space Flight compiled a list of the data that it required the Office of Space Sciences to supply on the environment around the Moon and on its surface – i.e. Brainerd Holmes's *Requirements for Data in Support of Apollo*. In view of the urgency to feed such information into the design of the Apollo vehicles, the Office of Space Sciences asked JPL whether Ranger could serve as the basis for an orbiter. JPL in turn asked the Hughes Aircraft Company to consider the possibility of a 360-kg orbiter which could be launched by the Atlas-Agena. Hughes replied that in order to meet this mass limit, the scientific payload could not exceed 27 kg, which was unrealistic in view of the activity to be pursued. JPL calculated that if the solid-rocket motor that Surveyor was to use in the initial phase of its descent to the Moon were to be used to augment the Agena in the translunar injection, it would be possible to increase the scientific payload to 57 kg, but this was still too little. Even although the development of the Centaur stage was running behind schedule, the Office of Space Sciences decided to proceed with the Centaur-based orbiter.[1] To meet the Apollo requirements, the orbiter would require to provide photography of potential landing sites capable of revealing protuberances and pits as small as 1 metre in size and slopes as shallow as 7 degrees. But even the stereoscopic views from the Surveyor orbiter's TV system would have a resolution no better than 10 metres. A photographic system employing film would be needed to meet the requirements of the Office of Manned Space Flight.

On 21 September 1962 Oran Nicks, Director of Lunar and Planetary Programs in the Office of Space Sciences, asked Lee R. Scherer to form a committee to evaluate proposals which had been submitted by the Space Technology Laboratories and the Radio Corporation of America for a 'lightweight' lunar orbiter compatible with the Atlas-Agena.

On 23 October Joseph Shea, the Deputy Director of the Office of Manned Space Flight, specified the relative priorities of the data that Apollo would require from the Office of Space Sciences. There was a greater need for the information which a soft-lander would provide, since this would feed into the design of the Apollo vehicles, whereas the information from an orbiter would not be required until later, in mission planning. Shea stressed that if funding was tight in Fiscal Year 1963, then the Office of Space Sciences should favour the lander over the orbiter.

Scherer reported to Nicks on the issue of an Agena-based orbiter on 25 October. The proposal by the Radio Corporation of America was for a Ranger bus to make a lunar flyby, dropping off a 200-kg package which would insert itself into orbit. The orbiter would be 3-axis stabilised and use a vidicon system (no doubt a development of the camera the company had provided for Ranger Block II) to provide pictures at a resolution of 130 metres in the wide-angle coverage and 30 metres in the narrow-angle coverage. The Space Technology Laboratories had envisaged an orbiter with a mass of 320 kg. It would have a monopropellant engine which was capable of firing several times. In addition to a midcourse manoeuvre and orbit insertion, this engine would permit changes to the orbit. One mission profile would be to enter a circular

[1]   The delay in the Centaur stage was in part due to problems with the configuration of its propellant tanks, but also because the Marshall Space Flight Center was busy with the Saturn launch vehicles. In early 1962, therefore, the Centaur had been transferred to the Lewis Research Center.

polar orbit at an altitude of 1,600 km and map the entire Moon, resolving objects as small as 18 metres in size. Alternatively, it could be placed into equatorial orbit at an altitude of 40 km to photograph that zone with a resolution of 0.5 metre. It would be spin-stabilised, and use a 'spin scan' camera of a design similar to that proposed by the RAND Corporation in 1958. It would use film to obtain a higher resolution than was obtainable using a vidicon. Scherer reported that only the proposal by the Space Technology Laboratories offered the prospect of meeting the requirements set by the Office of Manned Space Flight for imaging resolution, and he recommended that the company further refine the concept so as to enable the Office of Space Sciences "to establish the confidence needed [to consider] a flight program of this type, should it be deemed preferable to a Centaur-based orbiter". In fact, once the viability of an Agena-based reconnaissance orbiter had been established, this in itself undermined the case for pursuing the Surveyor orbiter.

On 26 October, Clifford Cummings, unaware of Scherer's study, wrote to advise Oran Nicks that JPL was about to conduct a study to refine the configuration of the Surveyor orbiter in order to specify how it would perform its mission. In his reply on 8 November, Nicks pointed out that the Office of Space Sciences was looking into the possibility of an Agena-based orbiter.

On 2 January 1963 Nicks asked Floyd L. Thompson, the Director of the Langley Research Center, to consider the possibility of his staff taking on the development of a lightweight orbiter. Thompson set up an internal feasibility study. After the Space Technology Laboratories had refined its concept, a review was held at Langley on 25 February involving representatives of the company, the Office of Space Sciences, the Office of Manned Space Flight, Langley and Bellcomm. Lee Scherer and Gene Shoemaker reported on a study they had undertaken for Nicks to determine how a lightweight orbiter might satisfy the photographic requirements of Apollo. Dennis Jones of Bellcomm reported an assessment made for Shea on the degree to which an orbiter might support the manned and unmanned exploration of the Moon. A second meeting on 5 March agreed that not only was a lightweight orbiter viable, it would also significantly support Apollo. Langley then sent a delegation headed by Clinton E. Brown to brief Robert Seamans and present the case for Langley taking on such a project; Seamans authorised planning to proceed.

In order to assist Langley draw up the request for proposals, in April 1963 the Office of Manned Space Flight refined its requirements. The *critical* needs were: (1) data on the radiation flux in lunar space over a typical 2-week period; (2) a summary and analysis of all efforts for short-term prediction of severe solar proton events; (3) measurements of particles capable of penetrating 0.01 cm and 0.1 cm of aluminium in an average peak 2-week period of micrometeoroid activity; and (4) photographic data capable of showing protuberances 3.5 metres tall and slopes of 15 degrees in an area of the lunar surface with a radius of 60 metres (to be provided by the autumn of 1965) and then equivalent data showing 50-cm protuberances and 8-degree slopes in an area with a radius of 1,600 metres. Other needs were: (1) measurements of the distribution of slopes greater than 15 degrees in areas of 3.5 metres radius; and (2) the greatest possible coverage of the zone within 5 degrees of the lunar equator with a resolution of 25 metres or better.

On 25 April Edgar Cortright put it to Homer Newell that since one successful orbiter could be worth "dozens of successful Ranger TV impactors", the three new Rangers which had recently been funded in order to obtain high-resolution pictures and gamma-ray and radar reflectance data on the Moon should be cancelled. Newell accepted this reasoning and passed the recommendation to Robert Seamans, who concurred on 12 July. Later in the year, the second batch of rough landing Rangers was also cancelled.

BOEING WINS

On 25 June 1963 Floyd Thompson went to Washington to define the terms of the request for proposals. In particular, he did not wish it to be stated that the spacecraft should be spin stabilised; he wished to see what the bidders proposed. It was agreed to say only that the primary requirement was photographic data at medium and high resolution in order to facilitate the selection of sites for Surveyor and Apollo landers. The secondary objectives were to provide information on the size and shape of the Moon and the properties of its gravitational field. Information would also be sought on conditions near the Moon, including the micrometeoroid flux and total exposure to energetic particles and gamma rays – the latter having been shown by Ranger 3 to exist. A key requirement of the photographic system was that it identify the altitude of the orbiter at the time of an exposure, the orientation of the line of sight (relative to lunar north) and the angle of the Sun to the surface. In particular, it was desired to be able to determine the location of any surface feature to an accuracy of 1 km.

On 23 August Lee Scherer presented the request for proposals to Oran Nicks and Edgar Cortright, who duly reviewed it with Robert Seamans. The Project Approval Document signed by Seamans on 30 August officially initiated Langley's first deep-space project. It was given the mundane name of Lunar Orbiter. The Lunar Orbiter Project Office was set up at Langley, with Clifford H. Nelson as Project Manager,[2] William J. Boyer as Operations Manager and Israel Taback as Spacecraft Manager. In Newell's office, Lee Scherer was appointed as Lunar Orbiter Program Director, Leon J. Kosofsky as Program Engineer and Martin J. Swetnick as Program Scientist.

On 30 August 1963 NASA invited bids from industry. In September the Lunar Orbiter Project Office established a Source Evaluation Board chaired by Eugene C. Draley of Langley. Five bids were received. The evaluations began in October and ran to late-November.

A key factor in the requirements was that, where possible, off-the-shelf hardware be used to minimise the development effort. The Hughes Aircraft Company, which was prime contractor for Surveyor and would have built the 3-axis-stabilised orbiter for that project, proposed a spin-stabilised spacecraft that would use a solid rocket

---

[2]   In October 1964 Langley recruited James S. Martin from Republic Aviation as Assistant Project Manager.

motor to enter lunar orbit. The Space Technology Laboratories submitted a refined version of its spin-stabilised design. The Martin Company, which supplied the Titan missile to the Air Force but had limited experience of spacecraft systems, offered a 3-axis-stablised design. The Lockheed Missile and Space Company, which had built the Agena as a 3-axis-stablised vehicle and integrated various payloads into it for the Air Force, including reconnaissance cameras, suggested that the Agena be adapted to operate in lunar orbit. Eliminating the need to develop a new vehicle satisfied the desire for off-the-shelf hardware, but the operational concept was flawed because it would require a lot of propellant to insert such a heavy rocket stage into lunar orbit. The Boeing Company's expertise was aircraft, but it wished to gain experience with spacecraft systems. It proposed a 3-axis-stabilised spacecraft with a mass of 360 kg that would enter lunar orbit using a liquid rocket (just developed by Marquardt as an attitude control thruster for the Apollo spacecraft) and be powered by solar panels. The Source Evaluation Board was particularly impressed by Boeing's plan to use a lightweight form of a photographic system developed by Eastman Kodak in 1960 for a reconnaissance satellite. The camera used two lenses in a configuration that would take wide-angle and narrow-angle frames simultaneously and interleave them onto a single strip of film.[3] The film would be developed and fixed using the 'semi-dry' Bimat process introduced by Kodak in 1961, as this obviated the complication of handling 'wet' chemicals in weightlessness.[4] The clinching argument in favour of Boeing was the proposal to use Kodak SO-243 fine-grain aerial film to obtain the required high resolution. This film had an exceedingly 'slow' rating of 1.6 ASA, whereas the other bidders intended to use 'fast' film. In the case of the spin-stabilised designs, a high-speed film was essential. But adding up the time spent flying to the Moon, the time spent in orbit preparatory to imaging, the 10 days spent imaging, and the time spent scanning and transmitting the film, a mission might last up to a month. During this time there was a fair chance of the particle radiation from a solar storm 'fogging' a high-speed film, and the heavy shielding to protect it would be prohibitive. Boeing's proposal to use slow film showed that the company had a better understanding than its competitors of the mission requirements. The Source Evaluation Board strongly recommended in favour of Boeing, and this was accepted. On 20 December 1963 James Webb announced that the contract would be awarded to Boeing of Seattle, Washington.

Boeing appointed Robert J. Helberg to manage the development of Lunar Orbiter. George H. Hage was Chief Engineer. Carl A. Krafft, the Business Manager, led the contract negotiations that began on 6 January 1964 and involved both Langley and the merged Office of Space Sciences and Applications. Boeing subcontracted Kodak to provide the photographic system, and the Radio Corporation of America for the communications system. In March, Boeing

---

[3]   In the case of Lunar Orbiter, the wide-angle images would be referred to as medium (M) frames and the narrow-angle images as high-resolution (H) frames.

[4]   In fact, Bimat was similar to the Polaroid process.

suggested that the photographic data be processed into pictorial format at Kodak in Rochester, New York, where there was already the necessary equipment, but NASA decided that the processing, handling and distribution of all scientific data provided by Lunar Orbiter should be done at Langley – in the case of photographic data by utilising equipment and technicians supplied by Kodak. Langley appointed Calvin Broome as Chief of the Photographic Subsystem Section.

The plan called for five Lunar Orbiter missions to be launched by Atlas-Agena D, with the first in either late 1965 or early 1966. They were to photograph the lunar surface from a perilune of 40 km. As in the case of Ranger at JPL, Langley would be responsible for overall systems integration of the spacecraft and the launch vehicle, as well as the necessary ground support, but, significantly, by this point NASA had gained control of both the procurement of launch vehicles and of launch operations. Because JPL had established the Deep Space Network to track and communicate with spacecraft, the Lunar Orbiters would be run from the Space Flight Operations Facility. In April 1964, Langley discussed this collaboration with Eberhardt Rechtin. This was the first time that JPL had provided another NASA centre with deep-space support, and so, in effect, a 'contract' had to be negotiated to define what JPL would do. But since trajectory design was closely related to the design of the spacecraft's communications system, and JPL had neither the manpower nor the computer time available to involve itself in this, the transit trajectory and operations in lunar orbit

On 16 April 1964 the Langley Research Center signed the contract with Boeing to develop the Lunar Orbiter spacecraft. It was sent to NASA headquarters for final review and endorsed by James E. Webb on 7 May.

would have to be planned by Langley and Boeing after JPL had educated Boeing's engineers in the capabilities and procedures of the Deep Space Network.

Langley and Boeing signed the detailed contract on 16 April 1964. It was sent to NASA headquarters for ratification. James Webb agreed on 7 May, and the formal contract was signed on 10 May.

MISSION OBJECTIVES

On 12 May 1964 the Office of Space Sciences and Applications announced how Lunar Orbiter would satisfy Apollo's requirements for maps of the Moon, as agreed with William B. Taylor of the Advanced Manned Missions Program Directorate of the Office of Manned Space Flight. The Manned Spacecraft Center in Houston was interested primarily in the near-side within 5 degrees of latitude of the equator, and had specified stringent requirements for accuracy of selenodetic and topographic data in the vicinity of selected landmarks to assist in navigation in orbit and landing site selection. The US Geological Survey was to produce a variety of maps based on Lunar Orbiter photography.

Oran Nicks suggested to Sam Phillips on 23 September 1964 that the Office of Manned Space Flight should make a study of how Lunar Orbiter could best support Apollo. This would aid the Lunar Orbiter Project Office in developing guidelines for mission planning. Bellcomm was asked to make this study, and on 25 January 1965 Douglas D. Lloyd and Robert F. Fudali submitted the report *Lunar Orbiter Mission Planning*. This discussed the relative merits of clockwise and anticlockwise orbits of the Moon aligned near the lunar equator. It was confirmed that to achieve the specified 1-metre resolution in the H frames the pictures could be taken from an altitude no greater than 46 km. A strategy of obtaining contiguous high-resolution coverage of multiple targets was recommended. To avoid the possibility of orbital instability as a result of such a low perilune, it was recommended that the initial inclination of the orbit should not exceed 7 degrees to the lunar equator (because gravity perturbations would tend to increase the inclination) and that the spacecraft should have sufficient propellant to perform corrective manoeuvres. Bellcomm followed up on 30 March with *Apollo Lunar Site Analysis and Selection*, which recommended that the Office of Manned Space Flight and the Office of Space Sciences and Applications form a Site Survey Steering Committee with responsibility for choice of measurements and their relative priorities and instruments, target selection, launch schedules, control of data handling, and methods of data analysis for the Lunar Orbiter and Surveyor missions. On 10 May Bellcomm further recommended that the Office of Manned Space Flight and the Office of Space Sciences and Applications create a joint Lunar Surface Working Group to coordinate mutual planning activities concerning site survey requirements and the means by which these should be satisfied.

In May the Surveyor/Orbiter Utilisation Committee was formed. It was chaired by Edgar Cortright, and its membership comprised senior representatives of these two programs and their project offices: Oran Nicks of Lunar and Planetary

Programs, Urner Liddel of Lunar and Planetary Science, Lee Scherer of the Lunar Orbiter Program, and Benjamin Milwitsky of the Surveyor Program, all of whom were from the Office of Space Sciences and Applications; Israel Taback of the Lunar Orbiter Project Office at Langley; Victor Charles of the Surveyor Project Office at JPL; Sam Phillips, the Apollo Program Director and Everett E. Christensen of Manned Operations, both at the Office of Manned Space Flight; and William A. Lee of the Apollo Spacecraft Project Office and William E. Stoney of Data Analysis, both at the Manned Spacecraft Center. The Committee was to coordinate the Surveyor and Lunar Orbiter projects for their mutual benefit and in support of Apollo. In July, the Apollo Site Selection Board was established in the Office of Manned Space Flight. Although the Surveyor/Orbiter Utilisation Committee would gather engineering and science information and *assess* proposals for Lunar Orbiter imaging coverage and for Surveyor landing sites, and later *recommend* landing sites for Apollo, the Apollo Site Selection Board chaired by Sam Phillips would make the *decisions*.

The Surveyor/Orbiter Utilisation Committee's first meeting on 20 August 1965 discussed four Lunar Orbiter mission options which had been developed by Langley and Boeing in response to Bellcomm's report. In order of priority they were: type 1, to photograph ten evenly distributed target areas near the equator, each of which would be covered stereoscopically with both M and H frames; type 2, to photograph four areas in order to 'screen' for possible Surveyor landing sites near the equator; type 3, to photograph using H frames an area containing a landed Surveyor in order to study its context; type 4, to obtain topographic data which would not otherwise be obtained. It was decided to start with the type 1 mission, in order to provide as soon as possible the data that was required by the Apollo planners. If the Office of Space Sciences and Applications had not been obliged to support Apollo, the preferred first mission would have been to enter a high circular polar orbit for a global survey at a resolution better than that obtainable using a terrestrial telescope and, significantly, to view the limbs from a vertical perspective.[5] In 1963, when the Office of Manned Space Flight began to specify its requirements for Apollo in terms of surface slopes, Gene Shoemaker had hired Jack McCauley to develop methods of photoclinometry. In June 1965 the Surveyor project asked McCauley to use this technique to suggest possible landing sites for their landers. He formed a small team and compiled a list of 74 sites. Owing to uncertainty in the accuracy of Surveyor's approach trajectory, the sites were specified in terms of 'target circles' 25, 50 and 100 km in radius. After factoring in vertical descent and illumination constraints, they selected only circles of 25 and 50 km radius. McCauley presented the final list to the Surveyor/Orbiter Utilisation Committee on 20 August. There were 24 sites with 50-km-radius circles on the maria, and seven in the highlands. There were also 13 'scientific' targets with 25-km-radius circles that would require greater landing accuracy.[6]

[5] As yet, the only images of the far-side had been provided by Luna 3 in October 1959 and Zond 3 in July 1965.

[6] In fact, all Surveyors except the last would be sent to sites on McCauley's list.

On 8–9 September 1965 Langley hosted a meeting which (in part) drew up lists of photographic targets judged compatible with Apollo, Surveyor and Lunar Orbiter constraints. James Sasser of the Apollo Spacecraft Project Office in Houston argued for distributed coverage which 'sampled' different types of terrain near the equator, although with the emphasis on apparently smooth areas. Lawrence Rowan of the US Geological Survey described an analysis based on a map produced by the Air Force Chart and Information Center on a scale of 1:1,000,000. This analysis identified the types of terrain available for 'sampling' by Lunar Orbiter: namely an ordinary mare, a dark mare, mare ridges, mare rays, crater rims, deformed crater floors, and several different types of terrain in the highlands. These discussions led to the 'A' mission plan which was formally presented to the Surveyor/Orbiter Utilisation Committee on 29 September. This called for a type 1 mission to inspect a number of areas in the 'Apollo zone' – defined as being within 5 degrees of the equator and 45 degrees of the central meridian – to assess their suitability for Apollo and Surveyor landings. It would start with test pictures taken in the high-perilune initial orbit of sites between 60°E and 110°E. Although not in the Apollo zone, these pictures would show a vertical perspective of the limb region in which landmarks would later be selected for Apollo orbital navigation. After the perilune had been lowered, ten sites, mostly in the zone, would each receive a single photographic pass timed to maintain a given angle of illumination as the terminator advanced westward. The targets would cover a variety of terrains, including the Flamsteed Ring in Oceanus Procellarum, which was the favoured site for the first Surveyor. In May, a team of photo-interpreters led by Lawrence Rowan had been created by the US Geological Survey to suggest sites for Apollo. Each site was subjected to a detailed analysis, drawing in data from all sources. This work continued through the Summer Study on Lunar Exploration and Science held 19–31 July 1965 in Falmouth, Massachusetts. Rowan presented a list of ten potential Apollo landing sites to the Surveyor/Orbiter Utilisation Committee on 29 September (just over a month after the Committee received McCauley's list of candidate Surveyor sites – some sites were on *both* lists). The meeting approved the 'A' mission proposal with nine primary (P) sites, including several that were not on the smooth maria.

The Planetology Subcommittee of the Space Sciences Steering Committee met on 21–22 October to discuss the 'A' mission plan. The meeting was chaired by Urner Liddel, who was a member of the Surveyor/Orbiter Utilisation Committee. Harold Masursky of the US Geological Survey explained how the methods of structural and stratigraphic geological mapping would be applied to the pictures supplied by Lunar Orbiter. Liddel then wrote to Oran Nicks on 5 November to emphasise the merit of developing a Lunar Orbiter Block II for a multifaceted scientific study of the Moon to obtain the data which would be required to plan 'advanced' Apollo missions.[7]

---

[7] In particularly, the Planetology Subcommittee called for the Lunar Orbiter Block II to undertake selenodesy, gamma-ray, X-ray, magnetometry, microwave and non-imaging radar studies from orbit.

On 14 December 1965 Langley Director Floyd L. Thompson (farthest, front row) and George E. Mueller (to his right) from the Office of Manned Space Flight at NASA headquarters are briefed on the 'A' mission scheduled for the first Lunar Orbiter spacecraft.

The Apollo Site Selection Board held its inaugural meeting on 16 March 1966. Although the only materials available were telescopic studies and their interpretation on the basis of close-up views of three sites provided by the Ranger project, several potential areas were identified in the expectation that it would prove possible to land the first Apollo mission at one of them.

On 4 April Leonard Reiffel, representing Apollo, informed Oran Nicks of another Apollo requirement. The original plan had been to store all the data returned by the Lunar Orbiter missions on film, but magnetic tape had a greater dynamic range and was more readily processed by computer, and NASA wished the process of analysis to be as automated as possible – in particular the photoclinometry by which the US Geological Survey was to measure the slopes. Nicks duly ordered that state-of-the-art recorders be purchased to enable the data to be written directly onto tape.

By the time of the Apollo Site Selection Board's second meeting on 1 June 1966, Surveyor 1 had landed on the Moon and the first Lunar Orbiter was soon to attempt to photograph it to provide a sense of context which would allow the 'ground truth' from the lander to be applied more generally.

THE SPACECRAFT

Except for a few minor problems, the Lunar Orbiter design phase was finished by mid-April 1965. The spacecraft stood 1.7 metres tall, was 1.5 metres in diameter at the base of its body, and massed 385 kg. Its interior had three decks. The lowest and largest deck contained the flight programmer, inertial reference unit, photographic system, Canopus star tracker, communications system, batteries and the transponder for radio tracking. The 'windmill' of four large solar panels were affixed to the rim of the lowest deck, together with two antennas. The span was 4 metres with the solar panels deployed. The low-gain antenna on a boom mounted on one side was for the uplink, and the 92-cm-diameter high-gain antenna opposite was to send engineering telemetry and photographic and other scientific data. Both antennas operated in the S-Band. The antennas increased the overall span to 5.6 metres. The middle deck had the Sun sensors and the 100-pound-thrust bipropellant engine that burned hydrazine and nitrogen tetroxide. There were four tanks (two for each propellant) each driven by a nitrogen-pressurised bladder. As the propellants were hypergolic there was no need for an igniter. The top deck formed a heat shield to protect the spacecraft from the efflux of the engine, and it had four attitude control assemblies mounted around its rim, each including a pair of cold-gas jets that delivered a thrust of 0.5 pound and drew nitrogen from the same tank as pressurised the main engine.

The solar panels had a total of 10,856 cells for 375 watts, and were to hinge down and hold a fixed position after the spacecraft had separated from the Agena. While the vehicle was briefly oriented with its solar panels facing away from the Sun, and also when passing through the Moon's shadow, it would draw upon a rechargeable 12-ampere-hour nickel-cadmium battery. Much of the communications system supplied by the Radio Corporation of America was off-the-shelf. The equivalent of Ranger's computer/sequencer was the flight programmer. It had a 128-word memory, and was capable of storing commands for 16 hours of autonomous picture-taking. It would supervise the duration of a main engine burn by monitoring accelerometers until the requisite velocity was achieved. The engine was gimballed to allow for asymmetries in the thrust and offsets in the centre of mass. The attitude control system comprised the Sun and Canopus sensors, inertial reference unit and gas thrusters. Attitude control was directed by the flight programmer in conjunction with the associated electronics control assembly. Gyroscopes in the inertial reference unit would enable the flight programmer to recognise when the desired attitude had been achieved, then hold that attitude. By default, the spacecraft would align its roll axis to the Sun so that its solar panels would be illuminated. A feature of the inertial reference unit was that it could be instructed to remember the current attitude, and after the vehicle had temporarily adopted a different attitude (for example for photography) the inertial reference unit would automatically fire the thrusters to restore the previous attitude.

Those parts of the spacecraft structure that would be in direct sunlight would have to endure a temperature of +120°C, whereas those in the shade would be at −160°C. The temperature control system had to maintain the internal systems between

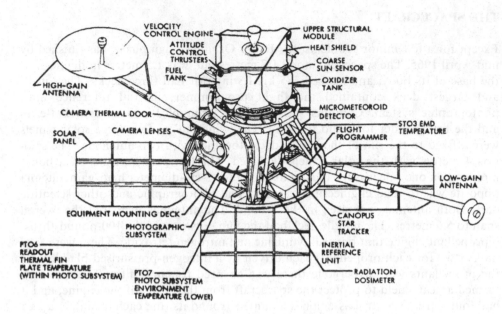

NOTE: SHOWN WITH THERMAL BARRIER REMOVED

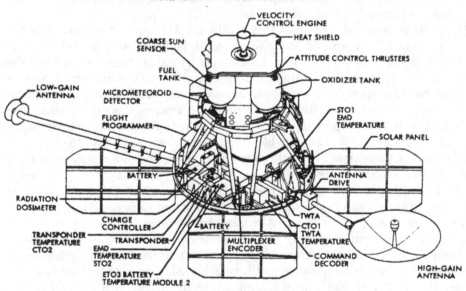

Details of the Lunar Orbiter spacecraft.

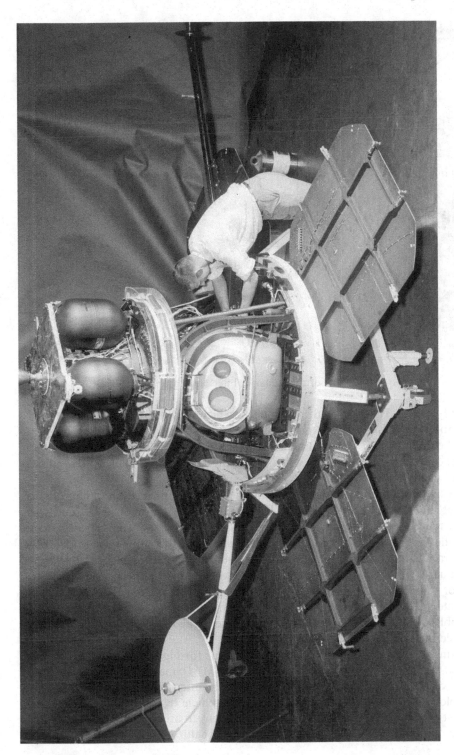

Working on a model of the Lunar Orbiter spacecraft.

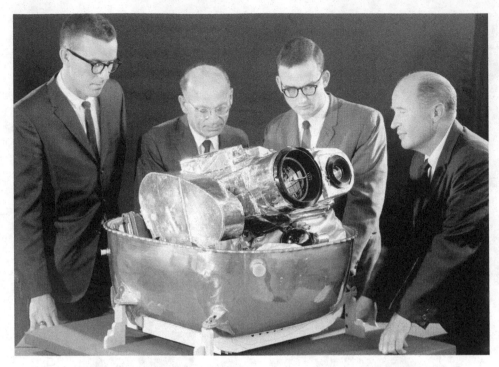

The two-lens camera of the Lunar Orbiter photographic system.

+2°C and +30°C. The base of the main deck, which would routinely face the Sun, had a coating of thermal paint. Apart from the lens of the Canopus sensor, the Sun sensors, the thermal door of the camera compartment and the heat shield on the top section, the remainder of the bodywork was protected by a blanket of aluminised mylar and dacron.

The 68-kg photographic system was contained in an ellipsoidal aluminium 'bath tub' about the size of a large suitcase. It was carried on the main deck, and was affixed by four legs to minimise conduction – the heat transfer was largely radiative, which made heat absorption and dissipation a slow and therefore smooth process. The camera was immersed in nitrogen in a pressure vessel, with its lenses observing through quartz windows. It was protected by a hinged thermal door that would be opened for a photographic session and closed immediately afterwards. The door was to eliminate light which might fog the film stored in the loopers between the various subassemblies, and it would also maintain the lenses at an even temperature in order to preclude thermal distortion of the optics.

The camera used a 610-mm-focal-length f/5.6 lens for high-resolution (H) frames and an 80-mm f/2.8 lens for medium-resolution (M) frames. The shuttering gave exposures of 1/25th, 1/50th or 1/100th second. At the intended perilune of 40 km the orbital velocity would be 1.6 km/sec, causing the vehicle to travel 64 metres during the slowest exposure. To enable the narrow-angle lens to attain the required 1-metre surface resolution, it was necessary to compensate for this motion. It was to be done

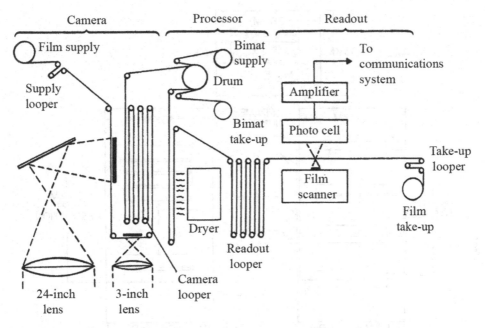

Detail of the Lunar Orbiter photographic system.

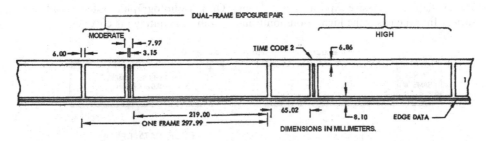

Each exposure of the Lunar Orbiter photographic system simultaneously took a regular medium-resolution frame and an elongated high-resolution frame, but the manner in which the film travelled through the camera meant that the pairs were interleaved.

using a velocity/height (V/H) sensor that made repeated circular scans of a portion of the image in the narrow-angle field in order to calculate the rate and direction of image motion, and then compensate for the changing perspective by driving a servo mechanism to vary the rate at which the film crossed the focal-plane shutter. Kodak subcontracted the V/H sensor and the narrow-angle shutter to Bolsey Associates. At an altitude of 40 km, the M frame would cover an area of 33 × 36 km at a resolution of about 7 metres, and the H frame 4.1 × 16.4 km at a resolution of 0.9 metre. Thus the H frame viewed 5 per cent the area of the M frame at 8 times the resolution. Because the two shutters were to operate together, each 'exposure' would comprise a pair of frames on the film, with the H frame corresponding to the central portion of

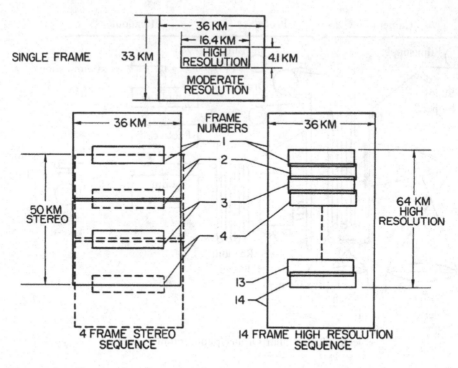

The overlap of successive exposures of the Lunar Orbiter photographic system could be varied. In some cases the high-resolution frames produced contiguous coverage.

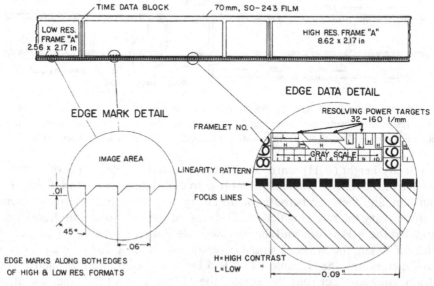

Calibration information was pre-exposed alongside the frames of the Lunar Orbiter photographic system's 70-mm film.

the M frame. The system could take 1, 4, 8 or 16 exposures in sequence. Contiguous coverage could be obtained by timing the shuttering to obtain overlapping frames at either the M or H scale. The known change in perspective from one exposure to the next in overlapping frames would enable stereoscopic analysis to be performed from a single photographic pass.

The specification for the photographic system stated that it must provide at least 194 frame pairs, but Kodak managed to accommodate a strip of film 70 metres long that was nominally capable of 212 exposures. The unperforated film had a thin band of pre-exposed data running along one edge that would be read out along with the adjacent image, and this contained a grey-scale and calibrated charts to indicate the resolving power. Film from the supply reel travelled through the focal plane of the optical imaging system and, after being exposed, was stored in a looper until it was fed to the processor.

For the Bimat development process, a gelatine layer of Kodak type SO-III film soaked in imbibant type PK-411 solution was briefly laminated against the film to develop and fix the negative image in one step – it had to remain in contact with the film for at least 4 minutes, after which it was peeled off and collected on a take-up reel. The processed film was dried by passing it over a heated drum and fed into a second looper. It could then either be scanned or passed through the scanner to a storage reel to be scanned later whilst being rewound. The scanner used a 'flying spot' to produce a signal whose strength represented the density of the emulsion. As there was no tape recorder onboard to store this data, it was transmitted in real-time as a frequency modulated analogue signal. The film was scanned at 287 lines per millimetre. The scanning produced a succession of rasters, each comprising a 2.54 × 65-mm 'framelet' (the calibration data beyond the edge of the image completed the 70-mm width) which took about 23 seconds to transmit. It took 43 minutes to scan a frame pair. Photography could be terminated at any time by commanding the Bimat to be cut and the film to be scanned as it was rewound, with the first picture read out being the last one taken.

In accordance with the requirements of the Office of Manned Space Flight, Lunar Orbiter would also measure radiation and micrometeoroid fluxes in order to assess the hazard to Apollo in cislunar space and in lunar orbit.

The radiation experiment was designed by a team headed by Trutz Foelsche at Langley, and built by Texas Instruments. It had two dosimeters. As conceived, the camera was to have exposed its entire film in contiguous coverage of a single area, but it was later decided to investigate a number of sites at different longitudes. In the 'concentrated' mode the entire film would have been shot in a single pass and then transmitted. However, in the 'distributed' mode the photography would occur over a 10-day period. The photographic system was shielded by aluminium at 2 gm/cm² around the film feeder and elsewhere at 0.2 gm/cm² to protect against a reasonable dose of radiation. As this shielding would be inadequate protection against a severe solar storm, the real-time measurements from the radiation experiment would enable the threat to the film to be assessed and, if necessary, the photography terminated so as to switch to scanning.

The micrometeoroid experiment was led by Charles A. Gurtler and William H.

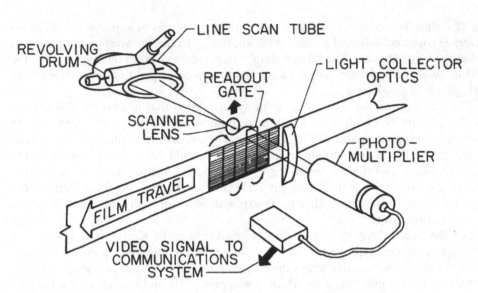

The scanner of the Lunar Orbiter photographic system produced a succession of narrow 'framelets' spanning the width of the film, including the calibration strip.

Kinnard at Langley. It comprised 20 detectors strung in a ring outside the thermal blanket of the middle deck. Each was a semi-cylinder which contained a pressure-sensitive microswitch that was held closed by the pressure of the gas in that volume. In the event of a cell being punctured, the fall in pressure would open the switch and generate a one-off signal. It would yield an estimate of the flux rate, but not the size and velocity of the micrometeoroids. Although crude, it was sufficient for the Office of Manned Space Flight.

The spacecraft's transponder would enable the Deep Space Network to measure both range and range-rate along the line of sight to the antenna on Earth. This information was essential to conducting the mission, and also offered a source of selenodetic data. On 4 May 1962 the Lunar Sciences Subcommittee in Newell's office had set up the Working Group on Selenodesy. This had discussed mission planning for the orbital version of Surveyor, in particular orbital trajectories around the Moon. At a meeting of the Planetology Subcommittee of the Space Sciences Steering Committee on 24 September 1963, Gordon MacDonald of the University of California at Los Angeles explained to representatives of the recently formed Lunar Orbiter Project Office how radio tracking data could produce valuable science. To date, the tracking of Ranger spacecraft had proved insightful, but Lunar Orbiter offered the prospect of surveying broad areas of the Moon. On 10 June 1964 Boeing and Langley discussed how this selenodesy experiment, led by William H. Michael at Langley, could be integrated into the mission design, and it was decided that the gravitational field would be surveyed during an 'extended mission' lasting anything up to a year after the photography was concluded.

As with Ranger, the Atlas-Agena for a Lunar Orbiter mission would use a parking orbit, and the Agena would make the translunar injection. In this case,

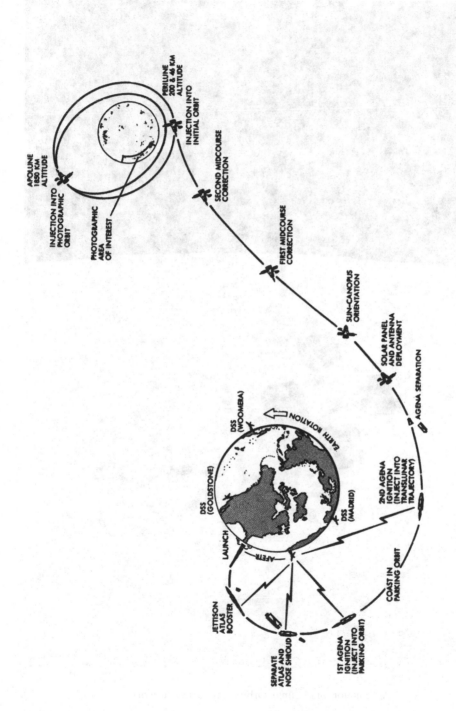

The typical sequence of events of a Lunar Orbiter mission.

A depiction of a Lunar Orbiter spacecraft at work.

however, the spacecraft would have the option of two midcourse manoeuvres to refine its aim for the point in space at which it would insert itself into lunar orbit. After radio tracking had determined the initial orbit, the spacecraft would make a series of manoeuvres to shape the orbit for the photographic activity.

10

On the surface

"MY GOD, IT LANDED!"

The assembly of Surveyor 1 started in the autumn of 1964 at the Hughes Aircraft Corporation in El Segundo, California. The preliminary tests revealed a number of design problems, which were overcome. It was then subjected to vibration tests in June 1965. After upgrades designed to improve its performance, the spacecraft was fully checked and in August it was installed in a vacuum chamber for solar-thermal tests during which it 'flew' several missions. These tests revealed various issues, and it took until 22 January 1966 to achieve acceptable performance. It arrived at the General Dynamics-Astronautics Combined Systems Test Stand near San Diego on 19 February for compatibility tests with its launch vehicle. The principal activities in this facility were to demonstrate the physical compatibility of the adapter between the spacecraft and the Centaur, and to conduct a simulated countdown and flight to the point at which the spacecraft was released to verify the operational connectivity of the entire vehicle. The Atlas had been installed horizontally in the test stand on 29 December, and on 31 January the Centaur had been stacked on the interstage support ring nearby in a vertical orientation. By 4 March the spacecraft had been mated and the aerodynamic shroud fitted. After a thorough systems readiness test on 7 March the spacecraft was retrieved and then flown to Florida aboard a C-133 Cargomaster on 13 March.

The space frame of the Surveyor spacecraft was made using thin-wall aluminium tubing. The base was an equilateral triangle, with a landing leg at each corner. Each leg had a fixed lower strut that was hinged onto the base, and an upper telescoping strut affixed further up the frame that incorporated an aircraft-style shock absorber. To fit into the shroud, the legs were hinged up against the frame. They would later be swung down into place by torsion springs.

The superstructure was asymmetrical, with the 'apex' being offset away from leg no. 1, and a mast affixed at this point supported the solar panel and the planar high-gain antenna. The mast comprised a fixed section and an upper section that could be rotated to simultaneously enable the solar panel to face the Sun and the antenna to

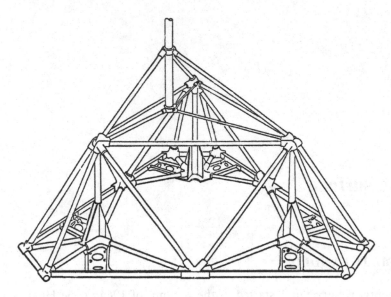

The asymmetric configuration of the Surveyor spacecraft's frame.

face Earth. For launch, the solar panel was stowed flat against the mast, but once the spacecraft had orientated itself to point the mast towards the Sun the panel would be raised 90 degrees. When face-on to the Sun, its 3,960 transducer cells would produce 85 watts. The high-gain antenna was not to be deployed until the vehicle was on the lunar surface. Communications in flight would be by small conical omni-directional antennas at the tips of two folding booms that were hinged to the frame and, as with the legs, were to be swung down by springs.

There were two compartments attached to the sides of the frame to provide active thermal control for sensitive electronics. The apparatus was on a thermal 'tray' that distributed heat throughout the compartment. Each compartment comprised an inner and an outer shell, between which was sandwiched an insulating blanket of 75 sheets of aluminised mylar. The upper surface of each compartment comprised a radiator of mirrored glass. Compartment 'A' maintained its interior between 5°C and 52°C, and housed two transmitters, two receivers, the main battery, the charge regulator, main power switch and ancillary apparatus. Compartment 'B' maintained its interior between –18°C and 52°C, and housed the central command decoder, central signal processor, engineering signal processor and low-data-rate auxiliary. Both boxes had sensors to report their temperatures by telemetry. They had heaters to maintain their trays above their minimum specified temperatures, and used bimetallically activated thermal switches to provide a conductive path to the radiators to remain below their maximum operating temperatures. All apparatus not housed in these compartments was mounted directly on the frame and relied on paint patterns and polished surfaces for passive thermal regulation.

The attitude would be controlled during the cruise by cold-gas thrusters fed by a single spherical tank with a capacity of 4.5 pounds of nitrogen under high pressure.

There was a pair of opposed thrusters on each leg. The thrusters on leg no. 1 were to provide roll control. The thrusters on the other legs fired perpendicular to the plane of the legs to achieve pitch and yaw control. It was inherent in the design that the legs would have been deployed prior to the spacecraft being released by the Centaur, to enable the spacecraft to use its thrusters to stabilise itself once free.

The solid rocket motor that would initiate the powered descent to the Moon was provided by the Thiokol Chemical Corporation, Elkton, Maryland. The 0.94-metre-diameter spherical casing was made of high-strength steel, insulated with asbestos and rubber to maintain the case at a low temperature while firing. It burned a case-bonded composite of aluminium, aluminium-perchlorate and poly-hydrocarbon, and required an igniter. The nozzle had a graphite throat and a laminated plastic exit cone. The thrust would vary between 8,000 and 10,000 pounds depending upon the pre-ignition temperature, which would vary with solar illumination and hence with the lunar phase at the spacecraft's arrival. Its task was to slow the approach from 6,000 miles per hour to approximately 250 miles per hour. Although the nozzle was partially submerged into the casing to minimise its projection from the base of the space frame, it still projected below the plane of the foot pads. It had to be jettisoned after burnout both to shed 'dead weight' and to enable the legs to make contact with the ground on touchdown. The casing was affixed to the base of the frame at three points, just inboard of the legs, and was to be released by explosive bolts. The mass of propellant in the retro would vary depending on the requirements of the approach and the overall mass of the vehicle, but in general would account for two-thirds of the spacecraft's launch mass.

The vernier propulsion system comprised three bipropellant engines supplied by the Reaction Motor Division of the Thiokol Chemical Corporation, Denville, New Jersey. It was to perform the midcourse manoeuvre, stabilise the vehicle during the retro-rocket's burn, and complete the descent. An igniter was not required, because the monomethyl hydrazine and nitrogen tetroxide propellants were hypergolic. Each vernier engine had its own fuel and oxidiser tanks, with the propellants being forced by bladders pressurised by helium from a common supply. The nozzle of the thrust chamber assembly was made of molybdenum, and each vernier could be throttled to deliver a thrust in the range 30 to 104 pounds. The engine on leg no. 1 was capable of swivelling to provide roll control, but the others were fixed.

The design policy for Surveyor was to minimise the complexity of the spacecraft by placing responsibility for decision-making on Earth. In turn, this required a large vocabulary of commands (256) and a large number of telemetry channels (100). In fact, it was the first spacecraft to be wholly reliant upon commands from the Deep Space Network. Ranger had been able to use an L-Band radio, but Surveyor required the greater bandwidth of the S-Band.

The spacecraft was interfaced to the Centaur by a 52-pin electrical connector on the base of the frame. This supplied power and enabled systems monitoring prior to launch, and once in space it carried pre-separation commands from the Centaur. The spacecraft's power subsystem comprised the solar panel, the main battery, auxiliary battery, charge regulator and associated electronics. The main battery comprised 14 silver-zinc cells, delivered 22 volts of direct current, and had a capacity of 3,800 watt-hours at a discharge rate of 1 ampere. It was the sole source of power after the feed from the ground was removed for launch. It would hand over to the solar panel once this was deployed in flight, then be recharged. The solar panel was to be the primary power source during the cruise and in daylight on the Moon, with the main battery backing it up at times of peak demand. Whenever there was surplus power from the solar panel, this would be made available to recharge the main battery. The silver-zinc auxiliary battery was in a sealed magnesium canister mounted externally. It was non-rechargeable, with a maximum capacity of 1,000 watt-hours, and was to provide backup for the main battery. As the motion of the Sun in the lunar sky was only 0.5 degree per hour, there was no need for an autonomous system to locate the Sun and maintain the solar panel facing it – after landing, the panel would be set and adjusted by Earth-command. And because from a vantage point on the lunar surface Earth remains essentially fixed in the sky, the high-gain antenna was to be controlled from Earth. This was in keeping with the design strategy of keeping the spacecraft as simple as possible.

The primary objectives of the first mission were to demonstrate the ability of the spacecraft to make a midcourse manoeuvre and then a powered descent to soft-land on the Moon using a variable-thrust rocket engine, as was planned for Apollo; and also to demonstrate that the Deep Space Network could maintain contact with the vehicle, both in flight and on the lunar surface. The secondary objectives involved obtaining engineering data on the spacecraft's systems at all phases of the mission.

Although Surveyor 1 was an engineering test model, and as such did not carry any scientific experiments, it did have a TV camera and a range of instrumentation. The tertiary objectives were to provide data on radar reflectivity during the descent; to measure the bearing strength of the surface in the act of landing; to take pictures of one of the foot pads and the adjacent surface, and subsequently of the broader lunar landscape; and to monitor spacecraft temperatures as the elevation of the Sun in the sky changed, including after sunset, to gain some insight into the thermal properties of the surface.

Launch windows were controlled (in part) by the illumination at the landing site, since the Sun had to be low in the east to give the vehicle time to achieve its primary surface activities prior to the heat of lunar noon. This defined the launch date. The translunar injection requirements defined the time of day. On this first mission, the Centaur was to make a single long burn to achieve translunar injection. A direct ascent was feasible on days when the Moon was south of the Earth's equator.[1] The available days of the month were determined by the attainable flight path angle of the Centaur at injection – i.e. the angle at which the vehicle travelled relative to the horizontal plane of the Earth below. Varying this angle compensated for the daily change in the position of the Moon as it pursued its monthly orbit of Earth. The

[1] This was because on a direct ascent the translunar injection point was necessarily near the latitude of the launch site, and for a launch from Florida this was north of the equatorial plane on a southerly heading, which meant that by the time the spacecraft reached lunar distance it would be south of the equatorial plane.

attainable range in the flight path angle was determined by the fuel in the Atlas-Centaur combination, because in a direct ascent both vehicles were involved in the path of the Centaur at injection. Also, any deviation from a horizontal flight path at injection increased the amount of propellant required for the manoeuvre. A primary constraint on the window was that the launch pad at the time of launch and the Moon at the time of the spacecraft's arrival had both to be in the plane of the Earth–Moon transfer orbit. With the launch site moving eastward with the axial rotation of the Earth, conditions were acceptable only once per day for a given plane. However, the window could be extended to several hours by altering the plane by varying the azimuth at launch between the allowed range of 80 to 115 degrees east of north. This azimuth constraint was imposed by range-safety considerations.

The time of flight of between 61 and 65 hours was determined by the constraint that the spacecraft reach the Moon while the primary Deep Space Network station at Goldstone had a line of sight. The approach velocity could not exceed the braking capability of the retro-rocket. Finally, for this first mission the approach to the Moon could be no more than a few degrees off local vertical. This required the target to be near the equator at a midwestern longitude. And since Surveyor was restricted to the Apollo zone, this meant a longitude of about 45°W. The chosen target was in the Flamsteed Ring, an ancient 110-km-diameter crater which had been 'inundated' by Oceanus Procellarum to such an extent that only the high points of its rim remained visible. Superimposed on its southern rim was the 10-km crater Flamsteed. The mare in the eastern half of the ring was considerably darker than was typical for Oceanus Procellarum. The fact that when viewed telescopically this dark patch appeared to be free of craters, rays and ridges made it attractive to the Apollo selectors. The target was a 60-km-diameter circle in the northern sector of the ring. The nearest sizeable crate (about 1 km in diameter) was 10 km east of the nominal aim point.

At the Cape, Surveyor 1 was checked out in Hangar AO, then subjected to a series of performance verification tests which concluded on 5 April. Meanwhile, the Atlas was erected on Pad 36A on 21 March and the Centaur was stacked on 31 March. On 7 April the spacecraft was mated with its launcher, and the ability of its radio system to communicate with the Deep Space Network facility at the Cape was verified. A 'plugs out' acceptance test on 26 April ran a simulated flight through to spacecraft release. Surveyor 1 was then returned to its hangar, and between 14 and 16 May its liquid propellants were loaded and the retro-rocket was installed. The related checks were finished on 24 May. The spacecraft was re-installed on its launcher on 26 May, and three days later subjected to a final readiness test that concluded at 03:52 local time on 30 May. In the expectation of the spacecraft being declared healthy, the countdown for the launcher had begun at 03:30. This ran without incident to T–90 minutes at 06:50 at which time a planned 1-hour hold began. When the countdown resumed at 07:50, the one for the spacecraft was integrated, and events progressed smoothly to launch at 14:41:01 GMT. Considering that this was the first operational mission for both a new launch vehicle and a new spacecraft, this was remarkable.

The boost phase was steered by the programmed inertial autopilot. In the case of an Atlas-Agena, the Atlas steered the sustainer phase using updates radioed by the

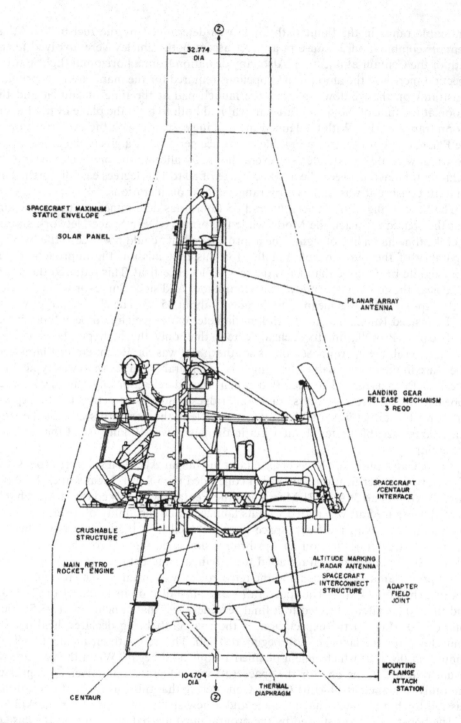

Detail of the Surveyor spacecraft inside the aerodynamic shroud of the launch vehicle.

Testing the fit of the shroud around the Surveyor spacecraft (left) and the ability of the Centaur stage to command the deployment of the spacecraft's legs and antenna booms.

ground on the basis of radar tracking, but for the Atlas-Centaur the inertial guidance system of the upper stage issued the steering commands. The Centaur had four side panels to insulate its cryogenic hydrogen tank from the aerodynamic heating of the climb through the dense lower atmosphere. The panels were jettisoned at an altitude of 92 km to shed 1,200 pounds. The fibreglass aerodynamic shroud was jettisoned at 120 km. The Atlas shut down after 240 seconds of flight, released the Centaur and fired eight retro-rockets to open the separation. The RL-10 engines of the Centaur were then chilled by helium at $-269°C$ in order to minimise vaporisation of liquid hydrogen at $-253°C$ as this was supplied to start the engines in flight. The Centaur ignited 9.6 seconds after its release by the Atlas at an altitude of 155 km and 340 km downrange. It continued until the guidance system sensed that a velocity of 34,496 ft/sec had been attained, which in this case was after a burn of 438.4 seconds ending at $T + 689.2$ seconds – at which time the propellant remaining was sufficient for only another 2 seconds of thrusting. At translunar injection the vehicle was at an altitude of 166 km and 3,600 km southeast of the Cape.

The Centaur used the electrical umbilical to command the spacecraft to deploy its legs and antenna booms and place its transmitter on high power. When pyrotechnics released the pins that had held the legs against the upper part of the frame, a torsion spring at each hinge swung that leg into position. Similar springs were to deploy the

On 30 May 1966 an Atlas-Centaur lifts off with the Surveyor 1 spacecraft.

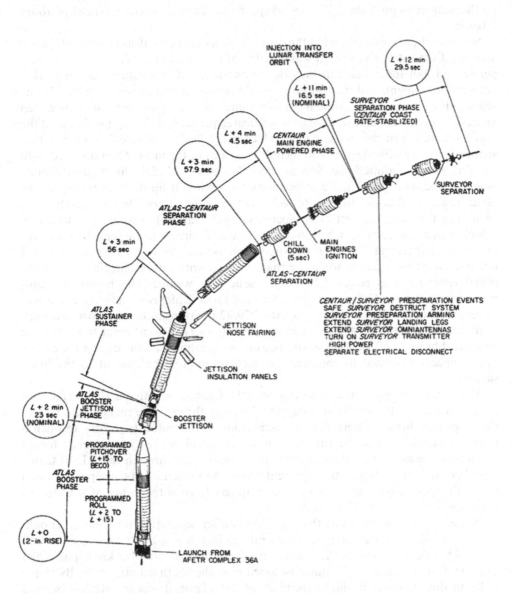

INJECTION INTO
LUNAR TRANSFER
ORBIT

L + 12 min
29.5 sec

L + 11 min
16.5 sec
(NOMINAL)

SURVEYOR
SEPARATION PHASE
(CENTAUR COAST
RATE-STABILIZED)

L + 4 min
4.5 sec

CENTAUR
MAIN ENGINE
POWERED PHASE

L + 3 min
57.9 sec

SURVEYOR
SEPARATION

ATLAS-CENTAUR
SEPARATION
PHASE

L + 3 min
56 sec

CHILL   MAIN
DOWN   ENGINES
(5 sec)   IGNITION

ATLAS-CENTAUR
SEPARATION

ATLAS
SUSTAINER
PHASE

JETTISON
NOSE FAIRING

CENTAUR/SURVEYOR PRESEPARATION EVENTS
SAFE SURVEYOR DESTRUCT SYSTEM
SURVEYOR PRESEPARATION ARMING
EXTEND SURVEYOR LANDING LEGS
EXTEND SURVEYOR OMNIANTENNAS
TURN ON SURVEYOR TRANSMITTER
HIGH POWER
SEPARATE ELECTRICAL DISCONNECT

JETTISON
INSULATION PANELS

ATLAS
BOOSTER
JETTISON
PHASE

L + 2 min
23 sec
(NOMINAL)

BOOSTER
JETTISON

PROGRAMMED
PITCHOVER
(L + 15 TO
BECO)

ATLAS
BOOSTER
PHASE

PROGRAMMED
ROLL
(L + 2 TO
L + 15)

L + 0
(2-in. RISE)

LAUNCH FROM
AFETR COMPLEX 36A

The sequence of operations of a direct ascent Surveyor launch.

omni-directional antenna booms. The Centaur then severed the electrical connection, and 5.5 seconds later – at 14:53:38 – fired pyrotechnics to enable three springs on the adapter to push the 2,193-pound spacecraft away at a relative speed of about 1 ft/sec.

Surveyor 1 promptly activated the cold-gas thrusters of its attitude control system and stabilised itself. The system was capable of overcoming a roll rate of 3 degrees per second, but the actual rate was barely one-tenth of this value. The separation occurred in sunlight, and the trajectory would remain continuously illuminated. The spacecraft automatically manoeuvred to acquire the Sun. This was a slow, two-step process involving angular rates of only 0.5 degree per second. First, it rolled until the Sun was detected by the coarse sensor, which had a 10-degree-wide field of view that spanned an arc of 196 degrees and included the axis of the mast. Then it yawed until the fine sensor acquired the Sun at 15:00:34. Next, at 15:03:20, explosive bolts unlocked the solar panel and a stepping motor rotated it up through 90 degrees to generate electrical power to start to recharge the main battery. On a southerly track such as in this case, the first Deep Space Network station to acquire the departing vehicle was Johannesburg in South Africa. It was desirable to inspect the telemetry to verify the spacecraft's state of health as soon as possible, as the transmitter could not operate at high power for more than an hour without overheating. When the initial telemetry was received, the only issue was whether the boom of omni-directional antenna 'A' (located above leg no. 1) had deployed – the sensor that was to report this event had not done so. At 15:20:42 the transmitter was switched back to its low power setting. A command was sent at 16:21:21 to attempt to deploy the antenna, but to no effect. This would not impair communications during the cruise, but engineers convened to consider the implications for the descent to the lunar surface.

Meanwhile, having delivered its payload, the Centaur initiated a 180-degree turn around using its attitude control engines. Midway through the manoeuvre, two of the 50-pound thrusters were fired perpendicular to the initial trajectory to open a lateral separation. Once the turn had been completed, valves were opened for 10 seconds to release the residual propellants – the oxygen through the RL-10 thrust chambers and the hydrogen through vent tubes – to further open the separation and to put the spent stage into an orbit with an apogee beyond the radius of the Moon's orbit and with a period of 11 days.

A mission specification was that by 5 hours after separation the Centaur should be at least 360 kilometres from the spacecraft. In fact, it was 1,000 km away and well clear of the field of view of the spacecraft's Canopus sensor. Also known as alpha Carinae, Canopus is a bright star positioned near the south celestial pole. By virtue of being almost perpendicular to the plane of the ecliptic, it was an excellent 'second reference' for a vehicle that maintained 3-axis stability by using the Sun – which by definition is in the plane of the ecliptic. The sensor had been developed by JPL for the Mariner interplanetary missions. Canopus acquisition was an Earth-commanded manoeuvre during which the spacecraft was to maintain its lock on the Sun. The roll to initiate the Canopus scan began at 18:53:39. After 220 degrees, the star appeared in the sensor's field of view but the automatic lock-on did not engage. The vehicle

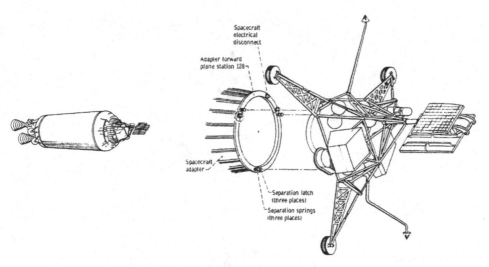

The configuration of the Surveyor spacecraft upon being released by the Centaur stage.

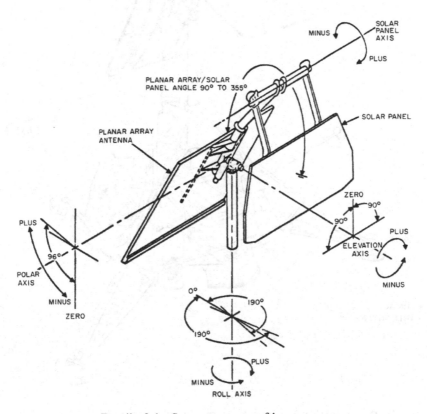

Detail of the Surveyor spacecraft's mast.

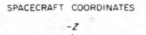

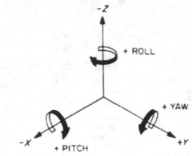

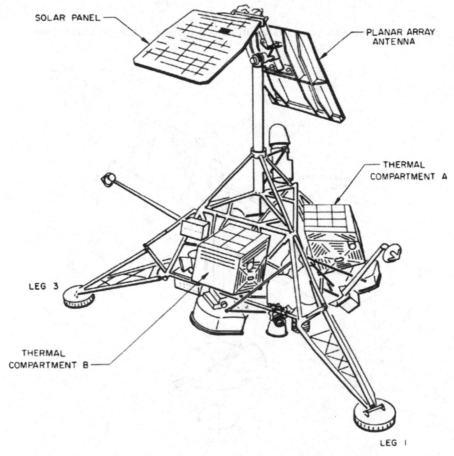

Detail of the Surveyor spacecraft, including its axes of rotation.

was allowed to roll a full 360 degrees, and when the star reappeared at 19:13:20 the lock-on was commanded by Earth. In the event of the spacecraft losing its cruise attitude, the flight control system would use the 3-axis strap-down gyroscopes of its inertial system in its attempt to recover.

Surveyor missions were managed by the Space Flight Operations Facility at JPL, which operated three support teams: (1) to determine the trajectory of the spacecraft, both in planning and in flight; (2) to monitor the telemetry on a continuous basis in order to evaluate the state of the spacecraft; and (3) to control spacecraft operations. At peak times, 300 people from various sources were involved in flight control.

In its early stage, the trajectory was an elliptical Earth orbit with a 160-km perigee and a 384,000-km apogee, but as it ascended the Moon's gravity would act to draw the vehicle's trajectory towards itself. Tracking showed that the translunar injection would result in an interception slightly west of the 45-km-diameter crater Hansteen, some 400 km southwest of the target. The flight plan included a manoeuvre to refine the aim. On 31 May the vehicle was to depart from its cruise attitude, and turn under inertial guidance to the attitude for this manoeuvre. The 'critical component' of this burn would refine the aim for the target (it would be 12.2 ft/sec in this case) and the remainder of the 66.7-ft/sec overall adjustment was to be directed to ensure that the speed relative to the Moon would be acceptable when the retro-rocket burned out at the end of the initial phase of the descent. At 06:08:25 the spacecraft switched its transmitter to high power, and its telemetry rate to high 2 minutes later. The roll was initiated at 06:30:13, the yaw at 06:34:48, and the desired orientation was attained at 06:36. Each rotation was individually commanded by Earth and specified in terms of the interval that would be required to subtend the desired angle at the nominal fixed turning rate of 0.5 degree per second. The six propellant tanks of the bipropellant vernier propulsion system had been pressurised at 300 psi for launch, and the helium tank at a nominal 5,175 psi. A pyrotechnic had been fired shortly prior to departing cruise attitude in order to open a valve to allow the helium to raise the pressure of the propellant tanks to 720 psi. A regulator would maintain the tanks at this pressure in readiness for the lunar descent. The midcourse manoeuvre began at 06:45:03. The three vernier engines were throttled to provide a constant acceleration of 0.1 times Earth's gravity – which meant 75 pounds of thrust from each engine – and were cut off after 20.8 seconds when an accelerometer aligned to the longitudinal axis of the inertial system sensed that the desired change in velocity had been attained. The effect was to move the arrival point to within 16 km of the centre of the target circle, which was about as accurate as was believed feasible at that time. Still under inertial guidance, the spacecraft initiated the reverse yaw at 06:52:55 and the reverse roll at 06:58:02, resuming an attitude which pointed its mast at the Sun at 07:00:55. It was decided to postpone Canopus reacquisition, in order to perform an engineering test by measuring the drift in the roll gyroscope over a 3-hour period.

A meeting 15 hours before Surveyor 1 was due to reach the Moon decided that because omni-directional antenna 'A' had failed to deploy, the pre-retro manoeuvre should be a roll-yaw-roll sequence, as this orientation would ensure that antenna 'B', located between legs no. 2 and 3, would not be masked by the spacecraft's structure during the descent. Since the spacecraft would lose solar power when it adopted this

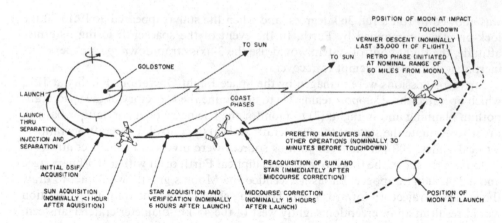

The typical sequence of events of a Surveyor mission.

attitude, it brought both batteries on line at 01:23:00 on 2 June, some 5 hours prior to initiating the descent. The pre-retro manoeuvre in which the vehicle departed from its cruise attitude began 38 minutes in advance of the initial phase of the descent. It started with a roll of + 89.3 degrees that was initiated at 05:36:46 and took 3 minutes to complete. The yaw of + 60.0 degrees at 05:41:47 lasted 2 minutes. Between them, these turns aligned the nozzle of the retro-rocket with the vehicle's velocity vector as this would be when that engine was ignited. The second roll of + 94.2 degrees that started at 05:45:17 and lasted just over 3 minutes was to provide omni-directional antenna 'B' with a clear line of sight to Earth both during the descent and operations on the surface until the lander could deploy its high-gain antenna.

A TV camera was carried inboard of leg no. 2, facing down along the longitudinal axis. Once the vehicle had completed the pre-retro manoeuvre, the plan had been for this approach camera with a 6.4 × 6.4-degree field of view to take 100 overlapping images, Ranger-style, between the altitudes of 1,800 km and 100 km. But owing to the complexity of the descent phase, it had been decided prior to launch that on this occasion the approach camera would not be operated.

The altitude marking radar was enabled at 06:12:58, just 2 minutes in advance of the predicted start of the descent. This pulsed-magnetron radar was aimed downward on the vehicle's longitudinal axis, and its singular task was to alert the flight control system when the slant range was 100 km. The engineers were reluctant to enable the radar any earlier, lest it malfunction and issue a spurious report. From this point, the spacecraft was on its own – it had to make the descent using its automated systems. The performance of the retro-rocket would depend on its pre-ignition temperature. Sensors had monitored its temperature during the cruise to enable the duration of its burn to be computed. With knowledge of the trajectory from tracking and the likely performance of the motor, it was simple to calculate the slant range at which the motor should be ignited in order for it to burn out at the desired altitude. The speed of the vehicle was known from tracking, so it was straightforward for the Space Flight Operations Facility to compute how long after

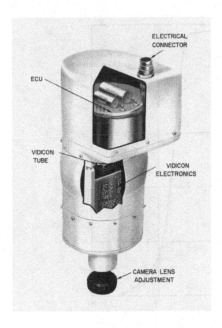

ELECTRICAL
CONNECTOR

ECU

VIDICON
TUBE

VIDICON
ELECTRONICS

CAMERA LENS
ADJUSTMENT

Details of the approach camera for the Surveyor
spacecraft.

receiving the 100-km 'mark' the flight control system ought to wait to initiate the
retro sequence – in this case, the calculated delay was 7.826 seconds. The initial
approach was at 6.1 degrees to the local vertical. The altitude marking radar started
to measure the slant range about 350 km out, and at 06:14:39.708 it provided its
mark. The vernier propulsion system was ignited after the programmed delay, and
the retro-rocket 1.1 seconds later – at which time the vehicle was travelling at 8,565
ft/sec. As the altitude marking radar now represented surplus mass, it had been
affixed to the aperture of the retro-rocket by friction clips to enable it to be readily
jettisoned by the igniter gas pressure.

The trajectory would angle down towards the surface as the vehicle slowed, but in
this phase of the descent the flight control system was to individually throttle the
verniers to hold the pre-ignition attitude by reference to the 3-axis gyroscopes of its
inertial system. Depending on the predicted velocity at retro burnout, the total thrust
for the three verniers would be between 150 and 200 pounds – the required value of
194 pounds was uplinked shortly before the altitude marking radar was enabled. In
fact, the retro had been so well aligned when it was inserted into the frame that the
verniers had little to do. A slight attitude transient coincident with retro ignition was
probably caused by omni-directional antenna 'A' swinging down and locking into
position.

As solid propellant was burned, the deceleration increased from four times Earth
gravity at ignition to ten times immediately prior to burnout. At 06:15:27.957 an
acceleration switch noted the peak thrust of 9,700 pounds sharply fall below a level
of 3,500 pounds. By definition this was burn termination, with a burn duration of 39
seconds. The flight control system used this signal to time the burnout activities. A
delay of about 9 seconds had to be allowed for the motor's residual thrust to tail off

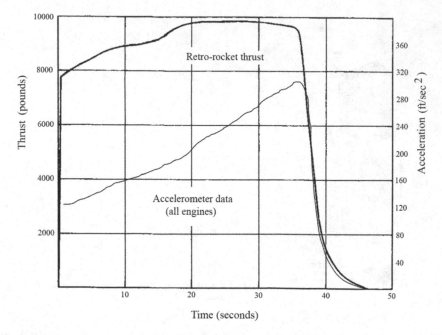

A plot of the thrust of the solid-fuel retro-rocket and the deceleration imparted by the combination of the retro-rocket and the vernier propulsion system during the initial phase of Surveyor 1's powered descent.

so that it could be safely jettisoned. Fully 1,244 pounds of the motor's 1,390-pound mass had been propellant. At 06:15:39.956 explosive bolts were fired to release the spent casing. At the same time, the verniers were throttled to deliver their maximum thrust for several seconds in order to ensure a clean separation. At burnout, the angle between the vehicle's thrust vector and velocity vector was 5.5 degrees, this offset having accrued during the braking burn.

The radar altimeter and Doppler velocity sensor (RADVS) was supplied by Ryan Electronics of San Diego, California.[2] One of its two subsystems was a single-beam coherent frequency modulated continuous-wave microwave radar altimeter with its beam pointing down the longitudinal axis to measure the slant range. The second subsystem was a three-beam coherent continuous-wave autodyne Dopper radar. The output of a single klystron was divided equally between the three transmitting horns, and the frequency shifts of the reflections from the lunar surface were processed to measure the lateral components of the vehicle's velocity relative to the lunar surface. Because these values were measured in terms of the spacecraft's coordinate system, they could be issued directly to the flight control system. The RADVS was activated during the retro burn. When its data was first used at 06:15:42.093, the slant range was 27,820 feet (and since the velocity vector at burnout was only slightly offset to

---

[2] Ryan was also to provide the descent radar for the Apollo lander.

vertical, the altitude was 27,800 feet) and the total velocity was 430 ft/sec (and since the vehicle had maintained its thrust along the velocity vector extant at the time of retro ignition, the 5.5-degree offset meant the longitudinal rate was 425 ft/sec). As the altimeter had locked on at a slant range of 36,000 feet, attitude control was switched from inertial to radar, and the vehicle immediately aligned its thrust axis with the instantaneous velocity vector to initiate the 'gravity turn' that would force the trajectory to vertical. The verniers were throttled to a total thrust of about 110 pounds to yield a constant deceleration at 0.9 times that of lunar gravity, so that as the vehicle fell it accelerated at a mere 0.1 times lunar gravity.[3] It reached the 'descent contour' in the range–velocity plane at 06:16:05.893, when the slant range was 18,000 feet and the speed was 442 ft/sec. A closed-loop algorithm then throttled the verniers to control the rate of descent to 'fly the curve'. By the 1,000-foot mark at 06:17:10.494, its trajectory was vertical with a sink rate of 103.1 ft/sec.

At 06:17:28.719, when the RADVS issued the 10-ft/sec mark at a height of 43 feet, the flight control system initiated 'attitude hold', slowed to a rate of 5 ft/sec and maintained this. The final task of the RADVS was to generate a mark when the foot pads were 12 feet above the surface. In fact, this was referred to as the 14-foot mark because the reference point for the radar system was 2 feet above the plane of the foot pads. At this point, the vehicle was to be descending at 5 ft/sec with no lateral components greater than 2 ft/sec. On receiving this signal at 06:17:34.169, the flight control system cut off the verniers to minimise erosion of the surface by their efflux. After falling in the weak lunar gravity for 1.53 seconds, the vehicle slammed down at 06:17:35.700 at a vertical rate of 11.6 ft/sec. The foot pads hit at intervals of 0.01 second – first pad no. 2, then no. 1 and finally no. 3. The horizontal rates were less than 1 ft/sec. Each foot pad was 30.5 cm in diameter and 12.8 cm tall, including the plate on its upper surface that incorporated the hinge to the leg. On impact, each pad pivoted to make firm contact, and the crushable aluminium honeycomb that formed its lower section absorbed some of the energy. In addition, the leg's 1.04-metre-long lower strut hinged upward and compressed the shock absorber in the upper strut. There was a cylindrical crushable block of aluminium honeycomb 18 cm in diameter and 20 cm in length affixed to the frame just inboard of each leg. As the 'stroke' on the shock absorbers was 11 cm, the maximum vertical offset of the legs was 43 cm, but when the rotation of the legs allowed the crushable blocks to hit the ground they absorbed the remaining energy. As the shock-absorbing struts re-extended to their original 95-cm lengths, the body of the vehicle was raised off the ground. In fact, the elasticity in the legs caused it to rebound to a height of about 6 cm, and then settle back. A strain gauge in the shock absorber of each leg provided an analogue record of the dynamics. The lander stood 3 metres tall, and had a maximum span across its legs of 4.27 metres. The inertial reference system indicated that it had settled within 1 degree of vertical.

---

[3]   The Moon's gravity imparted an acceleration of 5.3 ft/sec^2; hence 0.9 was 4.77 ft/sec^2, and 0.1 was 0.53 ft/sec^2.

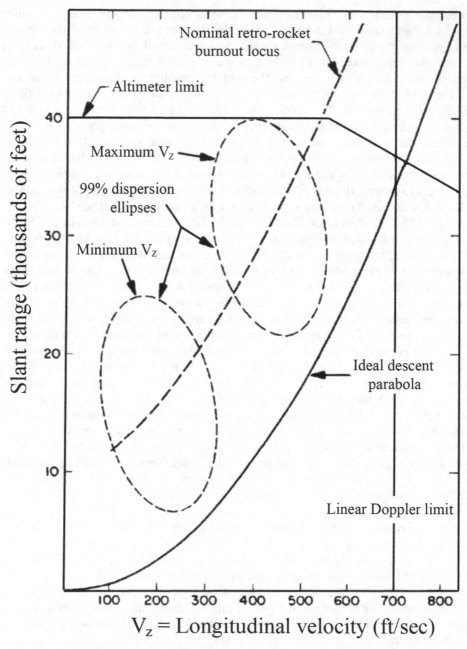

The longitudinal velocity of the Surveyor spacecraft at retro-rocket burnout was
designed to occur within a narrow slant-range band. Here the limits are shown as 99-
per-cent dispersions ellipses. Both cases presume the maximum propellant usage in the
midcourse manoeuvre. Following retro burnout, the vehicle must use its vernier
propulsion system to first acquire and then to follow the descent parabola.

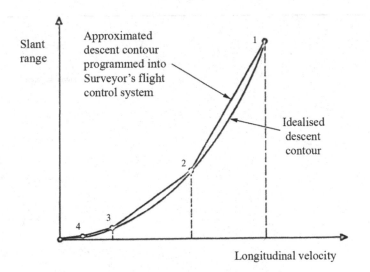

Although the ideal trajectory during the vernier phase of the Surveyor's powered descent was a parabola, the spacecraft was programmed to pursue a succession of linear approximations.

At JPL, the mood was tense. The commentator recited the diminishing altitude, ending with, "Touchdown!"

Gene Shoemaker, the PI of the TV team, exclaimed, "My God, it landed!" He had estimated the chance of a successful landing on the first attempt to be no better than about 10 per cent. Whereas Ranger 8 had enacted 11 commands in-flight from Earth, Surveyor 1 had received a grand total of 288 commands. Some engineers at JPL and Hughes had expected immediately to start an analysis of the telemetry to attempt to determine why the vehicle had crashed! An analysis of the trajectory at retro burnout indicated that the casing of the spent retro-rocket impacted at about 700 ft/sec, some 300 feet from the lander on a line just east of south. The mass of Surveyor 1 on landing was 645 pounds – although, of course, in the weak lunar gravity it weighed only one-sixth of this figure. Having come down on a particularly flat part of the Flamsteed Ring, the lander could reasonably be expected to sink from sight if Thomas Gold's idea that the maria were accumulations of fine dust was correct, but the continuing engineering telemetry established there to be no settling following the dynamics of the landing. The post-landing uplink included a command to lock the legs with the upper strut shock absorbers in the fully extended position.

The first task was to interrogate the lander to determine its status – not only was it in good condition, but omni-directional antenna 'A' had locked in position. The fact that it deployed in this way indicated that the pyrotechnic pin had indeed fired when the Centaur had commanded the boom to hinge. An investigation decided that the boom must have been misaligned sufficiently to allow friction to overcome the 0.25-pound force of the deployment spring, and that the dynamic load of the retro-rocket igniting had forced the boom to complete its swing and lock into position. Since the

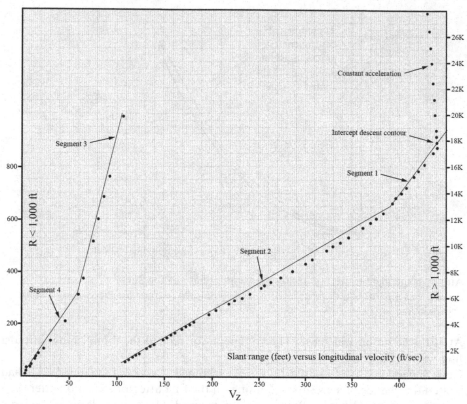

The actual descent of the Surveyor 1 spacecraft depicted in two sections, one for slant ranges above 1,000 feet and the other below 1,000 feet.

verniers were for the midcourse manoeuvre as well as the lunar descent, the lander had been loaded with sufficient propellant for the maximum midcourse manoeuvre and a nominal descent. If an inaccurate translunar injection had demanded a *greater* midcourse manoeuvre, then the additional propellant consumed early on would have required the descent sequence to be revised. In the case of Surveyor 1, however, all went well. The propellant load at launch was 184 pounds. The midcourse manoeuvre had consumed 16.5 pounds and the descent 119.7 pounds; so 47.5 pounds remained after landing. At 06:27 a command in the post-landing uplinked sequence vented the helium pressurant tank.

The 16-pound TV camera for use on the lunar surface was mounted with leg no. 2 to the right and leg no. 3 to the left. It consisted of a viewing mirror, filters, lens, shutter, vidicon and associated electronics. The mirror was an ellipse with axes 15 × 10.5 cm, and was supported on its minor axis. It had two drives, one for azimuth and the other for elevation. The azimuth range was almost 360 degrees and the elevation range extended from 65 degrees below to 35 degrees above the plane perpendicular to the principal axis of the camera. The hood rotated in azimuth with the mirror. The mirror assembly incorporated a wheel with a 'clear' aperture and a

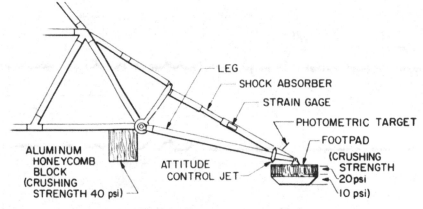

LEG
SHOCK ABSORBER
STRAIN GAGE
PHOTOMETRIC TARGET
FOOTPAD
(CRUSHING STRENGTH 20 psi 10 psi)
ALUMINUM HONEYCOMB BLOCK (CRUSHING STRENGTH 40 psi)
ATTITUDE CONTROL JET

Detail of the Surveyor spacecraft's leg.

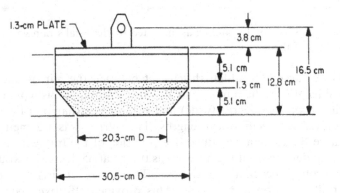

1.3-cm PLATE
3.8 cm
5.1 cm
16.5 cm
1.3 cm 12.8 cm
5.1 cm
20.3-cm D
30.5-cm D

Detail of the Surveyor spacecraft's foot pad.

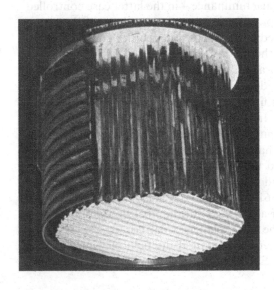

A crushable block of the type installed beneath the Surveyor spacecraft.

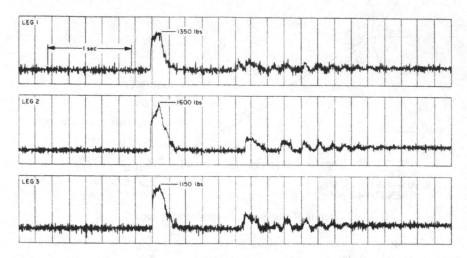

The dynamics of Surveyor 1's landing as indicated by the axial strain gauges on the shock absorbers of the legs.

trio of filters. A variable-focal-length lens set between the mirror and the vidicon provided focal lengths in the range 25 to 100 mm. Although this would normally be used at its limits for 25.3-degree (wide) and 6.43-degree (narrow) fields of view, it could be operated at specific focal lengths. The mirror was 1.2 metres above the ground. Because the camera was tilted over the side of the frame at an angle of 16 degrees to the lander's mast, if the mirror was tipped at 65 degrees it would provide a view of the ground beneath the camera. There was an iris to accommodate differences in illumination of the scene. This provided effective apertures ranging from f/4 to f/22, either in fixed increments that increased the aperture area by a factor of two or in proportion to the average luminance – in the latter case controlled by a sensor that used the secondary output of a beam-splitter to control the servo. The camera settings were to be controlled from Earth, but the iris could be set to automatically gauge the illumination of the field of view.

The focal plane of the optical system was on the vidicon tube, and the image was 11 mm square. In its 200-line mode the vidicon was scanned by an electron beam in 61.8 seconds, and a frame could be transmitted in 20 seconds by an omni-directional antenna at a bandwidth of 1.2 kilohertz. In its 600-line mode the vidicon scan took 3.6 seconds, and a frame could be transmitted in 1 second by the high-gain antenna at 200 kilohertz. In normal use, the mechanical focal-plane shutter was opened for 150 milliseconds and the vidicon retained the image until scanned and cleared. But for imaging in Earthlight the shutter could be opened and held while the vidicon was scanned every 3.6 seconds (600 lines) or 61.8 seconds (200 lines). As a variation on this mode, the vidicon could integrate for a specified duration before the shutter was closed – this would be used to observe the stars, planets, solar corona etc. A sensor had been installed to prevent the shutter from opening if the field of view were too bright, but this could be circumvented by Earth command. The filters were chosen to

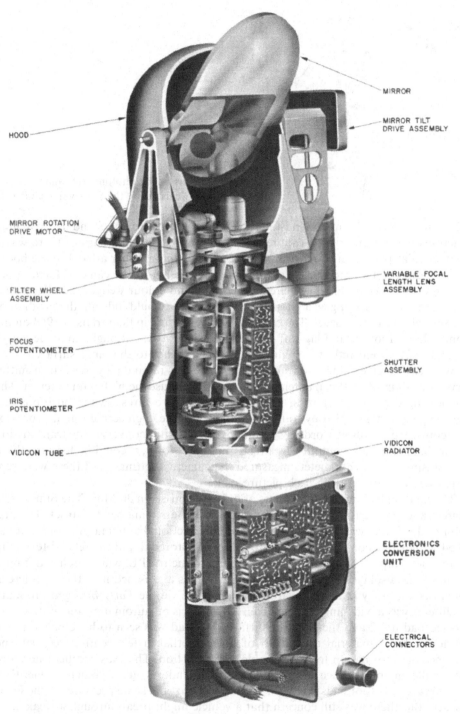

HOOD

MIRROR

MIRROR TILT
DRIVE ASSEMBLY

MIRROR ROTATION
DRIVE MOTOR

FILTER WHEEL
ASSEMBLY

VARIABLE FOCAL
LENGTH LENS
ASSEMBLY

FOCUS
POTENTIOMETER

SHUTTER
ASSEMBLY

IRIS
POTENTIOMETER

VIDICON TUBE

VIDICON
RADIATOR

ELECTRONICS
CONVERSION
UNIT

ELECTRICAL
CONNECTORS

Details of the lunar surface survey camera carried by the Surveyor 1 lander.

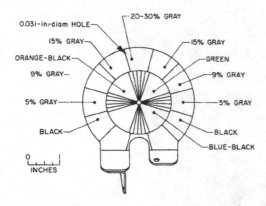

Details of the photometric calibration chart carried by the Surveyor 1 lander.

conform the overall camera-and-filter spectral response to the standard matching functions of colorimetry. To calibrate the camera on the lunar surface there was a small circular photometric chart on foot pad no. 2, and another affixed to the boom of omni-directional antenna 'B', on the frame to the left of the camera. Each target had grey-scale wedges arranged circumferentially and colour wedges radially, with a post in the centre serving as a gnomon whose shadow would indicate the position of the Sun relative to the target. The camera had been tested in field trials in 1964 on the Bonito Lava Flow near Flagstaff in Arizona, since it was thought that the rough ground there would reflect sunlight in a similar manner to the lunar surface.

With two receivers and two transmitters, there was redundancy. Each transmitter served one omni-directional antenna, but either could use the high-gain antenna. The engineering data (temperatures, voltages, currents, pressures, switch positions etc) were sequentially sampled by four commutators in the engineering signal processor and converted to 10-bit words by an analogue-to-digital converter for transmission to Earth. As the video output was already in digital form, it could be fed directly to the transmitter. Potentiometers measured the camera's settings, and these were sent as telemetry accompanying each picture.

The landing had occurred 57 hours after local sunrise on 30 May. The Sun was 28 degrees above the eastern horizon, and some 290 hours remained to sunset. The first 200-line TV frame was transmitted using omni-directional antenna 'B'. On receipt at 06:54, Goldstone converted the data to video and stored it and the related telemetry on magnetic tape. The video was relayed in real-time to JPL, which archived it onto 70-mm film, displayed it on closed-circuit monitors and issued it to the commercial networks, and they in turn forwarded it to Europe via the *Early Bird* geostationary satellite. It was a wide-angle frame taken with the iris on automatic gain control, and showed pad no. 2 and the adjacent surface. The pad was seen to have penetrated a surface of loose material to a depth of no more than a few centimetres, with no evidence of deformation to its crushable lower portion. This was excellent news for the Apollo engineers, since although the manned lander would be much heavier than the Surveyor, its foot pads had been scaled to impart the same pressure on the lunar surface. But there was still concern that a vehicle might break through a weak crust into a subterranean cavity.

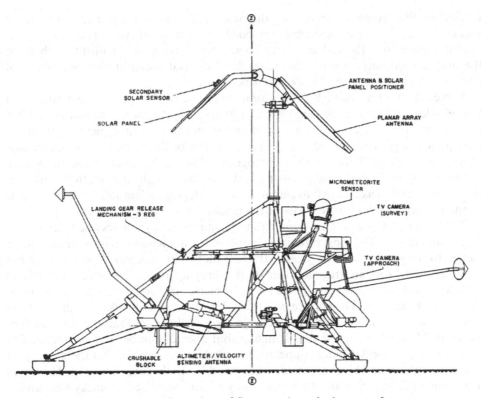

Details of the configuration of Surveyor 1 on the lunar surface.

After a total of 11 such pictures were taken over the next 50 minutes to inspect an arc between foot pads no. 2 and 3, the solar panel was commanded to face towards the Sun, and after it locked on at 08:56 the power was routed to recharge the battery. Next, the high-gain antenna was deployed. This was scanned until the strength of the signal received by Goldstone was maximised. With both the Sun and Earth located in the sky, it was possible to make a preliminary estimate of the lander's orientation on the surface – leg no. 1 was due west, and the survey camera mounted between the other two legs was on the eastern side. At 09:40 the camera was switched to its 600-line mode. By the time Goldstone lost its line of sight and handed over to Canberra, a total of 144 such pictures had been taken.

In 600-line mode, the resolution of the surface beside pad no. 2, some 1.6 metres from the viewing mirror, was 0.5 mm. It did not look like the 'frothy vacuum lava' that Gerard Kuiper had predicted. It appeared to be a blanket of fragmental material ranging in size all the way down to the limiting resolution. Some of the fine-grained material which had been disturbed by the foot pads formed a raised rim immediately adjacent to a pad, but the material displaced laterally was less than that which had been compressed in order to accommodate the foot pads. Rays of clumpy material that had been thrown out extended radially for several tens of centimetres. The clumps comprised weakly consolidated aggregates which had evidently been formed by the pressure imparted by the pads, and this indicated that the material was cohesive. The impact dynamics provided an estimate of the bulk density of the granular material at the surface as 1 to 1.5 gm/cm^3. Intriguingly, whilst the albedo of the undisturbed surface was 6 per cent, the disturbed material alongside the foot pads was typically one-third darker at 4 per cent.

Surveyor 1 took wide-angle frames for a panorama of the site – at this resolution a full 360-degree coverage required about 120 frames. Raymond Batson of the US Geological Survey had devised a procedure for mosaicking such pictures. Prints of the square images were pasted into their appropriate positions on the interior surface of a hollow sphere. This was then photographed in ten sections, the sections were flat-mounted side by side and photographed again as a single mosaic. In the process, the point of view was rectified to eliminate the 16-degree inclination between the axis of the camera and the vertical axis of the lander.

The landing site was a gently undulating plain pocked by craters and littered with fragmental debris. The most prominent crater in the immediate vicinity was about 11 metres to the southeast of the lander. It was 3 metres in diameter, had a distinct but irregular raised rim, an irregular inner slope averaging about 28 degrees, and was 60 cm deep. Most of the nearby craters up to 20 metres in size had either low rounded rims or were rimless. The surface was littered with 'coarse' blocks – with this term referring to their surface textures. The majority were brighter than the lunar surface material. Most were angular to subangular, but a small number were rounded. The majority of the angular blocks appeared to stand on the surface, but the major part of many of the round blocks seemed to be embedded in the loose material. There was a prominent 27-metre crater 60 metres away which possessed a blocky rim, and it had evidently excavated the coarse blocks in the lander's immediate vicinity. Some two dozen craters up to 100 metres in size were visible as narrow elliptical forms in the middle distance. A crater about 300 metres to the east had a prominent raised rim which had an exterior slope of about 12 degrees and stood 5 to 10 metres above its surroundings, with coarse blocks extending out from the crest of its rim to about one crater's diameter, the largest of which was about 1 metre in size. It was evident that these blocks had been excavated from a coherent substrate beneath the layer of fragmental material, and were single pieces of solid rock. Their faceted appearance implied they had broken at joints. Two larger craters further to the southeast had relatively low and inconspicuous rims. Overall, the characteristics of the craters implied that the fragmental debris layer extended to a depth of at least 1 metre. In terms of shape and distribution, the larger craters were similar to those seen by the Rangers immediately before they crashed, implying that this site was representative of a mare surface. The regional slope was determined by measuring the camera pointing angles of locations on the horizon and then using star sightings to obtain a true vertical in the celestial reference frame. This indicated the lander's vertical axis to be tilted to the southeast at 1.7 ($\pm$0.5) degrees. On the kilometre-scale, the area was essentially level, just as would be expected for a flow of low-viscosity lava which had flooded an ancient crater and solidified in place. The conclusion was that this site would be suitable for Apollo, which was welcome news because it established that there was at least one location available to that program.

Hills seen by Surveyor 1 projecting over the northeastern horizon.

With the camera's mirror positioned 1.2 metres above the ground, the horizon was about 2 km distant. Six features on the horizon in an arc spanning between northeast and northwest were clearly distant hills. The most prominent of these features was a ridge to the northeast. Ewen Whitaker identified them on a telescopic picture which had been taken using the University of Arizona's 61-inch NASA-sponsored reflector several months earlier when the illumination was similar, and triangulation showed the landing site to be well within the predicted 2-sigma ellipse. The M frames taken by Lunar Orbiter 1 in August 1966 confirmed these terrain identifications, and located the lander's position to within several kilometres. When Lunar Orbiter 3 photographed the general area in both M and H frames on 22 February 1967, the lander was found on frame H-183. Shadows cast by crater rims were perpendicular to the Sun line, but an object sitting on the surface cast its shadow down-Sun. At that time, the lander's mast-mounted panels were oriented to emphasise its shadow, and it was visible as a bright object with a long thin shadow.[4] During Lunar Orbiter 4's mapping mission in May 1967 the area was documented in frame H-143, and this enabled the position of the lander to be measured in relation to 13 craters whose selenographic coordinates were able to be measured on the global grid, thereby enabling the lander's position to be computed to an accuracy of ±0.01 degree, or roughly to within 600 metres. This established that it had landed within 15 km of the aim point.

On 3 June, Surveyor 1 sent sets of pictures using its three filters to enable a colour picture of pad no. 2 to be produced. It transpired that the only object to display real colour was the small circular photometric chart – the Moon seemed to be a gradation of grey, with black shadows. It took several days to accumulate the 1,000 frames for a single-filter panorama at narrow-angle resolution. In the process, detailed pictures were obtained of two nearby half-metre-sized coarse blocks resting on the surface. One, 5 metres to the southeast, was rounded on its upper side. Its texture hinted at the presence of vesicles in a fine-grained material. This suggested that it formed near the surface of a lava flow. The other block, 5 metres to the southwest, was angular with well-developed facets which were rounded at their corners and edges. It had a mottled appearance, with the lighter portions forming small knobs. The intersecting

---

[4]   Note that Surveyor 1 was the only lander in the series to be directly observed on the surface.

Frame H-143 taken by Lunar Orbiter 4 in May 1967 shows the Flamsteed Ring. The outline shows the area covered by the next illustration.

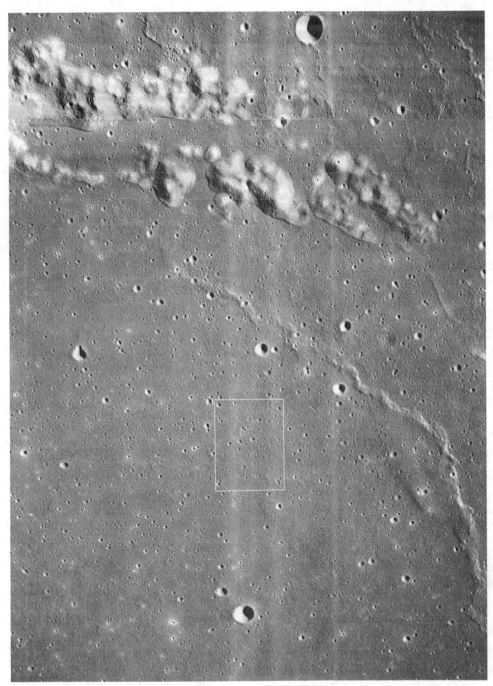

Frame M-183 taken by Lunar Orbiter 3 in February 1967 shows the area within the
Flamsteed Ring in which Surveyor 1 landed. The hills on the lander's horizon form part
of the ring. The outline shows the area covered by the next illustration.

Frame H-183 taken by Lunar Orbiter 3 shows the neighbourhood of the Surveyor 1 landing site. The outline shows the area covered by the next illustration.

A close examination of frame H-183 taken by Lunar Orbiter 3 revealed the Surveyor 1 lander (arrowed) casting a distinctive shadow on the lunar surface.

fractures suggested cleavage planes imposed by plastic flow under pressure. The fact
that the block was surrounded by small fragments of similar material suggested that
it broke apart on impact owing to a relatively low shear strength.

At 12:55 on 4 June, and again at 09:46 on 6 June, an experiment was performed in
an effort to determine whether firing a cold-gas attitude control thruster caused any
erosion of the lunar surface. The thrusters were mounted on the legs, just inboard of
the foot pads. The thruster on leg no. 2 was chosen because the ground under it was
well illuminated. One of the two nozzles was angled 'up' and the other 'down'. The
nozzle was 15 cm above the surface, and directed its 0.06-pound thrust at an angle of
72 degrees to the ground. It was pulsed for 20 milliseconds with a 30-millisecond
pause for a duration of 4.5 seconds, with this cycle being repeated several times. A
total of 30 pictures were taken prior to, during and after this activity. An intriguing
impression was observed where the jet would have impinged upon the surface, but
unfortunately it was not possible to prove beyond doubt that this had been caused by
the experiment. Although the crushable blocks on the underside of the frame
inboard of the legs were unobservable to the camera, it was decided to try to find out
if they had made contact with the ground. The ground beneath crushable block no. 1

A view of the cold-gas thruster on foot pad no. 2 of a Surveyor spacecraft in the
workshop. One jet points generally upwards and the other downwards. Note also the
hinged foot pad and the photometric calibration target on the leg.

was in shadow and there was no line of sight beneath no. 2, but there was an imprint visible under no. 3. The fact that the legs made contact essentially simultaneously implied that all three blocks would have left imprints. It was decided to fit later landers with auxiliary mirrors in order to improve the camera's view beneath the vehicle.

As a result of having revised the roll angle for the descent in order to maximise the strength of the signal from omni-directional antenna 'B', the orientation of the vehicle upon landing meant the camera spent a great deal of time in the shadow of the mast-mounted arrays, and so was able to operate at higher Sun angles than had been planned. This facilitated multiple wide-angle surveys to monitor the changing illumination. In particular, a 360-degree panorama was taken in colour to measure the spectral reflectance of the surface in the vicinity of the lander. The fact that the coarse blocks had a higher albedo than the surface made them very easy to survey at mid-day. As the landing site was 2.5 degrees south of the lunar equator, the highest that the Sun rose in the sky at local noon, at 06:17 on 7 June, was 87.5 degrees. The camera had to cease work for two days because it became too hot to operate, but it resumed activity when the elevation of the Sun declined to 78 degrees. On 10 June a wide-angle 360-degree panorama was taken to document the shadows lying in the opposite direction to previously, and beginning on 11 June a wide-angle 360-degree panorama was taken every 24 hours to monitor the lengthening shadows, including that cast by the lander itself. As the lunar day drew to a close, the vehicle recharged its battery and oriented its solar panel facing due west to ensure that when the Sun rose again there would not be a surge of power to the chilled battery.

Local sunset was at 15:12 on 14 June. A series of pictures taken with the solar disk just below the horizon recorded the corona extending out as far as 4 solar radii (measured from the centre of the solar disk) and the presence of a prominent coronal streamer. The final picture of this set, taken at 15:37, was a 4-minute exposure in Earthlight showing foot pad no. 2. During its first day of operations, the lander had transmitted 10,338 pictures. It reported the decline in temperature to determine the equilibrium heat loss from its thermally controlled compartments. After 53 hours of darkness it was commanded at 20:31 on 16 June to switch off its transmitter and to hibernate – only the two receivers and the command decoder were left on, because they would be required to awaken the lander if it survived the night.

An attempt was made to command the lander to reactivate its transmitter several hours before sunrise on 28 June, but in vain. The command was issued daily, and at 11:29 on 6 July, some 171 hours after sunrise, it responded. The first command was to turn the solar panel to recharge the exhausted battery. Although it could not be measured directly, the temperature during the lengthy lunar night was estimated to have plunged to –180°C. On 7 July the camera returned 24 pictures, showing that it was still working. A plan was drawn up for using the camera in the 3 days remaining before sunset, but before this could be put into effect it was pre-empted on 8 July by the temperature of the battery increasing anomalously. After an emergency plan was devised to make the best use of what were expected to be the lander's final hours of life, the battery began to return to its normal state. As an engineering test, on 8 July the approach camera was switched on to verify that it was functional – although no

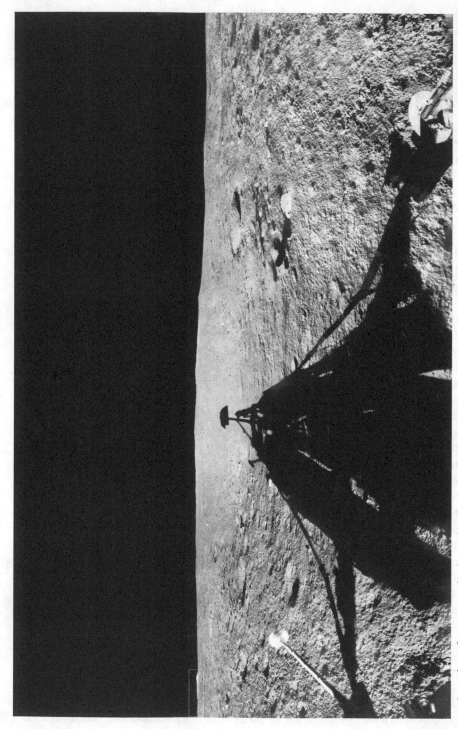

A section of a panoramic view taken by Surveyor 1 shortly before sunset, featuring the spacecraft's shadow. The view spans northeast to southeast, and the outline on the horizon marks the hills shown in an earlier illustration. (Courtesy of Philip J. Stooke, adapted from *International Atlas of Lunar Exploration*, 2007)

video was actually transmitted. In another test on 8 July, an attempt was made to fire the vernier engines but the power circuitry to activate the solenoids of the propellant valves failed, probably due to the chill during the night. Pictures taken on 12 July to record the state of the vehicle revealed that the glass of the radiator on top of one of the thermally controlled compartments had cracked. On 13 July Surveyor 1 began to take pictures at hourly intervals to monitor the lengthening shadows. By the time the operational phase of the mission was declared complete at sunset on 14 July, a total of 11,240 pictures had been taken.

Although Surveyor 1 continued to reply to engineering interrogations for several further lunations, the camera was not operated. During this extended mission, radio tracking of the lander on the surface provided data on the motions of the Moon. The last data returned was at 07:30 on 7 January 1967, during the eighth lunar day.

TUMBLING OUT OF CONTROL

Since Surveyor 2 was essentially complete at the time of the first mission, it was identical to its predecessor. Upon the surprising success of Surveyor 1, NASA opted not to postpone the second mission to install scientific instruments. It duly lifted off from Pad 36A at 12:32:00 GMT on 20 September 1966 on a direct-ascent trajectory with the objective of landing in Sinus Medii.[5] In general, this was much rougher-looking than the area in Oceanus Procellarum assigned to Surveyor 1. This time, it was intended to operate the downward-looking TV camera during the approach in order to gain a sense of perspective of the landing site.

The translunar injection trajectory would intercept the Moon just northeast of the crater Mosting on the western margin of Sinus Medii, 142 km from the centre of the 60-km target circle. The sequence for the 31.5-ft/sec midcourse manoeuvre began at T + 15h 42m with an interrogation to verify the readiness of the vehicle's systems. At 16h 12m the spacecraft initiated a roll of + 75.3 degrees, followed 5 minutes later by a yaw of + 11.5 degrees. This successfully oriented the vehicle for the burn. The burn was started on command at 16h 28m but vernier no. 3 failed to ignite. After the specified duration of 9.8 seconds, the system shut down. The asymmetric thrust had left the vehicle spinning about one axis at 1.22 revolutions per second, with this axis precessing in 12 seconds. This saturated the gyros in the minus pitch, plus yaw and minus roll directions. The flight control system set about regaining stability utilising the attitude control thrusters. After 7 minutes the precession had been cancelled. But after 14 minutes, with a residual spin around the main axis of 0.85 revolutions per second and with the jets having consumed half of the nitrogen supply, a command was sent from Earth to inhibit the system and save the remaining gas for use in the event that the problem involving vernier no. 3 could be overcome, at which time the verniers would be used to stabilise the vehicle.

[5]   In fact, this was site I-P-5 on Lunar Orbiter 1's target list.

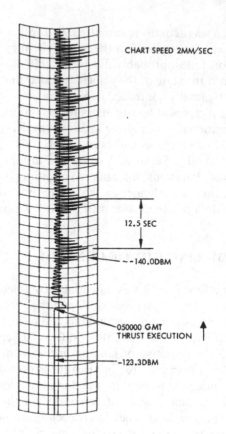

CHART SPEED 2MM/SEC

12.5 SEC

~ -140.0DBM

050000 GMT
THRUST EXECUTION

-123.3DBM

Deep Space Instrumentation Facility
automatic gain control variations served to
document the initial tumbling of the
Surveyor 2 spacecraft.

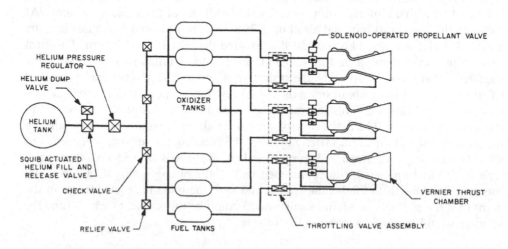

SOLENOID-OPERATED PROPELLANT VALVE

HELIUM PRESSURE
REGULATOR

HELIUM DUMP
VALVE

HELIUM
TANK

OXIDIZER
TANKS

SQUIB ACTUATED
HELIUM FILL AND
RELEASE VALVE

CHECK VALVE

VERNIER THRUST
CHAMBER

RELIEF VALVE

FUEL TANKS

THROTTLING VALVE ASSEMBLY

A simplified depiction of the Surveyor spacecraft's vernier propulsion system.

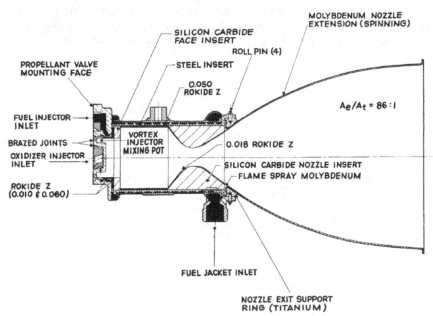

Detail of an engine of the Surveyor spacecraft's vernier propulsion system.

The solar panel was unable to provide power while the vehicle was tumbling, so time was of the essence. It was apparent that verniers no. 1 and 2 had delivered the specified thrust during the manoeuvre, but no. 3 had delivered no thrust at all. It was decided to command a 2-second firing in an effort to clear vernier no. 3. This was done at 18h 56m and again at 19h 18m, but without success. In case the fault was a stuck flow regulator valve, the system was commanded to pulse five times for a duration of 0.2 seconds at 5 minute intervals starting at 31h 12m and then attempt another 2-second firing – again in vain. This sequence was repeated at 26h 28m, 37h 29m, 38h 45m and 39h 45m – each time with no effect. At 41h 11m an attempt was made to fire the engine at a 'harder' start and a higher thrust for 2 seconds. Because the other two verniers were participating in these tests, by this point the spin rate had increased to 1.54 revolutions per second. At 43h 13m a new sequence was initiated in which the engines were pulsed five times for 0.2 second with 1 minute between firings, as a preliminary to a 20-second firing. Although this time the temperature of vernier no. 3 increased somewhat, the engine did not respond properly.

Although the spacecraft was tumbling out of control, it was decided to undertake a series of tests to obtain engineering data on its subsystems, concluding at 45h 02m with a command to trigger the retro sequence. Contact was lost 30 seconds into the retro-rocket's burn. The inert vehicle would have struck the Moon several hundred kilometres southeast of Copernicus.

The post-flight investigation by propulsion engineers of JPL, NASA, Hughes and Thiokol decided that there had been no combustion in vernier no. 3 at the attempted midcourse manoeuvre, and that although fuel had flowed into the engine the oxidiser had not. As it was not possible to determine the root cause of the failure, a number of revisions were introduced for Surveyor 3 designed to provide better diagnostics of the vernier propulsion system, both during pre-flight testing and in flight.

11

The Apollo zone

RECONNAISSANCE FLIGHTS

Only 28 months elapsed between James Webb approving the Boeing contract for Lunar Orbiter and the first mission. It had been hoped to launch on 11 July 1966, but Eastman Kodak was late in delivering the photographic system and the launch was rescheduled for 9 August. At the Cape, the Air Force made available Hangar S for the project. The photographic system was installed on 1 August, and the next day the spacecraft was mated with its Atlas-Agena D on Pad 13. At this point, a Deep Space Network facility at the Cape verified the spacecraft's communication system.

The countdown on 9 August was scrubbed at T–7 minutes owing to an anomaly with the Atlas, but liftoff occurred at 19:26:01 GMT on 10 August. The spacecraft's launch mass was 387 kg – which was at the top end of the launcher's capability for a deep-space mission. The Agena made a 154.5-second burn to achieve insertion into parking orbit at an altitude of 190 km at 19:34:44. It reignited at 20:02:35 to perform the 89-second translunar injection manoeuvre.

Following its release at 20:06:48, Lunar Orbiter 1 deployed its solar panels and antennas, then acquired the Sun for the first step towards adopting its cruise attitude. Some 6 hours into the translunar coast, the spacecraft was ordered to roll in order to locate the star Canopus, but it failed to lock on. The sensor was being distracted by sunlight reflecting off the vehicle's structure – a possibility that really ought to have been 'designed out'. The vehicle was commanded to turn to point the sensor at the Moon to provide a second point of reference to initialise the inertial system.

The midcourse manoeuvre of 38 m/s was achieved by a 32-second burn started at 00:00 on 12 August. When its interior began to overheat, the vehicle was instructed to swing its main axis 36 degrees off-Sun. This thermal problem came as a surprise, because Boeing had subjected the design to thorough testing in the vacuum chamber at its Kent Test Facility in Seattle. The Canopus sensor was finally able to lock on at 13:50 on 13 August.

The trajectory was so accurate that the vehicle arrived within 10 km of the desired orbit insertion point. The command for the insertion manoeuvre was sent at 15:23 on 14 August, after a cruise of 92 hours as against 66 hours for Ranger. After adopting the requisite attitude, the spacecraft began the burn at 15:34, firing its engine for 579 seconds in order to slow down by 790 m/s and enter a 190 × 1,860-km orbit that was inclined at 12.2 degrees to the lunar equator and had a period of 3 hours 37 minutes. As viewed from an imaginary vantage point above the Moon's north pole, the orbit was anticlockwise. Shortly after insertion, the vehicle disappeared around the Moon's trailing limb.

On 15 August Lunar Orbiter 1 read out the pre-exposed test frames on the leader of its film strip. These had been read out during ground trials at Goldstone to verify the functionality of the scanning and communication system. This in-flight readout was to confirm that the entire apparatus, both in space and on Earth, was functioning properly. By now, the thermal problem had become acute. The paint on the base of the vehicle was meant to absorb heat while exposed to sunlight and radiate it back to space while in the shadow of the Moon, but it seemed the pigment was deteriorating. However, on 18 August an electrical issue had the beneficial side-effect of easing this overheating problem.

The phase of the Moon was 'new' on 16 August; 'first quarter' would occur on 23 August and 'full' on 31 August. As the alignment of the spacecraft's perilune would remain fixed relative to the stars, the longitude of perilune would migrate westward at a rate of 12 degrees per 24 hours as the Moon travelled around Earth. This meant that the spacecraft could photograph a succession of sites along the equatorial zone with the Sun at essentially the same elevation in the sky. The Canopus sensor would routinely recalibrate the inertial system to ensure that the camera was accurately aimed.

After advancing the film to its start point, the camera snapped its first pictures on 18 August. It viewed Mare Smythii, near the equator on the eastern limb. Although outside the Apollo zone, this area was of interest for selecting landmarks to assist in Apollo orbital navigation. In this case, the spacecraft was at an altitude of 246 km, its velocity relative to the surface was 6,400 km/hour, the shutter setting was 1/50th second, and the frame pairs were taken at 10-second intervals. These frames were processed and 5 hours later were scanned travelling forward through the system. The M frames were of excellent quality, but it seemed that the focal-plane shutter of the narrow-angle optics was out of synchronisation with the V/H sensor, smearing the H frames. To investigate the problem, it was decided to take additional pictures using different shutter speeds, both with and without the V/H sensor operating. The results confirmed the sensor to be inoperative. A further test was made in which the output voltage of the sensor was increased in the hope of engaging the shutter, but this did not work. An analysis of the telemetry showed that the logic control circuitry of the focal-plane shutter was susceptible to electromagnetic interference that made it fire at the wrong point in the motion-compensation cycle, and this was a problem which could not be overcome in flight.

Since the H frames were essential to addressing Apollo's requirements and they would *not* be able to be provided, Jack McCauley and Lawrence Rowan of the US Geological Survey recommended that the high perilune be retained in order to

Assembling the Lunar Orbiter 1 spacecraft.

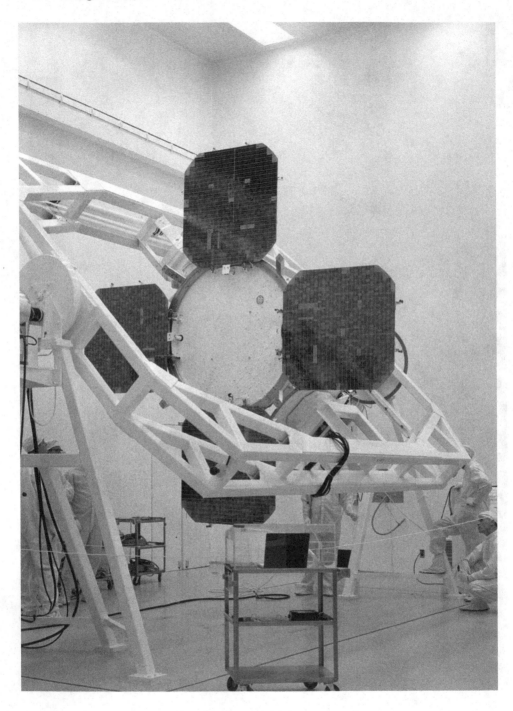

Testing the Lunar Orbiter 1 spacecraft.

Preparing to fit the aerodynamic shroud on the Lunar Orbiter 1 spacecraft, wrapped in its thermal blanket.

A press conference by Lunar Orbiter managers on 1 August 1966 in the run up to the first Lunar Orbiter mission.

On 10 August 1966 an Atlas-Agena lifts off with the Lunar Orbiter 1 spacecraft.

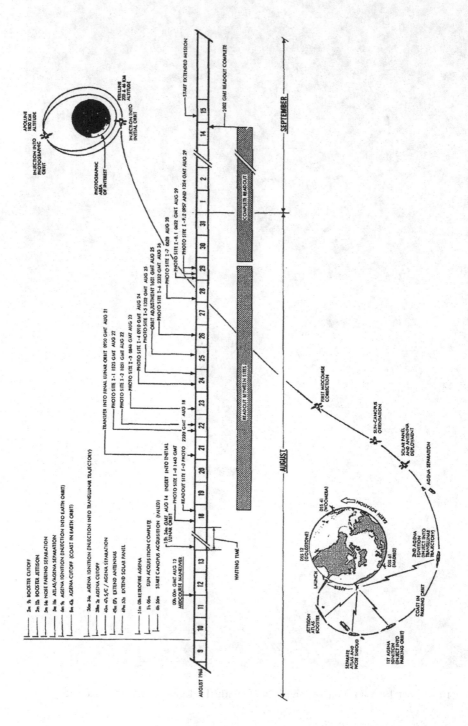

The Lunar Orbiter 1 mission timeline.

obtain wide-area *mapping* at a resolution of about 25 metres, to improve coverage of the majority of the Apollo zone at a definition better than that of a telescope. Doing so would enable other candidate sites to be re-evaluated prior to drawing up the target lists for subsequent Lunar Orbiter missions. The Bellcomm advisors were in agreement. But Clifford Nelson, the Project Manager at Langley, and his Boeing counterpart, Robert Helberg, argued that if the perilune were lowered then the increased rate of motion presented to the V/H scanner might induce the shutter to synchronise, in which case the mission would be able to proceed as planned. And so it was decided to make a 22.4-second, 40-m/s burn at 09:50 on 21 August to lower the perilune to 58 km.

As the spacecraft was approaching apolune, shortly prior to the perilune-lowering manoeuvre, it took a picture of the far-side. At an altitude of 1,497 km there was no need for V/H motion compensation, and when the H frame was read out it proved to be of excellent quality, confirming that the optics were correct. At this point, it was discovered that the Bimat strip was sticking to the processing drum. The plan had been to obviate this problem by not letting the Bimat remain immobile for more than 15 hours, and to shoot a frame-pair once every four orbits with the thermal door shut if no target was scheduled. The remedy was to reduce the time that the Bimat was immobile. The increased number of frames for this maintenance task meant revising the photographic mission. The most important targets would still receive the planned 16-exposure blocks, but others would get only half this number.

Lunar Orbiter 1 was to photograph nine primary targets in the southern part of the Apollo zone, seven secondary targets and the far-side from high altitude. At this stage in the reconnaissance process, however, each primary target spanned an area in which there might prove to be *tracts* which could be nominated as potential Apollo landing sites.

The primary targets were labelled by Roman numeral 'I' (for the first mission), 'P' for primary, and an Arabic number in a sequence progressing generally from east to west. In this scheme, the Mare Smythii area was I-P-0.

In terms of classifications:

I, mare
        A, average mare
        B, dark mare
        C, ridged mare
        D, rayed mare
II, upland
        A, highland basins
        B, subdued uplands
        C, upland plains
        D, sculptured highlands
III, craters
        A, well formed
        B, subdued

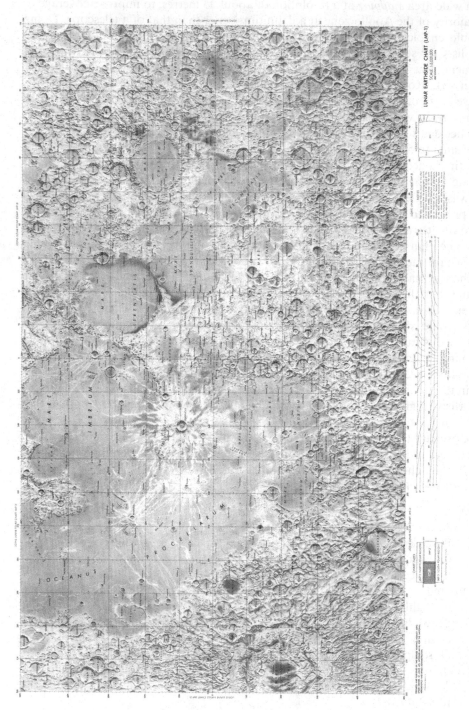

A map of the near-side of the Moon on a scale of 1:5,000,000 derived from imagery produced by the Lunar Orbiter missions.

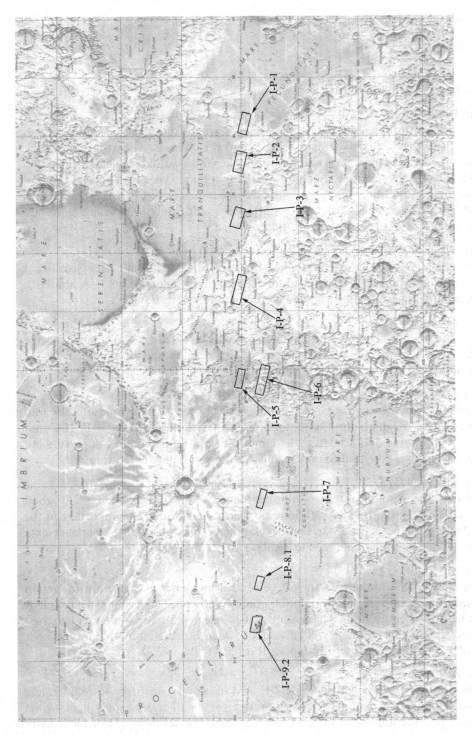

The Apollo zone annotated with the primary photographic targets assigned to the Lunar Orbiter 1 mission.

IV, structural features
    A, ridges
    B, domes
    C, rilles

the planners had the following to say about the primary sites:

Site I-P-1 (0.9°S, 42.3°E)
This site is within the lowlands of the western extreme of Mare Foecunditatis, and a portion of the eastern extremity of the central highlands. The inclusion of dark mare, moderately light mare and uplands makes this a valuable terrain calibration area. The 1-metre relative roughness of the two mare types is of special interest. The possible detection of generic relationships in the contact between the uplands and dark mare is of particular importance scientifically. The mare units are potential Apollo landing sites. (Rating A)

Site I-P-2 (0.2°S, 36.0°E)
A highland site bordering the southeast part of Mare Tranquillitatis. Significant terrain calibration data is expected for the mare and upland units II-A and II-B. Potential Apollo landing sites may be revealed here. (Rating B)

Site I-P-3 (0.3°N, 24.9°E)
This site in southwestern Mare Tranquillitatis is crossed by a ray. Data on the small-scale roughness and morphology of this area should be obtained. It is a potential Surveyor and Apollo landing site. (Rating B)

Site I-P-4 (0.0°, 12.9°E)
This site is located in the central highlands between Mare Tranquillitatis and Sinus Medii. High-resolution photography of terrain units II-A, II-B, II-C and II-D will provide data to define the 1-metre-resolution roughness of the upland areas. This is a potential Surveyor and Apollo landing site. (Rating A)

Site I-P-5 (0.4°S, 1.3°W)
Located in the southwestern portion of Sinus Medii, this is an especially good example of a smooth mare with low subdued ridges, which are important in the evaluation of the mare origin and development. This has high potentiality as a Surveyor and Apollo landing site. (Rating B)

Site I-P-6 (4.0°S, 2.9°W)
This area in the northern sector of the central highlands is important to terrain calibration by offering high-resolution photography of upland unit II-D as well as the deformed crater floor that is a previously selected Surveyor landing site. It is expected that this coverage will be valuable when bearing-strength data becomes available for the major terrain types. (Rating A)

Site I-P-7 (3.8°S, 22.8°W)
This site located between the craters Fra Mauro and Lansberg is a moderately good example of a mare with low sinuous ridges, small craters and a light ray. It should provide important information regarding the development of older mare

surfaces, and their characteristic morphology. It is a previously selected Surveyor landing site. (Rating B)

Site I-P-8.1 (3.0°S, 36.5°W)
This site in the southeast part of Oceanus Procellarum is a superior example of a relatively straight mare ridge, and is of particular importance in the definition of 1-metre-resolution roughness. It is an opportunity to investigate the generic processes concerned with the development of this mare morphology. It is a highly rated Surveyor landing site. (Rating A)

Site I-P-9.1 (2.3°S, 43.4°W)
This is the Surveyor 1 landing site. The location was changed on the basis of later refinement of the position of the lander, relabelled I-P-9.2a and I-P-9.2b, and scheduled for two successive orbits.

The first photography of a site in the Apollo zone was a 16-exposure sequence of I-P-1 taken on 22 August. The possibility of photographing Earth on the limb of the Moon had been discussed a year before the spacecraft was launched, but it was not made a mission requirement. On 22 August NASA representatives suggested that it be attempted, even though it would require the vehicle to adopt an unusual attitude. Robert Helberg, the Boeing manager, considered the impromptu manoeuvre to be an unnecessary risk because immediately after taking the picture the vehicle would pass beyond the limb and would have to re-establish its normal attitude out of contact with Earth. The company was understandably reluctant, because a large part of the bonus of its 'incentive' contract depended on completing the primary mission. Floyd Thompson, Clifford Nelson and Lee Scherer were in favour. Boeing relented when NASA agreed to compensate it if the vehicle were lost as a result of the experiment. At an altitude of 1,198 km, climbing towards apolune on 23 August, Lunar Orbiter 1 approached the trailing limb of the Moon as viewed from Earth. It turned to point its main engine perpendicular to the line of the limb, and then rolled until the camera's oblong H frame paralleled the limb. The shutter was fired at 16:36:28.6 GMT. The wide-angle view showed Earth as a tiny crescent against the limb, and also provided a magnificent view of the far-side crater Tsiolkovsky. The H-frame was much more dramatic, showing 'Earth set' against the cratered lunarscape. This was prominently featured by newspapers. A somewhat less impressive picture of Earth was taken at 07:15:00.9 on 25 August.

A 3-second, 5.4-m/s burn at 16:01 on 25 August lowered the perilune to 40 km to increase the motion of the image across the narrow-angle shutter, but the V/H sensor remained inoperative. At 13:23 on 29 August the spacecraft began the photographic sequence for I-P-9.2b, the final target on its list. The Bimat strip was cut at 18:14 on 30 August, and the full readout by rewinding the film began at 20:42. This process was finished at 21:18 on 16 September. Of the total of 205 frame-pairs, 38 had been obtained in the initial orbit. All nine primary targets had been photographed at lower altitude. Despite the loss of the high-resolution pictures of the primary targets, the mission was judged a successful engineering test of a new spacecraft which had produced some useful imagery.

On 23 August 1966 Lunar Orbiter 1 took this picture of Earth about to set behind the limb of the Moon. As the first view of our planet from lunar distance it was a sensation.

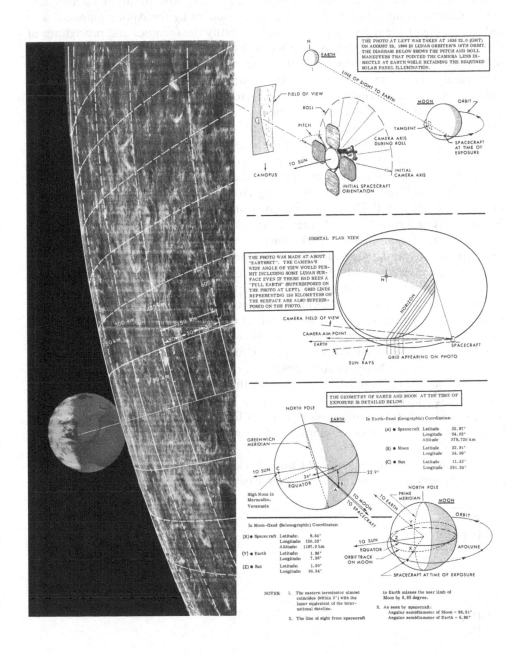

The geometry of Lunar Orbiter 1's historic picture of Earth near the Moon's limb.
(Courtesy of the Lunar Orbiter Image Recovery Project, Ames Research Center, 2008)

Its primary mission over, Lunar Orbiter 1 flew on to gather additional selenodesy, micrometeoroid and radiation data – all eagerly sought by the Apollo planners. The radio tracking of the spacecraft also served to certify the facilities and procedures of the Manned Space Flight Network at lunar distance.[1] By 28 October the condition of the vehicle had deteriorated – its battery was failing as a consequence of persistent overheating, its transponder was dropping out intermittently, its inertial reference unit was becoming unreliable, and the nitrogen for its attitude control thrusters was almost exhausted. To lose control of the communications system would risk radio interference with its successor, so on 29 October Lunar Orbiter 1 was deliberately crashed onto the far-side. In its 8 weeks in space, the micrometeoroid experiment had reported no hits at all. Because of the persistent thermal problems, the white paint for the base of the vehicle was modified for the next mission.

The Lunar Orbiter Photo Data Screening Group of specialists from JPL, Langley, the Manned Spacecraft Center, Boeing, Bellcomm, the Army Map Service, the Air Force Chart and Information Center, NASA headquarters and the US Geological Survey studied Lunar Orbiter 1's M frames to test the hypothesis that terrain which appeared smooth at the resolution of a telescope was probably smooth in finer detail. They also estimated the cratering of the different types of terrain, first by counting the craters down to the limiting resolution and thereafter using the photogrammetry technique. And, of course, photoclinometry was used to measure slopes. With Gene Shoemaker busy with Surveyor, the analysis at the US Geological Survey was led by Jack McCauley. For a spacecraft making a powered descent directly from translunar trajectory the target was a 60-km-diameter circle, but an Apollo lander descending from lunar orbit was expected to be able to be more accurate. But because pin-point accuracy could not presumed, at least not for the early missions, Apollo targets were ellipses with their major axes aligned along the line of approach. The US Geological Survey produced terrain and geological maps for smooth-looking areas which were large enough to accommodate ellipses of sizes corresponding to differing degrees of landing accuracy.

Taking into account the 'ground truth' provided by Surveyor 1 at I-P-9.2, the nine primary targets were rated in terms of increasing roughness as follows:

I-P-9.2
I-P-3
I-P-1
I-P-7
I-P-8.1
I-P-5
I-P-6
I-P-4
I-P-2

---

[1] The Manned Space Flight Network was operated under the direction of the Goddard Space Flight Center in support of the Manned Spacecraft Center.

Table 11.1 – Lunar Orbiter 1 orbital manoeuvres and photography

Date	Event/Site	Frames	
14 August	Orbit insertion		
18 August	I-P-0	22	(5–24)
19 August	far-side	1	(28)
20 August	far-side	1	(30)
21 August	Perilune cut to 58 km		
22 August	B-2	2	(48–49)
22 August	I-P-1	16	(52–67)
22 August	I-P-2	16	(68–83)
23 August	I-P-3	16	(85–100)
23 August	Earth	1	(102)
23 August	B-5	1	(103)
24 August	I-P-4	8	(105–112)
24 August	B-7	2	(113–114)
25 August	B-5	2	(115–116)
25 August	Earth	1	(117)
25 August	I-P-5	16	(118–133)
25 August	Perilune cut to 40 km		
25 August	B-8	2	(134–135)
26 August	I-P-6	8	(141–148)
27 August	B-9	1	(149)
27 August	B-10	2	(150–151)
27 August	B-11	4	(153–156)
28 August	I-P-7	16	(157–172)
29 August	I-P-8.1	8	(176–183)
29 August	I-P-9.2a	16	(184–199)
29 August	I-P-9.2b	16	(200–215)
30 August	Cut Bimat and start full readout		
16 September	Finish readout		

Notes: (1) Engineering exposures have been omitted. (2) The B-sites were previews of targets being considered for the forthcoming 'B' mission.

On 6 May 1966 representatives of Lunar Orbiter, Surveyor, Apollo and Bellcomm had met at Langley in order to plan the 'B' mission. Whereas the inclination of the first spacecraft's orbit was chosen to facilitate vertical photography of targets in the Apollo zone south of the equator, the inclination of the second spacecraft's orbit was to inspect targets north of the equator. As planning for this mission began before the first mission flew, the targets were drawn from those chosen by the Surveyor/Orbiter Utilisation Committee from the list submitted by Jack McCauley in August 1965 on the basis of telescopic studies. A total of 13 primary and 17 secondary targets were selected, each of which was to receive one pass. The secondary areas were not under consideration for the early Apollo landings, but were considered to be of 'scientific interest'. The photographic system had to advance at regular intervals to prevent the Bimat from sticking. In accordance with Boeing's procedure, for Lunar Orbiter 1 the film had been advanced with the thermal door closed. But Tom Young at Langley and Ellis Levin at Boeing had devised a procedure for doing this with the door open, and a schedule had been written to allow this maintenance function to provide useful pictures. The 'B' plan was approved by the Surveyor/Orbiter Utilisation Committee on 1 June.

In late September, however, the planners convened to review the target list in the light of the results of the first mission. One factor that influenced Apollo planning was Ranger 7's pictures of bright rays from Copernicus and Tycho crossing Mare Nubium. If (as was suspected) all rays were highly cratered, then this would argue against such areas being considered as landing sites. To test this hypothesis, Lunar Orbiter 1 had imaged a faint ray crossing Mare Tranquillitatis that originated from Theophilus, a prominent crater in the highlands to the south. There was no H frame coverage to confirm it, but the M frame suggested that the cratering was insufficient to rule out this site (I-P-3) as a potential Apollo site. It was decided to revise the 'B' mission to further study this issue by inspecting rays crossing Oceanus Procellarum between the craters Copernicus and Kepler. On 29 September the Surveyor/Orbiter Utilisation Committee, having reviewed the crater densities and slopes derived from Lunar Orbiter 1's M frames, recommended that eight areas be carried over to Lunar Orbiter 2 for study at higher resolution. Apart from the deletion of the high-perilune photography of the eastern limb, in purely operational terms the 'B' mission was to be a repeat of its predecessor.

After serving as the backup for the first mission, Lunar Orbiter 2 had been stored. Once modified, it was mated with its launch vehicle on 31 October and a countdown test that concluded on 3 November certified it ready for flight.

Lunar Orbiter 2 was launched at 23:21 GMT on 6 November 1966. After coasting in parking orbit for 14 minutes, the Agena performed the translunar injection burn. Parts of the spacecraft's surface had been painted black to reduce the opportunity for reflected sunlight to dazzle the Canopus sensor, and at 08:21 on 7 November this locked on without incident. At 19:30 on 8 November the spacecraft fired its engine for 18 seconds for a 21-m/s midcourse manoeuvre. The modified paint on the base of the vehicle eliminated the thermal problem that had marred the first mission. The 612-second, 830-m/s burn at 20:26 on 10 November put the spacecraft into an initial orbit of $196 \times 1,871$ km that was inclined at 12.2 degrees to the lunar equator and had a period of 3 hours 38 minutes. At 22:58 on 15 November a 17.4-second 28-m/s burn lowered the perilune to 50 km. The photography began on 18 November. The phase of the Moon was 'new' on 12 November; 'first quarter' would occur on 20 November and 'full' on 28 November.

Several modifications had been made to the photographic system. In particular, an integrating circuit had been added to the focal-plane shutter control logic to ensure that an output signal was not an electrical transient but a genuine command pulse; and a filter had been added to the 20-volt power line in order to minimise electromagnetic interference and spurious triggering of the circuits. An early check showed that the V/H sensor was operating correctly. As the spacecraft's perilune migrated west with the terminator, on 20 November it photographed site II-P-5 in Mare Tranquillitatis in search of Ranger 8's impact point. On the next revolution it

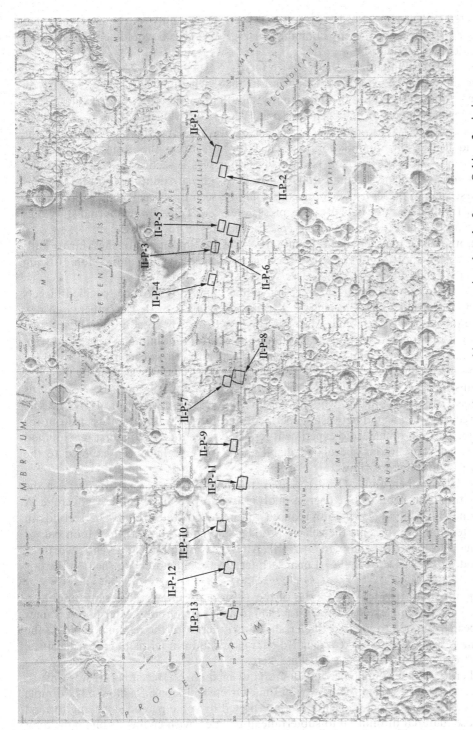

The Apollo zone annotated with the primary photographic targets assigned to the Lunar Orbiter 2 mission.

inspected the nearby I-P-3 target that had been rated highly on the basis of Lunar Orbiter 1's M frames, thereby obtaining the H frames which would be required for a more detailed evaluation.

The most striking image of any mission to date was taken on 24 November as one of the maintenance exposures. Douglas Lloyd of Bellcomm had suggested early in planning that at this point the opportunity should be taken to obtain an oblique view of Copernicus, and this had been agreed as II-S-12. The vehicle was at an altitude of 46 km, and the 100-km-diameter crater was some 240 km away to the north of the ground track. The illumination was ideal for discerning the nature of the local topography – on the horizon beyond the crater was a section of the Carpathian Mountains, whose peaks stand 1,000 metres above the surrounding terrain. This was the first view to present the Moon as it would be seen by an astronaut in low orbit. When the picture was issued to the press, the *New York Times* journalist Walter Sullivan described it as "one of the greatest pictures of the century".

Another such maintenance exposure, II-S-15, was an oblique of the Marius Hills in the western portion of Oceanus Procellarum, which gave the impression of being volcanic domes.

The photography concluded on 25 November, after all 30 of the specified targets had been documented. The high-gain transmitter failed on 6 December, one day before the readout was to end. Although this resulted in the loss of three M and two H frames of target II-S-1 in the eastern part of Mare Tranquillitatis, these had been taken at the start of the mission and their loss was not serious because some of them had been read out as part of the verification process as the film was running forward. The mission provided a number of high-quality pictures of the far-side. As regards the rays crossing the maria, while the results showed these were not always heavily cratered, II-S-11 located between Copernicus and Kepler was certainly ruled out for Apollo. The micrometeoroid experiment reported three hits. As the first mission had reported no hits, it was speculated that Lunar Orbiter 2 had been exposed to the annual Leonid meteor shower.

On 8 December Lunar Orbiter 2 fired its engine for 62 seconds to increase the inclination of its orbit to 17.5 degrees in order to improve the selenodesy coverage. A 3-second burn on 14 April 1967 reduced the period of the orbit by 65 seconds in order to minimise the time the vehicle would spend in darkness as the Moon passed through the Earth's shadow on 24 April. After providing almost a year's worth of selenodesy, on 11 October 1967 Lunar Orbiter 2 was deliberately crashed on the far-side.

The mission provided 184 frames of the 13 primary targets. The screening team reconvened at Langley on 5 December 1966 to assess the results. After the M frames had been examined to identify the terrain units in terms of the regional geology, the H frames were used to make a detailed characterisation in terms of crater densities, slope frequency distributions, blockiness etc.

When the Apollo Site Selection Board convened on 15 December, it was able to discuss the pictures from the first two Lunar Orbiter missions. Shallow pits ('dimple craters') on Ranger pictures had been interpreted by some people as having formed by loose material draining into subterranean cavities, with chains of pits indicating

A portion of the picture taken by Lunar Orbiter 2 on 24 November 1966 showing the central peak complex and terraced wall of Copernicus, with a mountain range beyond. This unprecedented oblique view was described as "one of the greatest pictures of the century".

An oblique view taken by Lunar Orbiter 2 on 25 November 1966 showing the hills and flow fronts in western Oceanus Procellarum near the crater Marius.

Table 11.2 – Lunar Orbiter 2 orbital manoeuvres and photography

Date	Event/Site	Frames	
10 November	Orbit insertion		
15 November	Perilune cut to 50 km		
18 November	II-P-1	16	(5–20)
18 November	II-S-1	4	(21–24)
18 November	II-S-2a	4	(25–28)
18 November	II-S-2b	4	(29–32)
19 November	II-S-3	1	(33)
19 November	II-S-4	1	(34)
19 November	II-P-2	8	(35–42)
19 November	II-P-3a	8	(43–50)
19 November	II-P-3b	8	(51–58)
19 November	II-P-4	8	(59–66)
20 November	II-P-5	8	(67–74)
20 November	II-S-5	1	(75)
20 November	II-P-6a	8	(76–83)
20 November	II-P-6b	8	(84–91)
21 November	II-S-6	1	(92)
21 November	II-S-7	1	(93)
21 November	II-S-8	1	(94)
21 November	II-S-9	1	(95)
22 November	II-P-7a	8	(96–103)
22 November	II-P-7b	8	(104–111)
22 November	II-S-10	1	(112)
22 November	II-P-8a	8	(113–120)
22 November	II-P-8b	8	(121–128)
22 November	II-P-8c	8	(129–136)
23 November	II-S-11	1	(137)
23 November	II-P-9	8	(138–145)
23 November	II-P-10a	8	(146–153)
23 November	II-P-10b	8	(154–161)
24 November	II-S-12	1	(162)
24 November	II-P-11a	8	(163–170)
24 November	II-P-11b	8	(171–178)
24 November	II-P-12a	8	(179–186)
24 November	II-P-12b	8	(187–194)
25 November	II-S-13	1	(195)
25 November	II-S-14	1	(196)
25 November	II-P-13a	8	(197–204)
25 November	II-P-13b	8	(205–212)
25 November	II-S-15	1	(213)
25 November	II-S-16	1	(214)
25 November	II-S-17	1	(215)
26 November	Cut Bimat and start full readout		
6 December	Readout halted		

fractures. There was concern that if the maria were lava flows, then by terrestrial analogy they might contain lava tubes which could collapse under the weight of an Apollo lander. Lawrence Rowan of the US Geological Survey ventured that in this respect younger-looking mare areas (such as II-P-2) appeared to pose a greater risk than older-looking mare areas (such as II-P-6). It had been possible to fit 23 Apollo target ellipses into clear-looking patches in the Lunar Orbiter pictures to date. After these had been screened by counting small craters and measuring slopes, eight were selected for further study.

On 5 January 1967 the Surveyor/Orbiter Utilisation Committee approved the plan for the third mission. As its predecessors had reconnoitred all the assigned targets, Lunar Orbiter 3 was to provide additional data for those which the screening process had deemed to be the most promising. The spacecraft was identical, but the operational plan was more sophisticated. At the western end of the Apollo zone, launches at different times of year favoured passing either north or south of the equator. In the case of the first mission, the orbit was inclined at 12 degrees to the lunar equator in order to photograph sites lying south of the equator from a vertical perspective, and for the second mission the orbit was inclined the other way for sites north of the equator. To enable Lunar Orbiter 3 to cover the entire latitude range, its orbit was to be inclined at an angle of 21 degrees. It was given 12 primary and 32 secondary targets. It was to obtain H frames of the best-looking areas that had been imaged by Lunar Orbiter 1 as M frames but not by Lunar Orbiter 2 as H frames. In particular, it was to provide pictures to facilitate stereoscopic analysis to compile terrain maps with 3-metre contours, in order to chart the topography on the line of approach from the east over which an Apollo lander would pass during its powered descent to a target.[2] This marked a switch from reconnoitring areas to certifying specific targets. Some of the secondary frames were oblique views of Apollo sites, taken looking west to show them as they would appear to astronauts preparing to make a descent. It was felt that three Lunar Orbiter missions in the equatorial zone should be sufficient to select targets for the first few Apollo landings. The Manned Spacecraft Center also wanted further radio tracking in lunar orbit to investigate anomalous gravitational effects which had been revealed by tracking the first two missions.

Lunar Orbiter 3 was launched at 01:17:01 GMT on 5 February 1967. Translunar injection was at 01:36:56. The spacecraft was released at 01:39:40 and deployed its appendages. Some 7 hours into the flight, it locked onto Canopus to adopt its cruise attitude. At 15:00 on 6 February the spacecraft executed a 4.3-second, 5.1-m/s midcourse manoeuvre to achieve the orbit insertion point for the desired inclination. At 04:20 on 7 February, an optional refinement was deleted. At 21:54:19 on 8 February the spacecraft began the 542-second burn to enter an initial orbit of

---

[2]   In fact, stereoscopic analysis of the Lunar Orbiter pictures proved difficult due to the manner in which they were scanned in narrow strips for transmission, as this gave the impression of the surface as being corrugated .

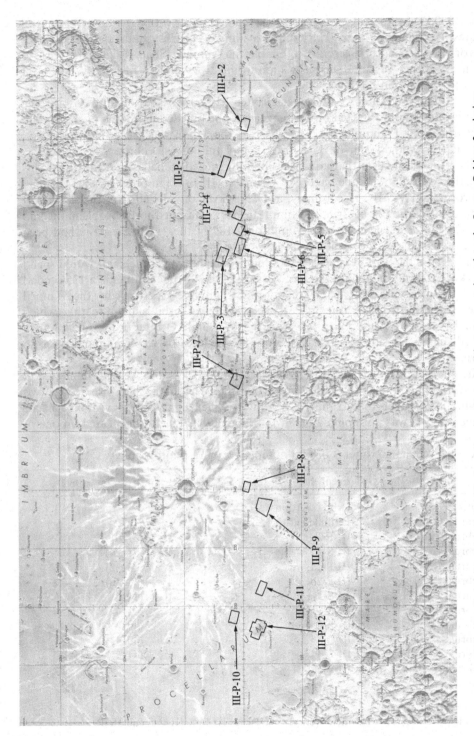

The Apollo zone annotated with the primary photographic targets assigned to the Lunar Orbiter 3 mission.

$210 \times 1{,}850$ km inclined at 21 degrees to the lunar equator with a period of 3 hours 25 minutes. As Lunar Orbiter 2 was still providing selenodesy data, the Deep Space Network tracked both spacecraft to enable the Manned Space Flight Network to obtain experience of simultaneously monitoring two vehicles in lunar orbit, as it would for an Apollo mission.

At 18:13:26 on 12 February, Lunar Orbiter 3 lowered its perilune to 55 km. The photographic mission started on 15 February – the first sequence was of the Apollo target in eastern Tranquillitatis. The phase of the Moon was 'new' on 9 February; 'first quarter' would occur on 17 February and 'full' on 24 February.

When the mechanism that advanced the film started to behave erratically, it was decided to cancel an oblique picture of Grimaldi (III-S-32) in order to start the read out process a day early. Accordingly, at 06:36 GMT on 23 February the spacecraft was told to cut the Bimat strip. But film transport continued to be problematic, and when the motor burned out on 4 March only 182 of the 211 frame-pairs had been transmitted; some of the earliest were lost.[3]

On 12 April the orbit was revised to minimise the time the spacecraft would spend in darkness during the lunar eclipse of 24 April. On 30 August it made a 125-second burn to circularise its orbit at 160 km, in order to provide experience for the Manned Space Flight Network in tracking a spacecraft in an orbit similar to that which would be used by Apollo. On 9 October 1967 Lunar Orbiter 3 was deliberately crashed on the far-side – two days before its predecessor was commanded to do likewise.

The photographs of III-P-12 were to locate Surveyor 1, and were accumulated in four overlapping blocks on successive revolutions. The lander had oriented its two mast-mounted panels to maximise its shadow and hence improve its visibility on the surface. After it was pin-pointed on an H frame, it was able to be located on one of

Table 11.3 – Lunar Orbiter 3 orbital manoeuvres and photography

Date	Event/Site	Frames	
8 February	Orbit insertion		
12 February	Perilune cut to 55 km		
15 February	III-P-1	16	(5–20)
15 February	III-S-1	4	(21–24)
15 February	III-P-2a	8	(25–32)
15 February	III-P-2b	4	(33–36)
15 February	III-S-2	1	(37)
15 February	III-S-3	1	(38)
16 February	III-S-4	1	(39)
16 February	III-P-3	4	(40–43)
16 February	III-P-4	8	(44–51)
16 February	III-P-5a	8	(52–59)

[3]   The lost frames degraded the coverage of III-P-3, III-P-4, III-P-5 and III-P-6.

Table 11.3 *cont.*

Date	Event/Site	Frames	
16 February	III-P-5b	8	(60–67)
16 February	III-P-6	4	(68–71)
17 February	III-S-5	1	(72)
17 February	III-S-6	1	(73)
17 February	III-S-7	4	(74–77)
17 February	III-S-8	1	(78)
17 February	III-S-9	1	(79)
17 February	III-S-10	4	(80–83)
18 February	III-S-11	1	(84)
18 February	III-S-13	1	(85)
18 February	III-P-7a	8	(86–93)
18 February	III-P-7b	8	(94–101)
18 February	III-S-14	1	(102)
18 February	III-S-15	4	(103–106)
19 February	III-S-16	1	(107)
19 February	III-S-17	4	(108–111)
19 February	III-S-18	4	(112–115)
19 February	III-S-19	4	(116–119)
19 February	III-S-21	1	(120)
19 February	III-S-21.5	1	(121)
19 February	III-S-22	1	(122)
20 February	III-S-20	1	(123)
20 February	III-P-8	8	(124–131)
20 February	III-S-23	4	(132–135)
20 February	III-S-24	1	(136)
20 February	III-P-9a	8	(137–144)
20 February	III-P-9b	8	(145–152)
20 February	III-P-9c	8	(153–160)
21 February	III-S-25	1	(161)
21 February	III-S-26	1	(162)
21 February	III-P-10	8	(163–170)
21 February	III-S-27	1	(171)
21 February	III-S-28	1	(172)
21 February	III-P-11	8	(173–180)
22 February	III-P-12b.2	4	(181–184)
22 February	III-P-12a	16	(185–200)
22 February	III-P-12b.1	4	(201–204)
22 February	III-P-12c	8	(205–212)
22 February	III-S-29	1	(213)
22 February	III-S-30	1	(214)
23 February	III-S-31	1	(215)
23 February	Cut Bimat and start full readout		
4 March	Readout interrupted		

Hyginus Rille by Lunar Orbiter 3.

Table 11.4 – Apollo site reconnaissance

Mission	No. sites	No. exposures
LO-1	9	136
LO-2	13	184
LO-3	18	162
Total	40	482

Note: Some sites were assigned to more than one mission.

the M frames taken by Lunar Orbiter 1 – although only in hindsight. The scientific targets included oblique perspectives of the 32-km-diameter crater Kepler (III-S-26) in Oceanus Procellarum and the 11-km-diameter crater Hyginus in Sinus Medii and the associated rille (III-S-6) that was suspected by some people of being of volcanic origin and was under consideration as a site for an advanced Apollo mission.

As soon as the Lunar Orbiter 3 readout ended, the screening team reconvened at Langley to assess the results. Now that the Surveyor 1 site was known, the landscape observed by that vehicle provided the 'ground truth' needed to test the validity of the process of interpreting overhead pictures – this was a major part of the rationale for devoting so many frames to seeking the lander.

As a result of the first three Lunar Orbiter missions 32 individual sites clustered in eleven groups in the Apollo zone had been comprehensively photographed. These, and nine less intensively imaged sites, were designated 'Set A' for the selection of the early Apollo landings sites.

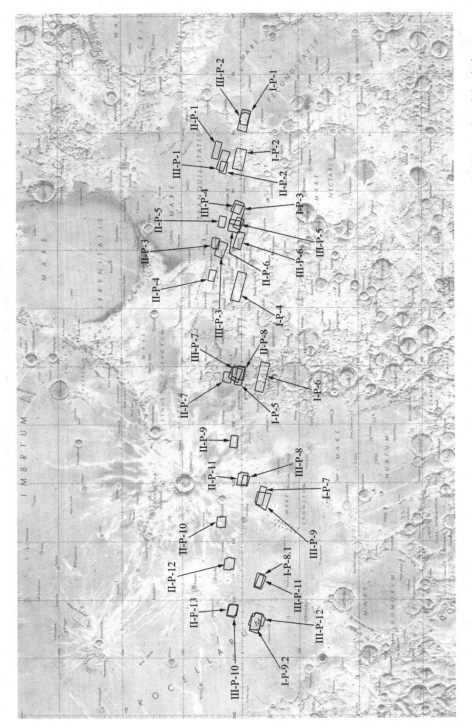

The Apollo zone annotated with the primary photographic targets jointly assigned to the Lunar Orbiter 1, 2 and 3 missions.

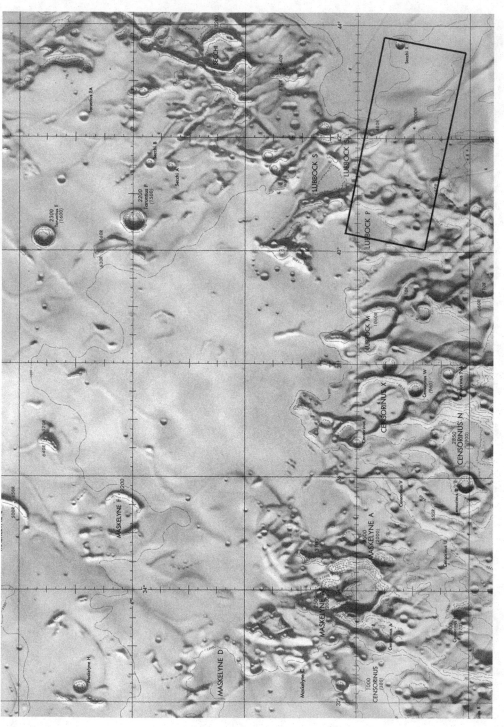

Lunar Orbiter photographic target I-P-1.

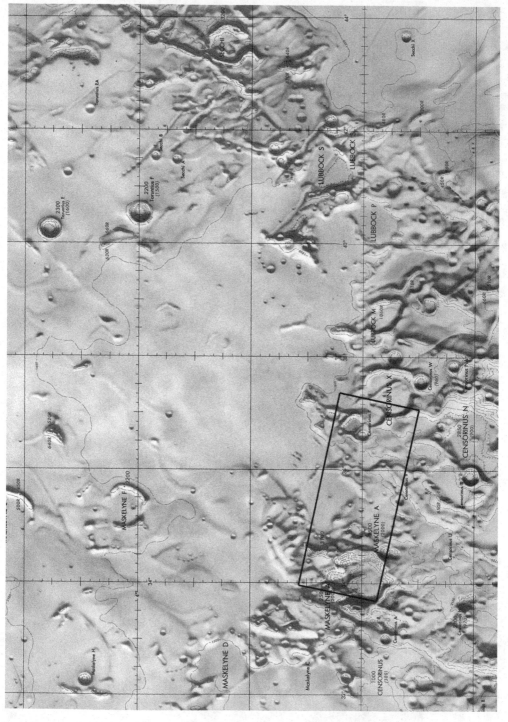

Lunar Orbiter photographic target I-P-2.

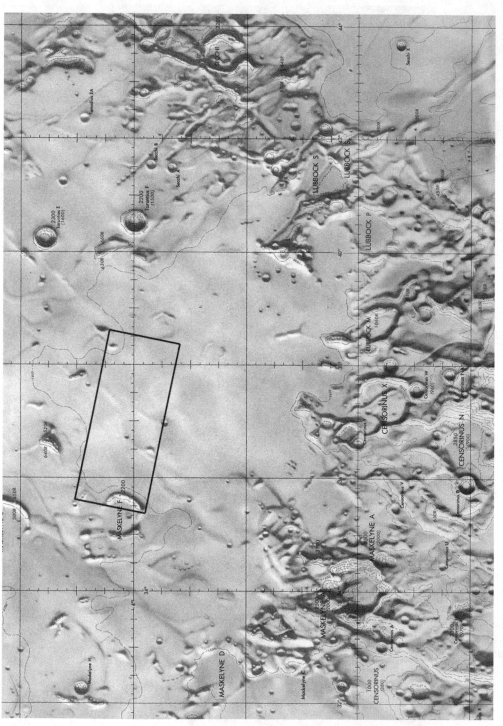

Lunar Orbiter photographic target II-P-1.

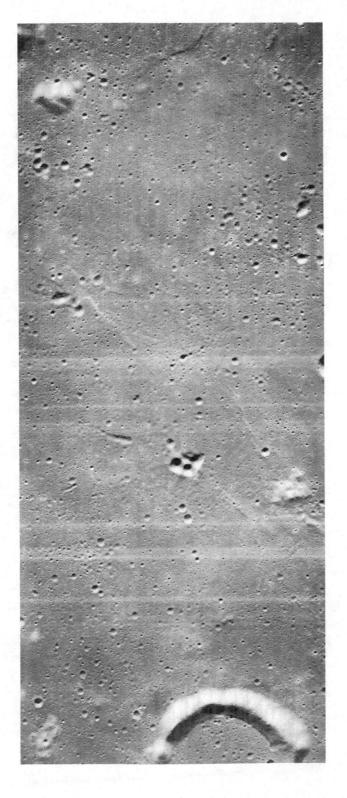

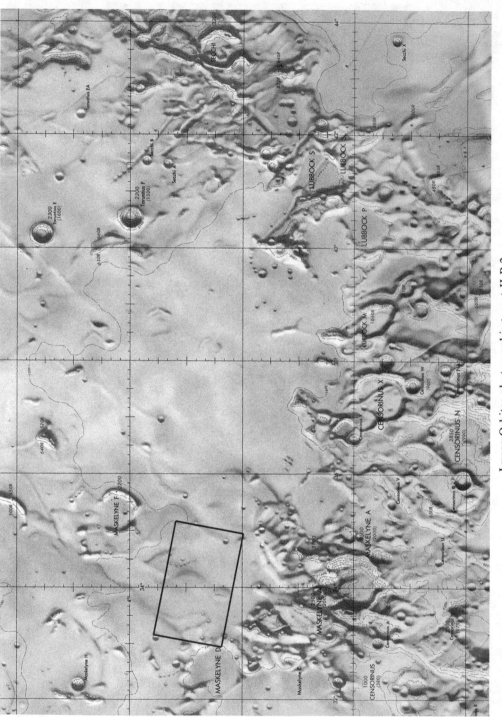

Lunar Orbiter photographic target II-P-2.

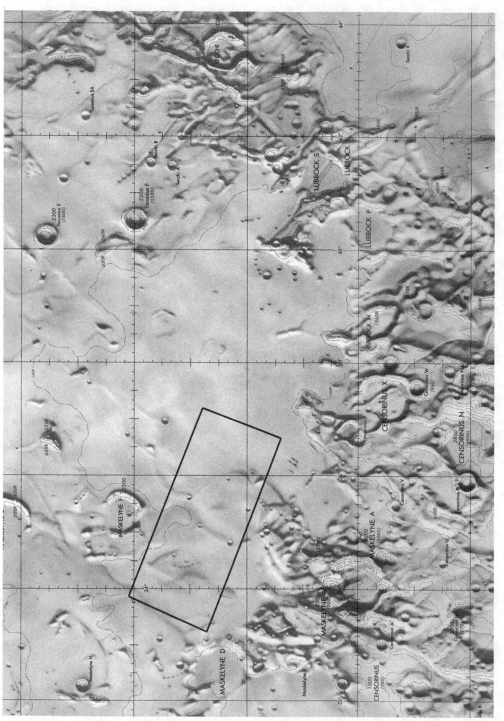

Lunar Orbiter photographic target III-P-1.

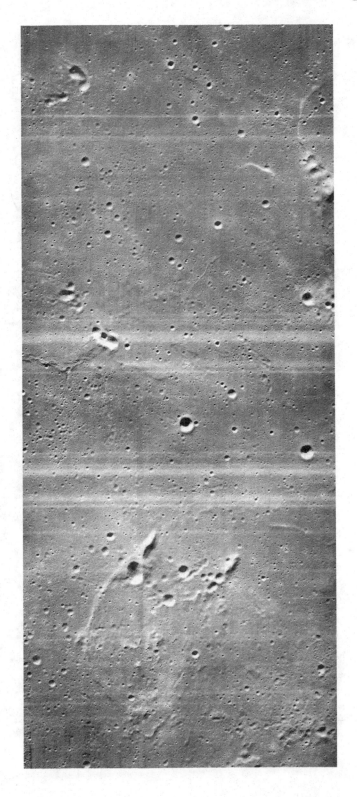

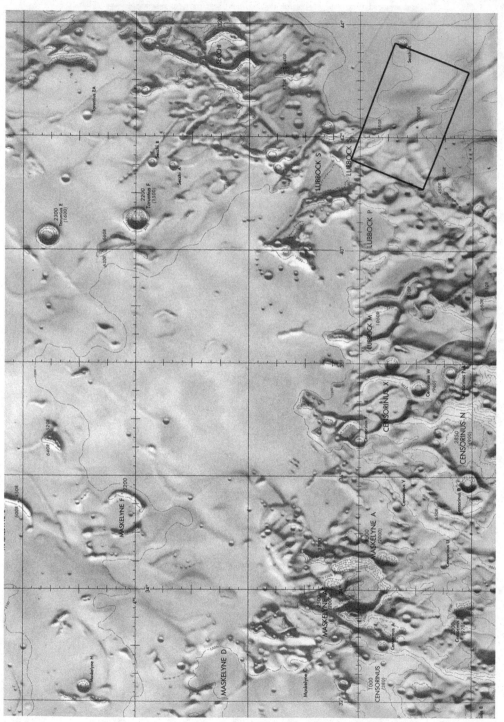

Lunar Orbiter photographic target III-P-2.

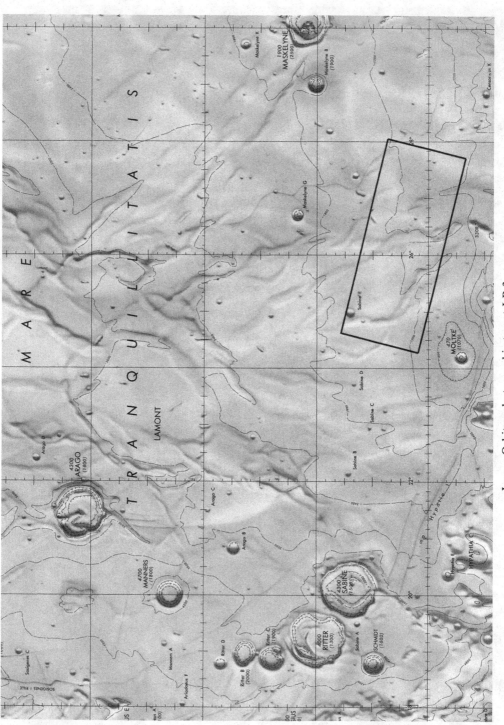

Lunar Orbiter photographic target I-P-3.

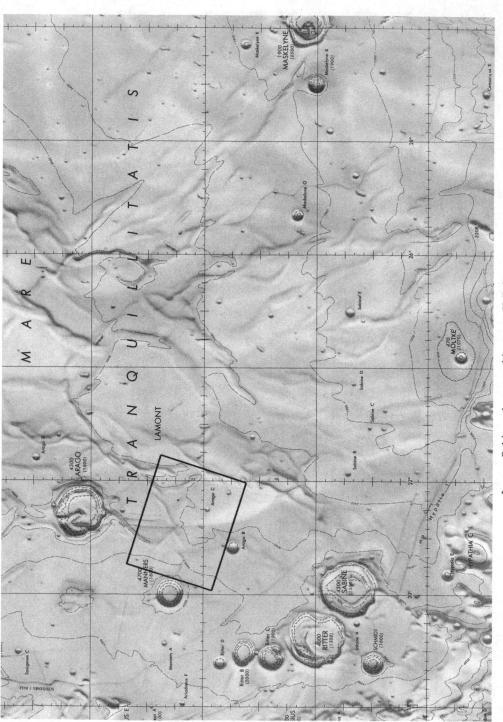

Lunar Orbiter photographic target II-P-3.

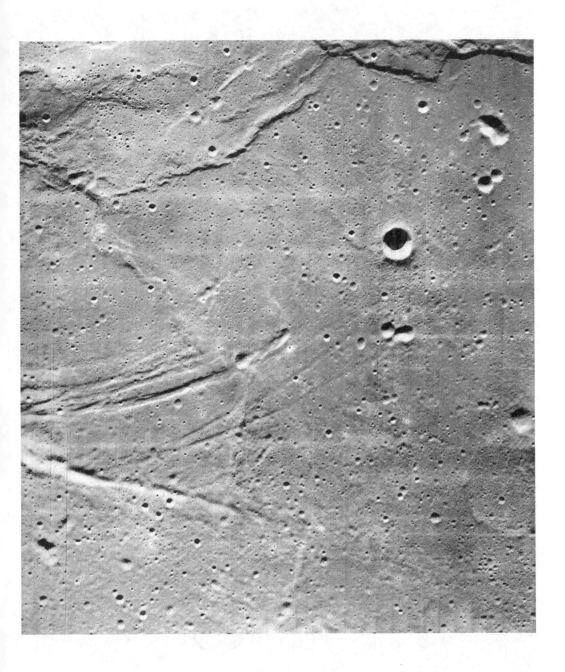

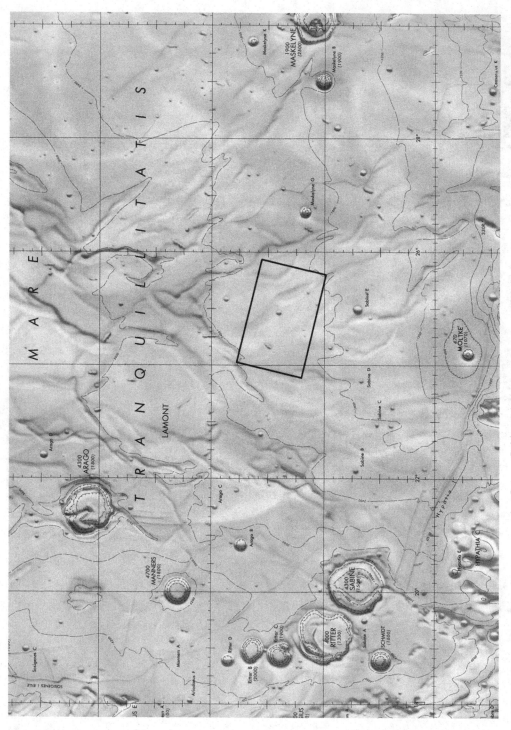

Lunar Orbiter photographic target II-P-5.

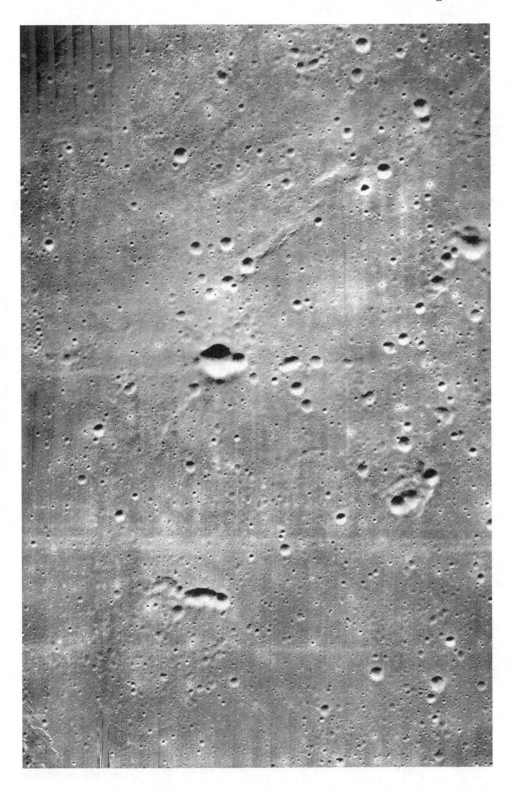

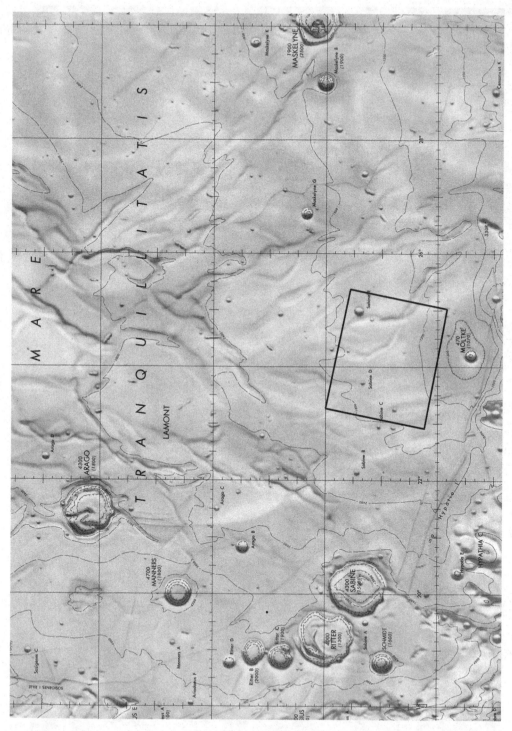

Lunar Orbiter photographic target II-P-6.

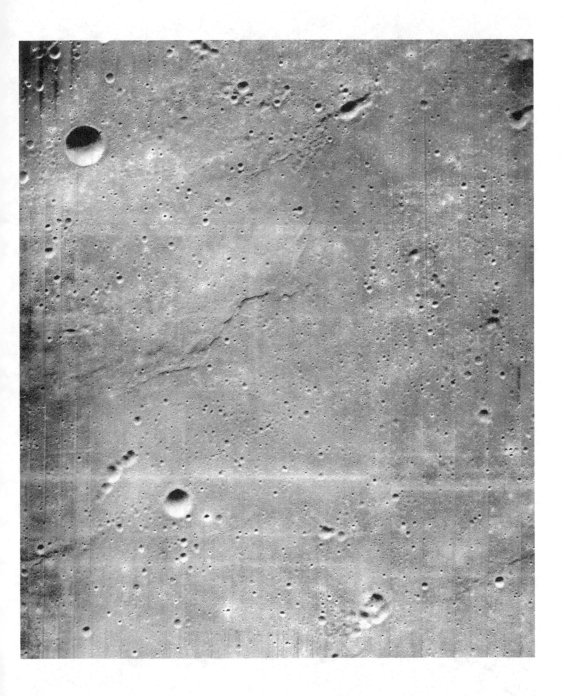

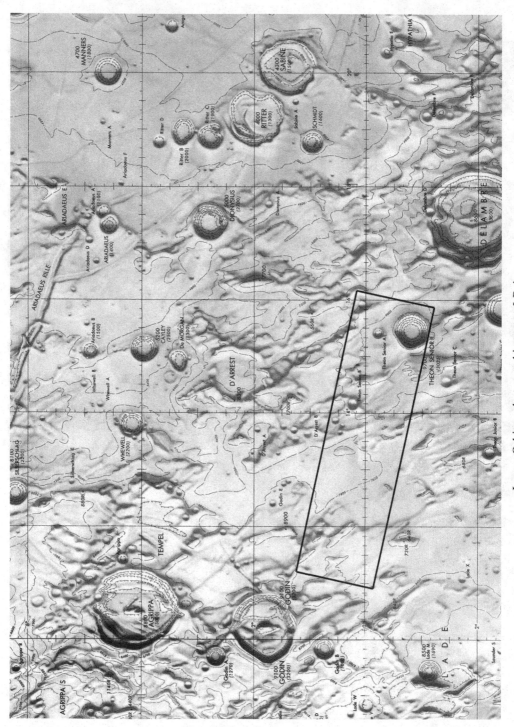

Lunar Orbiter photographic target I-P-4.

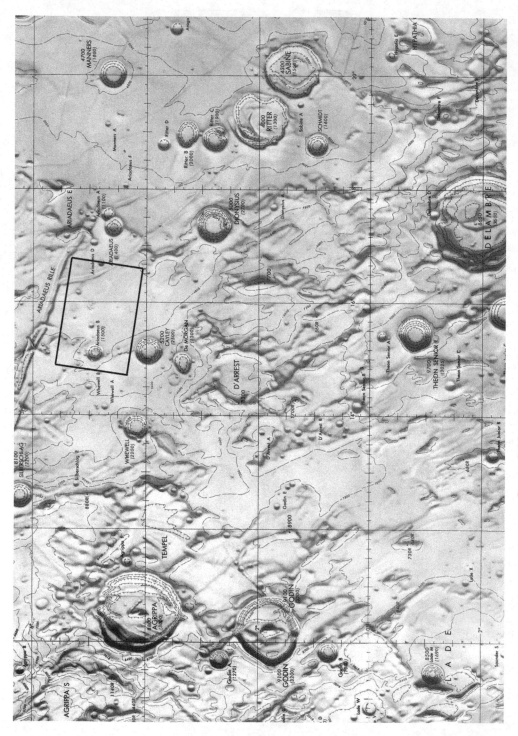

Lunar Orbiter photographic target II-P-4.

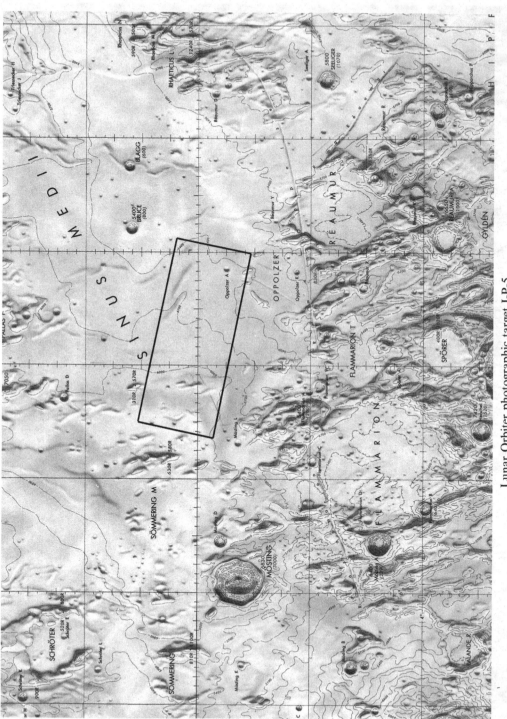

Lunar Orbiter photographic target I-P-5.

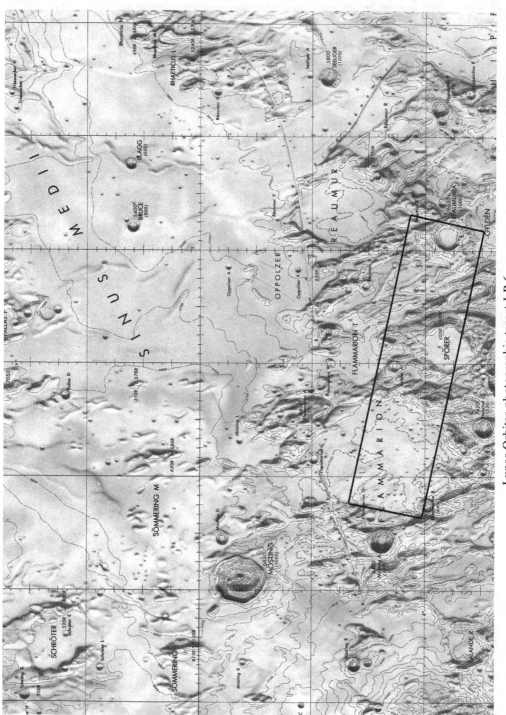

Lunar Orbiter photographic target I-P-6.

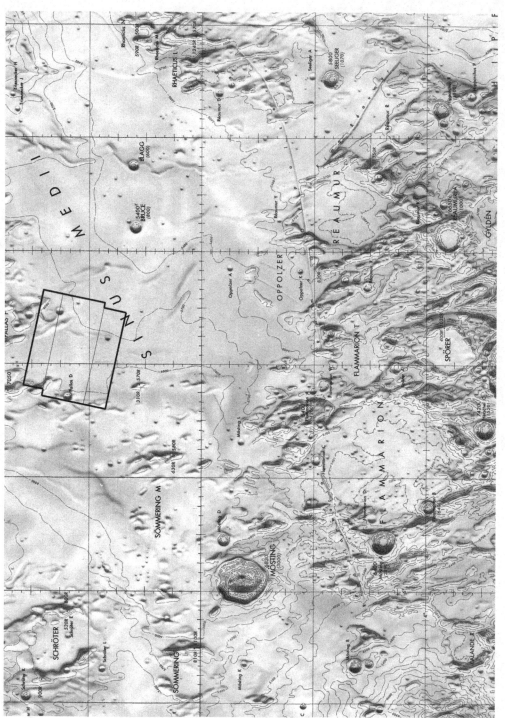

Lunar Orbiter photographic target II-P-7.

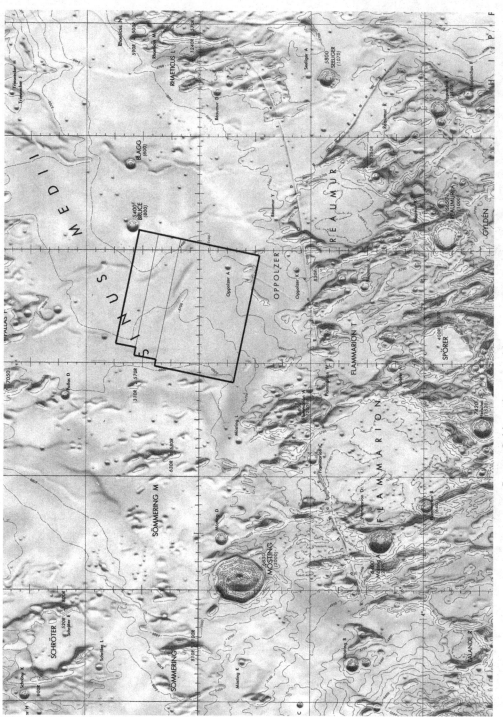

Lunar Orbiter photographic target II-P-8.

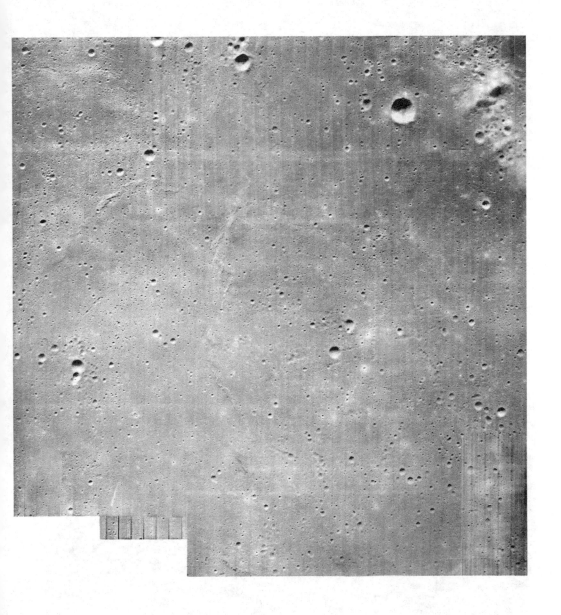

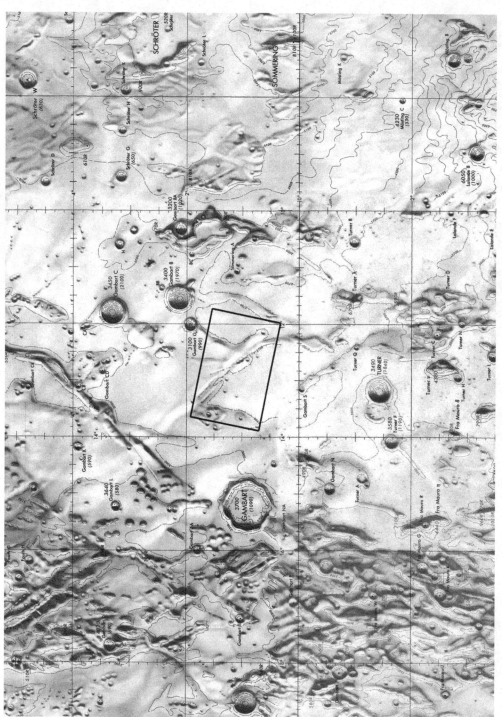

Lunar Orbiter photographic target II-P-9.

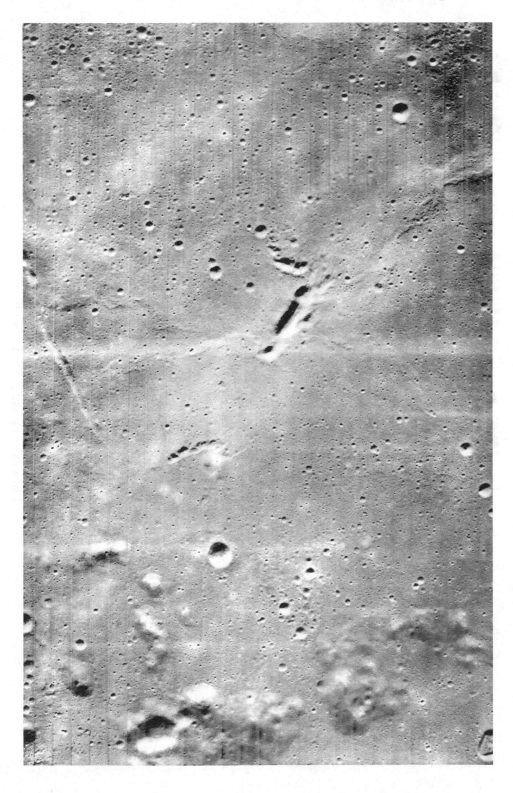

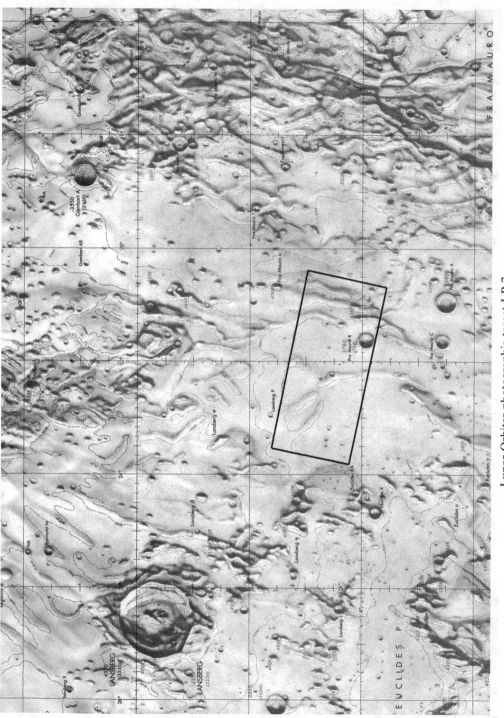

Lunar Orbiter photographic target I-P-7.

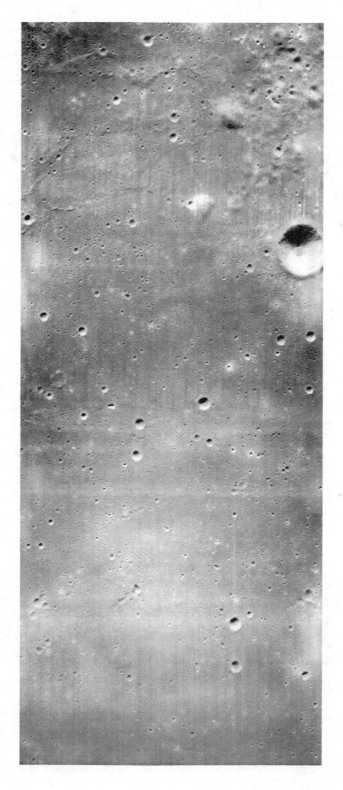

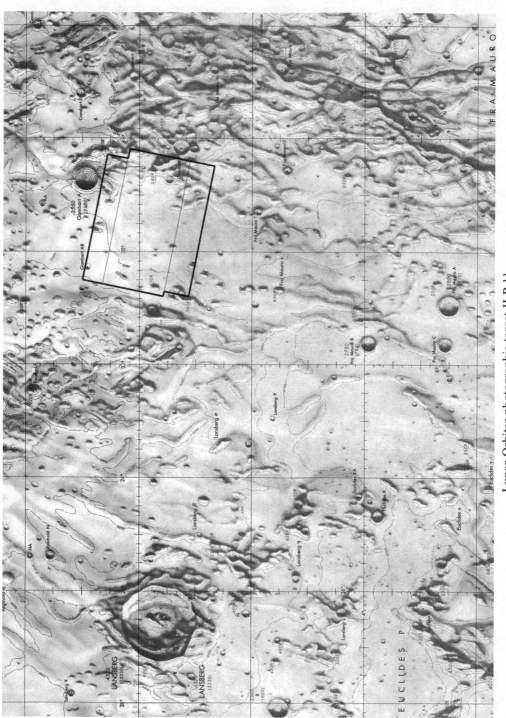

Lunar Orbiter photographic target II-P-11.

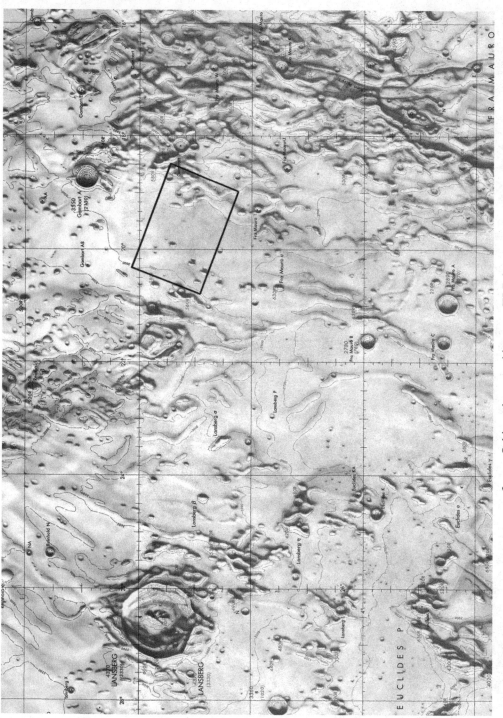

Lunar Orbiter photographic target III-P-8.

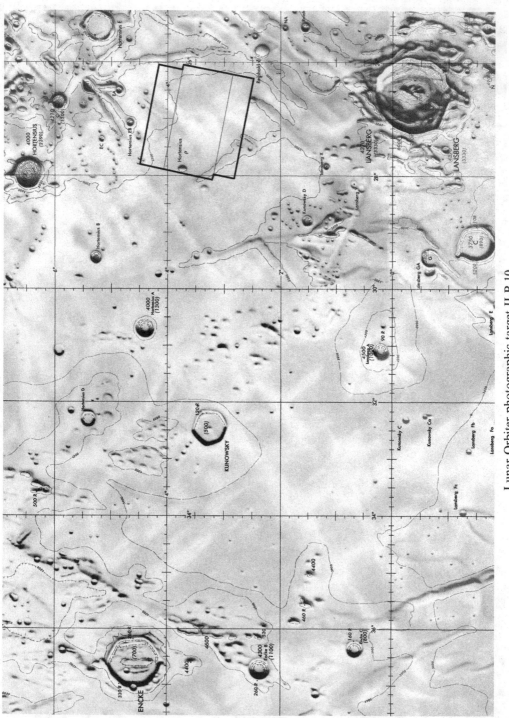

Lunar Orbiter photographic target II-P-10.

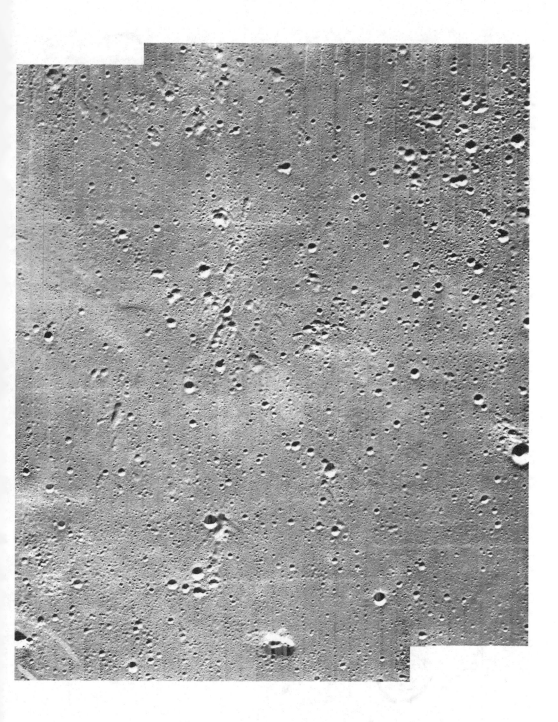

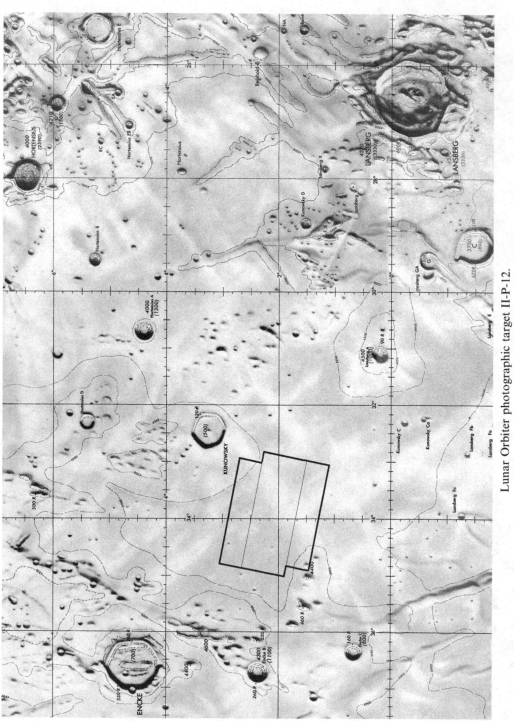

Lunar Orbiter photographic target II-P-12.

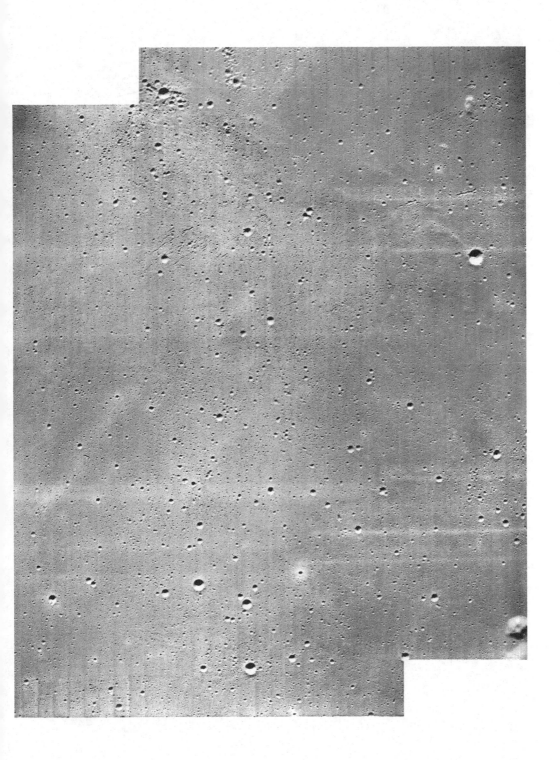

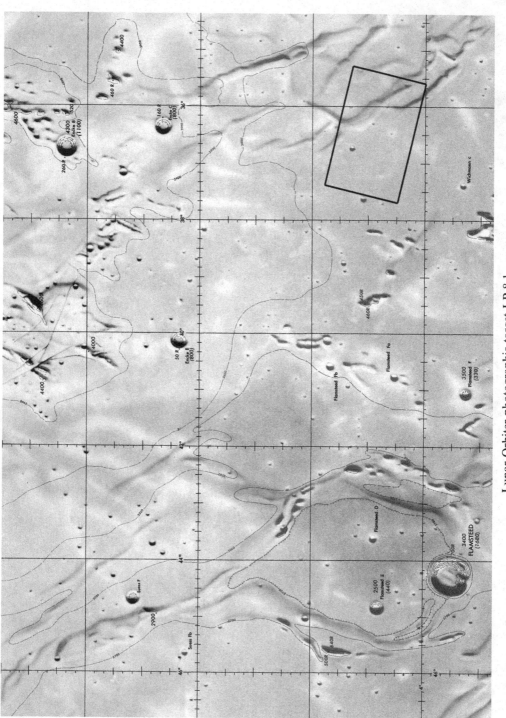

Lunar Orbiter photographic target I-P-8.1.

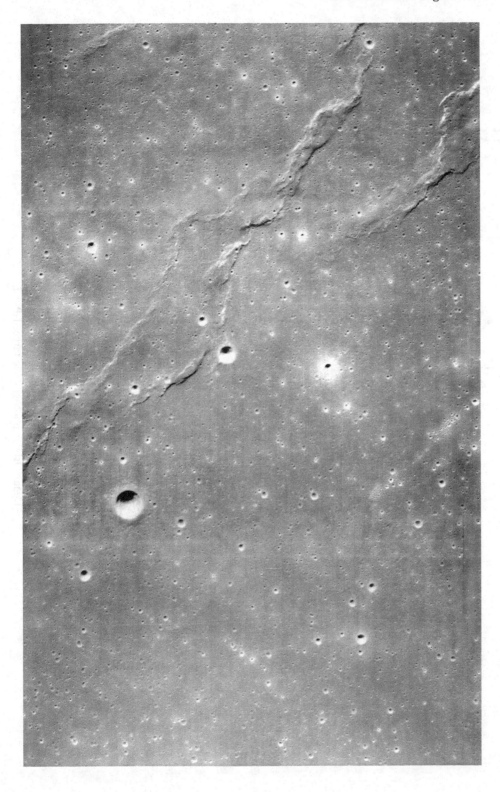

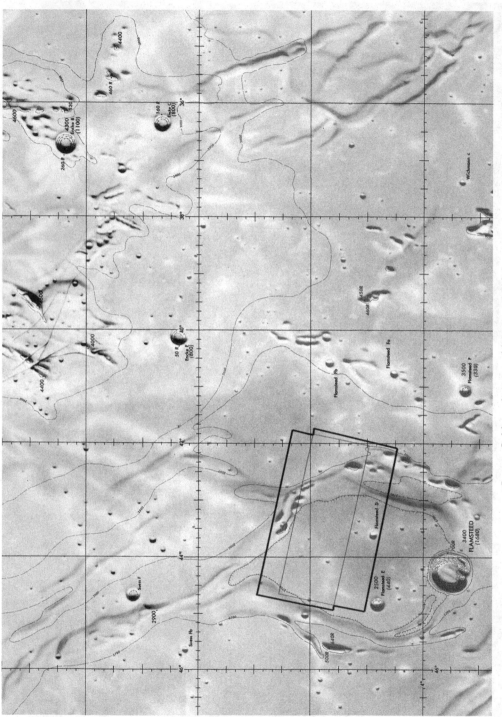

Lunar Orbiter photographic target I-P-9.2.

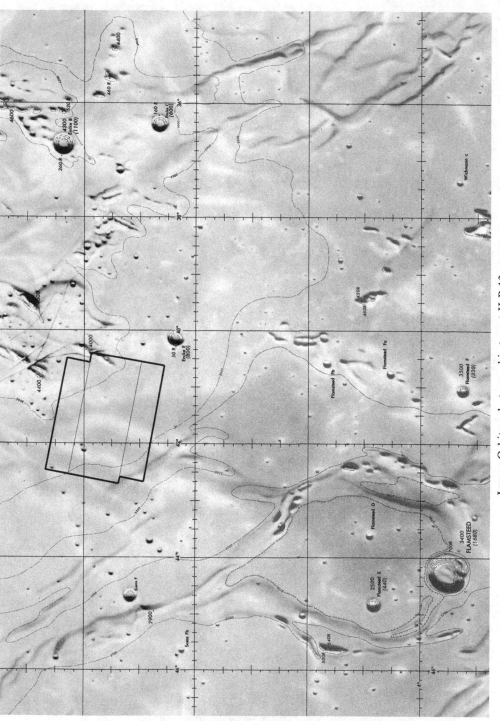

Lunar Orbiter photographic target II-P-13.

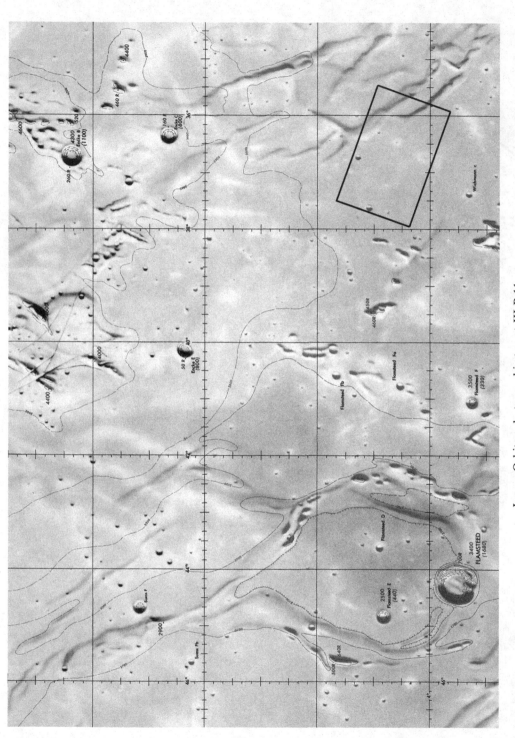

Lunar Orbiter photographic target III-P-11.

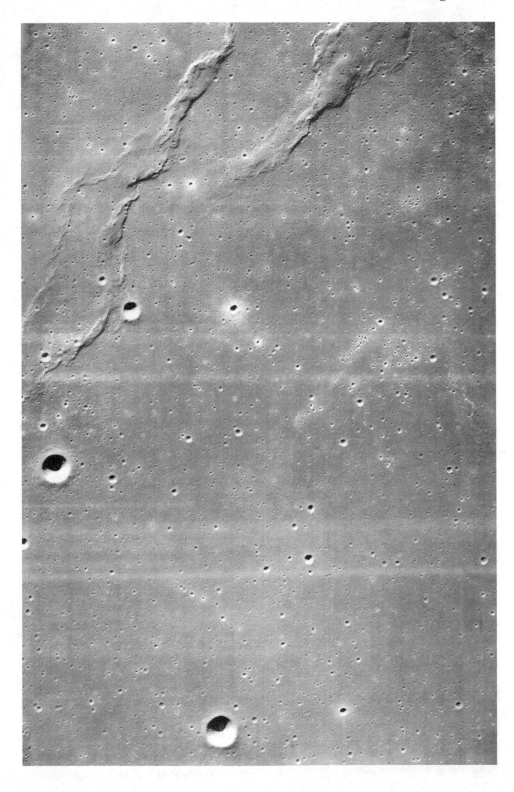

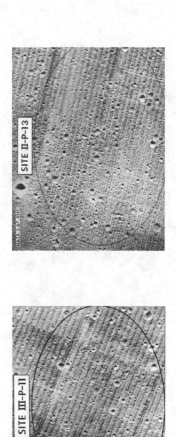

By design, the ellipses selected as potential targets for the first Apollo landing were bland areas.

APOLLO SITE SHORT-LIST

When the Apollo Site Selection Board met on 30 March 1967 the Apollo officials announced that whilst they would seek further Lunar Orbiter data, that from the first three missions satisfied "the minimal requirements of the Apollo program for site survey for the first Apollo landing". By now nine mare sites in the Apollo zone were deemed to be suitable as 'prime sites' for the early Apollo landings: one in Mare Foecunditatis, two in Mare Tranquillitatis, one in Sinus Medii and five in Oceanus Procellarum. These were designated 'Set B'.[4] For each such site, the US Geological Survey produced geological maps at scales of 1:25,000 and 1:100,000 to supplement the 1:1,000,000 regional maps. It was noticed that the sites on the eastern maria had high densities of large but shallow craters, and the sites on the western maria were generally flatter but rougher in detail. The astronomers had long ago noted that there was a difference in spectral hue, with the eastern maria being bluish and the western maria reddish.

On 15 December 1967 the Apollo Site Selection Board convened at the Manned Spacecraft Center to refine the target list for the first Apollo landing. All of the sites of Set B were acceptable in terms of their approach routes. However, as a landing in Mare Foecunditatis would not allow sufficient time after rounding the eastern limb for radio tracking to verify the lander's trajectory prior to powered descent, this site was discarded. Five sites were short-listed as 'Set C'. It was decided that three of these must be selected as options for the first landing mission, forming a prime site and two backups spaced in lunar longitude to accommodate successive 2-day delays in launch. It was recognised that the need for the crew to familiarise themselves with three sites would increase their training burden, but there would be no impact on the surface activities because the first landing was not to include a mapped traverse. In east to west sequence, the five sites were II-P-2, II-P-6, II-P-8, III-P-11 and II-P-13. Whilst it was clear that the prime site would be in the eastern hemisphere, the meeting did not specify whether it should be II-P-2 or II-P-6.

On 26 September 1968 the Set C ellipses were 'stretched' from 5.3 × 7.9 km to 5.0 × 15.0 km to allow for uncertainties in the Moon's gravitational field that might cause a lander to come in either 'short' or 'long' of the designated aim point. On 3 June 1969 the Set C sites were renamed Apollo Landing Site (ALS) 1 through 5 respectively.

[4]   They were I-P-1 in Mare Foecunditatis, II-P-2 and II-P-6 in Mare Tranquillitatis, II-P-8 in Sinus Medii, and II-P-11, III-P-9, III-P-11, III-P-12 and II-P-13 in Oceanus Procellarum.

12

Scratching the Moon

A BOUNCY LANDING

The 6-month hiatus after Surveyor 2 was not due to concern over the loss of that spacecraft, but to the wait for the restartable version of the Centaur. The two-burn configuration traded payload capacity against the hardware to restart the engines and the cryo-propellants that would be vented while coasting in parking orbit, but it offered Surveyor launches in winter months,[1] considerably lengthened the launch windows, and increased the flexibility in selecting the arrival time for optimal illumination at the landing site. The Centaur stage demonstrated its restart capability by a launch on 26 October 1966, thereby completing its test program.

The target for Surveyor 3 was a 60-km-diameter circle in the southeastern part of Oceanus Procellarum, centred 120 km southeast of the crater Lansberg. Its attraction was that although it was crossed by a ray from Copernicus 370 km to the north, at telescopic resolution it was sparsely cratered. The smooth patch of mare material was broken about 20 km to the west by rough hummocky terrain and isolated hills, and was bounded to the east by low ridges. It had been photographed at medium resolution by Lunar Orbiter 1 as target I-P-7, and at high resolution by Lunar Orbiter 3 as III-P-9c. These pictures revealed the presence of sub-telescopic craters in the target circle – with one, just over 1 km in diameter, situated very near the aim point. As a smooth-looking patch of Oceanus Procellarum, it bore a similarity to the Surveyor 1 landing site in the Flamsteed Ring, some 650 km to the west.

---

[1] Whereas in summer the Moon reaches its 'full' phase south of the equator, in winter it does so north of the equator, and since for the early Surveyors the landing sites were well to the west of the lunar meridian with arrival soon after local sunrise in winter months the translunar injection had to be made from south of the Earth's equator. The restartable Centaur facilitated this by using its first burn to achieve a parking orbit and, once south of the equator, using its second burn to head for the Moon.

Surveyor 3 lifted off from Pad 36B at 07:05:01 GMT on 17 April 1967. The Atlas jettisoned its booster section at T + 142 seconds and the sustainer engine shut down at T + 238. Once free of the Atlas, the Centaur established the desired circular parking orbit at an altitude of 160 km – with insertion at T + 569 seconds. The coasting phase would vary between 4 minutes and 25 minutes, depending on the geometry of the translunar injection – in this case it was to be 22 minutes 9 seconds. While coasting, the Centaur first fired two 50-pound-thrust hydrogen peroxide thrusters to settle the remaining propellants in their tanks, then continuously fired two 3-pound thrusters to maintain this condition. It had two clusters of 3.5 and 6-pound thrusters to control its attitude, and while maintaining its longitudinal axis to the local horizontal it rolled at a rate of 0.17 degree per second in order to even out solar heating of its payload and vented any propellant boil-off.

Owing to the predawn launch, the Centaur emerged from the Earth's shadow at 07:21:25. About 40 seconds prior to the translunar injection, the 50-pound thrusters fired again to guarantee that the propellants would enter their feed pipes. The main engines were shut down when the inertial guidance system sensed that the requisite velocity had been achieved – in this case at 07:38:49, after a 108-second burn. As on earlier missions, after it had configured and released the spacecraft, the spent stage performed the separation manoeuvre. Once free, Surveyor 3 stabilised itself and then adopted its cruise attitude. The midcourse manoeuvre at 05:00:03 on 18 April lasted 4.3 seconds and the 13.8-ft/sec change in velocity was entirely devoted to achieving the 'critical component' required to reduce the divergence from the centre of the target circle from the initial 480 km down to a mere 5 km.

The pre-retro manoeuvre in which the spacecraft departed from its cruise attitude involved starting a yaw of –158 degrees at 23:23:30 on 19 April and a pitch of –76.8 degrees at 23:30:17 to align the thrust axis with the velocity vector. The final roll of –64 degrees initiated at 23:34:35 was to optimise the RADVS. The initial approach was at 23.6 degrees to the local vertical. This would require a significant gravity turn during the vernier phase of the descent to force the trajectory to vertical.

The altitude marking radar was enabled at 23:59:33, and issued its 100-km slant-range mark at 00:01:12.829 on 20 April. The delay to the initiation of the braking manoeuvre was specified as 5.090 seconds. The verniers ignited precisely on time, and the new 'high-impulse' retro-rocket 1.1 seconds later – at which time the vehicle was travelling at 8,618 ft/sec. The acceleration switch noted the peak thrust of 9,550 pounds fall to 3,500 pounds at 00:02:00.587, giving a burn duration of 40.0 seconds. After allowing the thrust to tail off, the casing was jettisoned at 00:02:12.429. At burnout, the angle between the vehicle's thrust vector and velocity vector was 21.1 degrees.

When the RADVS-controlled phase of the flight began at 00:02:14.642, the slant range was 36,158 feet (and because the velocity vector at burnout was offset to vertical, the altitude was 32,900 feet) and the total velocity was 483 ft/sec (and since the vehicle had maintained its thrust along the velocity vector extant at the time of retro ignition, the longitudinal rate was 462 ft/sec). The altimeter had locked on at a slant range of 43,700 feet, only to drop out again. So when the RADVS was given control it aligned the thrust axis along the velocity vector extant at retro burnout

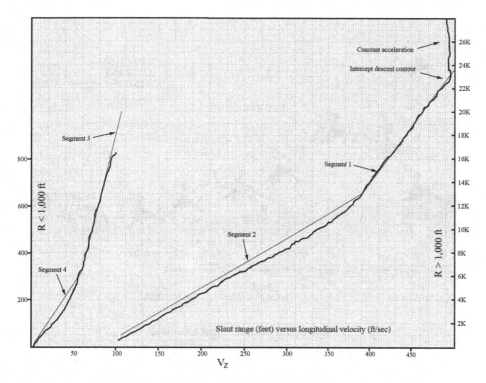

The descent of the Surveyor 3 spacecraft depicted in two sections, one for slant ranges
above 1,000 feet and the other below 1,000 feet.

and flew with the verniers at 0.9 lunar gravity, very slowly accelerating as it
descended. When the altimeter locked on again at 00:02:15.786, attitude control was
switched from inertial to radar, and the thrust axis was swung in line with the
instantaneous velocity vector to initiate the gravity turn. On intercepting the 'descent
contour' at 00:02:33.816, the slant range was 22,300 feet and the speed was 495 ft/sec.
By the 1,000-foot mark at 00:03:53.023, the vehicle was descending almost vertically
with a sink rate of 103.3 ft/sec. When the 10-ft/sec mark was issued at a height of 46
feet at 00:04:10:623, it seemed to be home and dry.

But at 00:04:13.275, at a height of 30 feet, one of the three angled radar beams lost
its lock on the surface. As the flow of data to the closed-loop computer abruptly
ceased at 00:04:13.387, the control system reverted to its inertial guidance system to
maintain its attitude and throttled the verniers to cancel out 0.9 of lunar gravity. But
because the RADVS was no longer operative, it was unable to issue the 14-foot mark
intended to cut off the verniers!

At 00:04:18.050 the vehicle touched down with a vertical rate of about 6 ft/sec.
Although it was level at this time, the ground was sloping down to the west,
causing leg no. 2 to make contact first. In response to the tilt induced by the other
two legs touching down, the flight control system – which was in attitude-hold
mode and did not realise that it was on the ground – increased the thrust of

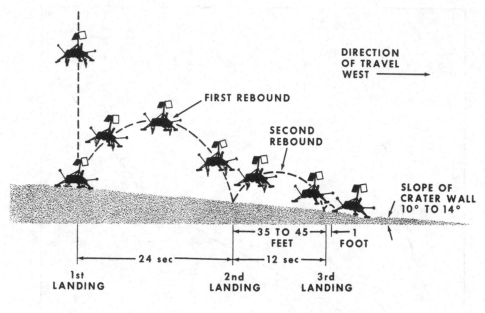

Details of Surveyor 3's tricky landing.

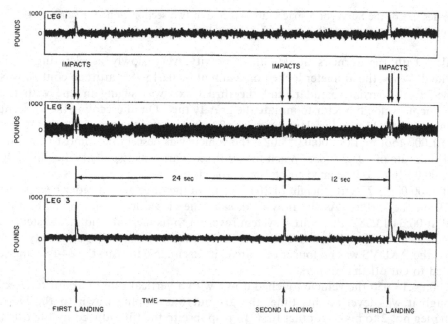

The axial forces on the shock absorbers of the three landing legs of Surveyor 3 from first touchdown to finally coming to rest.

verniers no. 1 and 3 to re-establish a level attitude, and this additional thrust caused the vehicle to lift off.

After peaking at about 38 feet, the vehicle made second contact at 00:04:42.030 some 50 feet west of the initial point, this time at a vertical rate of 4 ft/sec. Just as previously, the slope caused leg no. 2 to touch down first and in its effort to hold its attitude the vehicle lifted off again. The engines were cut off by a command from Earth at 00:04:53.907 – at which time the vehicle had peaked at a height of 11 feet and was at 3 feet and falling. Since a portion of the thrust had been aimed laterally at each liftoff, this had built up a horizontal component, with the result that when the vehicle struck the surface its vertical rate was only 1.5 ft/sec but it had a horizontal rate of 3 ft/sec. The elasticity in its legs caused it to rebound several inches and hop another 18 inches further downslope before it settled at 00:04:54.420, some 36 feet west of its second point of contact. The gyroscopes indicated the lander to be tilted towards the west at an angle of about 12.5 degrees from vertical.

An investigation concluded that the most likely cause of the RADVS dropping out as it neared the surface was that its logic ordered a 'break lock' as one of the beams crossed a field of rocks – to a microwave radar, angular rocks would have appeared much as broken mirror fragments would to a searchlight. The circuitry was designed to make the radar tracking circuits select the strongest signal if several were present. It was essential to ignore antenna 'side lobes' when the radar was preparing for the gravity turn. As Surveyor 3 made the final vertical descent, the scintillating side lobe had obliged the system's logic to break its lock. As the probability of losing lock on the main beam during the vertical phase of the descent was negligible, it was decided that this problem would be eliminated on future missions by having the flight control system inhibit this side-lobe rejection logic upon receiving the 1,000-foot mark.

The first 200-line picture was received 58 minutes after landing, and a total of 55 wide-angle pictures were obtained in this mode. For this mission, a small visor had been added to the hood of the camera to prevent direct sunlight from penetrating the optical train in the hope of reducing the glare that had afflicted Surveyor 1 when the Sun was above 45 degrees of elevation. Surveyor 3 arrived approximately 23 hours after local sunrise, and in the orientation in which the vehicle landed the camera was on the eastern side with the Sun 11 degrees above the horizon. There was therefore surprise that many of the preliminary pictures were partially or completely obscured by a veiling glare. It was concluded that either engine efflux or fine particles stirred up by the engines during the 'hot' landings had coated part of the camera's mirror such that when that part of the mirror was directly illuminated by sunlight the view of the lunar surface was obscured. In addition, any scene that included terrain which strongly reflected sunlight was similarly degraded. Later, intermittent sticking of the mirror in both its azimuth and elevation motions implied that dust had penetrated its mechanism. The hood rotated in azimuth with the mirror, and the mirror could be rotated in elevation to seal the hood, but the engineers had been reluctant to start off in that configuration in case the mirror failed to open. A better hood was already in development. As events would show, the camera's operational issues would impair the imaging schedule and the glare would degrade the results. A telemetry problem meant that scanning for the Sun and Earth could not start until 06:32. This issue had appeared at the time of the second contact in the protracted arrival. It proved to be a signal processing failure. The fact that the inoperative RADVS lost its high-voltage supply at the same time implied that the signal processing problem was the result of an electrical arc. After a detailed study of the performance of the system identified a number of short circuits, a work-around was devised to minimise the impact on the surface activities. Meanwhile, the Sun and Earth acquisition was completed at 08:15, and the first 600-line picture was taken at 08:42. By handing over in succession, the Deep Space Network stations at Goldstone in California, Canberra in Australia and Madrid in Spain maintained continuous communication with the lander.

An analysis of the early pictures determined that the horizon was 5 degrees higher than it would have been if the lander were on a level plain – indicating that it was in a shallow depression. From the tilt, it was inferred to be on the eastern interior wall of a medium-sized crater.

In-flight tracking could locate the landing site only to within a few kilometres, but the crater in which the vehicle had settled was able to be identified by comparing the landscape observed at ground level with the overhead view of frame H-154 taken by Lunar Orbiter 3 – although obviously the lander was not present at the time that this picture was taken. This showed that Surveyor 3 was within 2.8 km of the aim point. The area was in frame H-125 taken by Lunar Orbiter 4 in May 1967, but because that mission was mapping from high altitude the resolution was insufficient to show the lander – nevertheless, the resulting refinement of the selenodetic grid enabled the coordinates of the site to be measured to an accuracy of better than ± 0.01 degree in each ordinate.

Once the crater in which Surveyor 3 landed had been found in overhead imagery, its diameter was measured at about 200 metres. It was actually the largest of a tight cluster of craters arranged in a pattern which would later be dubbed the Snowman. Photoclinometry of H-154 suggested that the crater was about 20 metres deep, that there was a smooth transition from the concave floor, that the slopes of the interior walls averaged 10 to 15 degrees, and that the rim was low and gently convex. There was an inflection in the profile from concave to convex about half way between the centre and the rim crest, in both radial and vertical directions. The 'ground truth' of Surveyor 3 offered a means of checking the automated photoclinometry of overhead imagery – in particular, the depth of 20 metres was seen to have been overestimated by about 5 metres.

The overhead view resolved about 100 small craters scattered over the floor, inner slopes and rim of the main crater. These ranged in size from 25 metres down to the effective limiting resolution of 1 metre. Most had gentle interior slopes and rounded rims, but a few had steep interior slopes and sharp rims. Since blocks were of higher albedo than the surface material it was possible to discern blocks down to half a metre in size, and it was evident that most were related to three of the largest craters superimposed on the main crater. By taking bearings on features and relating these to the overhead perspective, the location of the lander could be pin-pointed to within 0.5 metre – it was almost half way between the centre and the rim crest. Because it was at the inflection of the slope, its foot pads were about 7 metres below the rim

A photograph by the 61-inch reflector of the Lunar and Planetary Laboratory of the University of Arizona showing the part of Oceanus Procellarum to the southeast of Lansberg (top left corner) to which Surveyor 3 was assigned. Note the hummocky terrain to the west and the wrinkle ridges to the east. The outline shows the area covered by the next illustration.

crest and about 7 metres above the centre of the cavity. In fact, the eastern rim of the crater proved to be beyond the camera's horizon in the upslope perspective. In an exercise analogous to field surveying, the wealth of detail within the crater enabled a topographic chart of its interior to be compiled – it was an excellent example of how lunar orbiters and landers could work together. The tilt of the lander was measured by a variety of methods and estimated at 12 degrees, inclined almost due west. The fact that this was several degrees steeper than the local slope of 10 degrees was the result of foot pad no. 1, which was on the downslope side, having come to rest in a small depression.

Whereas Surveyor 1's verniers had cut off as intended at a height of 12 feet and – as hoped – had not disturbed the surface, the fact that Surveyor 3's verniers had kept firing through two touchdowns offered an opportunity to investigate how an intense gaseous plume affected the lunar surface material. Although the imprints of the first contact were unidentifiable owing to the highly foreshortened view of the rim of the crater and the problem with the camera made it difficult to look for erosional effects beneath the lander, the site of the second contact was conveniently positioned about

Frame M-154 taken by Lunar Orbiter 3 on 20 February 1967 showing the general target for Surveyor 3. The outline shows the area covered by the next illustration.

A portion of Lunar Orbiter 3 frame H-154 showing the specific area in which Surveyor 3 landed. The outline shows the area covered by the next illustration.

A close up of Lunar Orbiter 3 frame H-154 showing the crater in which Surveyor 3 landed, with the inferred position of the lander indicated by the arrow. (The lander was not present at the time, however.)

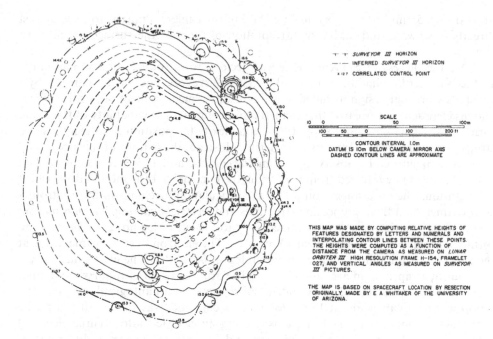

⊤ ⊤ *SURVEYOR III* HORIZON

—·— INFERRED *SURVEYOR III* HORIZON

x 127 CORRELATED CONTROL POINT

SCALE

CONTOUR INTERVAL 1.0m
DATUM IS 10m BELOW CAMERA MIRROR AXIS
DASHED CONTOUR LINES ARE APPROXIMATE

THIS MAP WAS MADE BY COMPUTING RELATIVE HEIGHTS OF
FEATURES DESIGNATED BY LETTERS AND NUMERALS AND
INTERPOLATING CONTOUR LINES BETWEEN THESE POINTS.
THE HEIGHTS WERE COMPUTED AS A FUNCTION OF
DISTANCE FROM THE CAMERA AS MEASURED ON *LUNAR
ORBITER III* HIGH RESOLUTION FRAME H-154, FRAMELET
027, AND VERTICAL ANGLES AS MEASURED ON *SURVEYOR
III* PICTURES.

THE MAP IS BASED ON SPACECRAFT LOCATION BY RESECTION
ORIGINALLY MADE BY E A WHITAKER OF THE UNIVERSITY
OF ARIZONA.

A contour map of the crater in which Surveyor 3 landed. The contours were drawn by interpolating between control points derived by the photographic trigonometry method. The probable vertical accuracy is ±0.5 metre.

11 metres away. However, the imprints were not able to be identified until 26 April, when the Sun was at an elevation of 77 degrees and no longer causing glare in the camera. These showed that pad no. 2 actually made two contacts, half a metre apart, and the position of a linear streak about 2 to 8 cm deep and 1 metre long suggested erosion by vernier no. 2 before the vehicle lifted off again. In addition, as the lander made its final touchdown each pad produced an overlapping cluster of imprints. The imprint adjacent to pad no. 2 was able to be studied in detail as a bonus for the soil mechanics team. Its floor was depressed about 5 cm below the level of the adjacent undisturbed surface. On the downslope side of the imprint there was a raised rim of disturbed material whose coarser texture suggested aggregation due to the force of impact, and there were lumps ranging out to a distance of 0.5 metre. The disturbed material was significantly darker than the undisturbed surface, and the fact that this was the case at both sites visited to date raised the prospect of it being characteristic of the maria. On observing that the compressed fine-grained material on the floor of the imprint bore the waffle pattern of the underside of the pad impressed through the aluminium foil cover sheet, engineers measured a number of foot pads on Earth and established that this 'detail' was on a scale of 40 to 80 microns. For the lunar surface material to preserve this pattern, a considerable fraction of it must have been particles finer than 60 microns and an appreciable fraction finer than 10 microns – this serendipitous study refined the measurement of

"less than 0.5 mm" which was made by tilting the camera's mirror to view almost directly downward and taking high-resolution pictures of the surface beneath the camera.

After the success of Surveyor 1, it was decided to introduce the soil mechanics surface sampler that had been intended for later Surveyors, possibly for carriage by a rover. The original design included sensors for direct measurement of position, force and acceleration, but because the telemetry and commanding capability of the initial form of the lander could not support this complexity the position measuring system, strain gauges and accelerometers were deleted. Instead, the actions of the arm would have to be monitored by observing it using the camera, and a limited amount of data would be able to be inferred from measuring the current that the motors drew whilst in operation. The experiment comprised the articulation mechanism, the electronics compartment and their supporting structures and electrical cabling. The mechanism was mounted on the space frame immediately to the left of leg no. 2, in the position formerly occupied by the approach TV camera. The electronics compartment was at the same level and almost midway between legs no. 2 and 3. The electromechanical mechanism comprised an arm and a scoop. The arm consisted of tubular aluminium cross members which could be extended and retracted in a pantograph fashion. The scoop was a container about 13 cm long and 5 cm wide, was rigidly affixed to the arm and its door was opened and closed using an electric motor. Three electrical motors operated through drive trains to extend and retract the arm and to rotate it independently in azimuth and elevation. It was spring-loaded, extended by having an electric motor unreel a metal tape, and retracted by reeling in the tape. The arm had a maximum extension of about 1.5 metres, but was unable to access the ground immediately below its mount. It could be elevated to raise the scoop about 1 metre off the ground. On this mission the azimuth arc of the mechanism subtended 112 degrees, ranging from the left edge of pad no. 2 towards pad no. 3 – although because the legs were spaced at 120-degree intervals it stopped short of pad no. 3. In all, the sampler had an arcuate operating area of 2.2 square metres. It was controlled from Earth, but because there was no onboard memory for complex sequences it had to be operated one command at a time. The fact that it used the same radio channel as the camera meant their use had to be interleaved. The principal investigator was Ronald F. Scott, an engineer at Caltech who had worked on the Surveyor rover proposal, but the hardware had been designed and built by Hughes. Of the overall mass of 15 pounds, the mechanical assembly accounted for 8.4 pounds.

The soil mechanics surface sampler, which was also referred to as the 'scratcher', could manipulate the lunar surface material in a number of ways. A trench could be made by opening the door of the scoop to expose its blade, driving the scoop into the ground and retracting the arm. The scoop could hold up to 100 cubic centimetres of loose granular material, or a small rock fragment. The mechanism was designed to dig to a depth of 0.45 metre, providing that the material permitted this. An impact test involved raising the scoop and disengaging the elevation motor by releasing the clutch to allow a torsional spring to assist lunar gravity in drawing the scoop down to disturb the surface. If dropped on top of a rock, the scoop's blade could serve as a rudimentary geological hammer. There was a 2.5 × 5.1-cm strip on the lower edge of

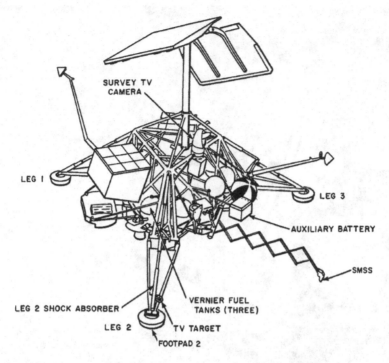

SURVEY TV
CAMERA

LEG 1

LEG 3

AUXILIARY BATTERY

SMSS

LEG 2 SHOCK ABSORBER

VERNIER FUEL
TANKS (THREE)

LEG 2

TV TARGET

FOOTPAD 2

The configuration of the Surveyor 3 lander.

the door in order to place a flat face on the lunar surface. A static bearing test would involve placing the scoop, door closed, directly above the target and then driving the scoop down until the motor stalled, with the current providing a measure of the force applied. The arm could also be manipulated to push rocks aside in order to inspect either the underside of the rock or its imprint on the surface.

The checkout of the soil mechanics surface sampler started at 10:00 on 21 April, shortly before the end of Goldstone's second session. After a pyrotechnic was fired to release the mechanism, JPL engineer Floyd I. Roberson commanded the arm to extend. The picture taken to confirm this showed that the arm had not advanced as far as expected. The command sequence was repeated, and the next picture showed that the arm was in the desired position. He then put it through a series of actions to verify that it could move in azimuth and elevation, checking its progress at each step by TV. This done, the arm was drawn back.

On 22 April the arm was swung to the middle of its operating area, and at 05:15 made its first bearing test of the lunar surface. The scoop was raised, the arm was swung to the right, and the scoop, door open, was driven into the surface at 09:14, after which the arm was retracted in order to scrape its first trench. Next the arm was swung left, beyond the bearing test position. After making a shallow scrape, it was raised and repositioned to make a second scrape on the same line. This time the motor stalled after just 10 cm – evidently it was more difficult to scrape an already existing trench. Meanwhile, the camera had suffered a difficulty moving in azimuth

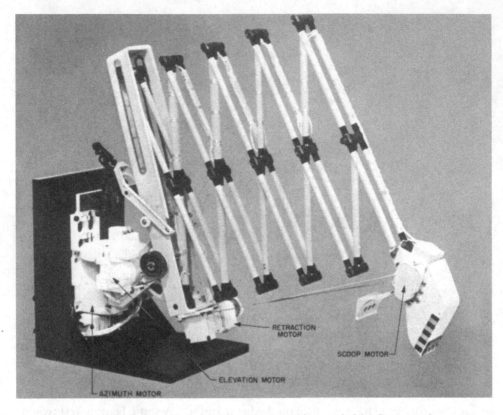

A model of the soil mechanics surface sampler carried by Surveyor 3.

that limited its ability to support the sampling activity. Work on the second trench resumed on 23 April with a third scrape being made along the same line. Arm work was suspended on 24 and 25 April owing to the heat. Because the latitude of the site was 3 degrees, the maximum solar elevation was 87 degrees. The arm had been left at the inner end of the second trench. Having noticed what appeared to be a rock at that position, the team decided to scoop it up on 26 April, but in the process of doing so the object crumbled. The arm was swung as far right as it could traverse and the sample was deposited on the upper surface of foot pad no. 2 so that the camera could inspect the clump of fine-grained material in colour at high resolution.

On 27 April the arm swung slightly left, away from foot pad no. 2, and conducted a second and third bearing test. It then moved a little further left and scraped a third trench involving 26 retraction steps, with a wide-angle picture being taken after each step and later sequenced to produce a 'stop-motion' movie. On 28 April the scoop picked up a small bright object from near the most recently made trench. This was added to the material dumped onto foot pad no. 2 for inspection, but there was loose material in the scoop from the trench and on falling from the scoop this covered the white object. When the scoop was dragged across the pile to expose the object of interest, it was observed to have darkened. Next, the arm made two parallel scrapes

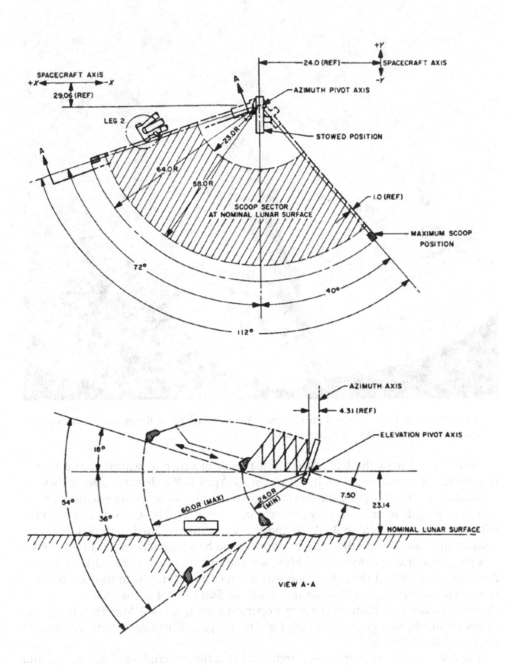

The operating area available to Surveyor 3's soil mechanics surface sampler.

A picture taken by Surveyor 3 on 28 April 1967 showing the soil mechanics surface sampler positioned between trenches no. 1 and 2.

successively offset to the left of the third trench in order to widen it, and then a bearing test was performed on its floor. On 29 April half a dozen impact tests were conducted in an arc beyond the recent trench, with the scoop being released from a variety of heights in order to vary the force of the impact. The arm was swung to the left of its operating area on 30 April and the scoop manipulated to draw a partially buried bright object onto the surface – it proved to be a fragment of hard rock, and it was photographed in colour. On 1 May two additional scrapes were made to deepen the second trench and then the scoop was dropped four times with its door open to loosen the floor prior to a final scrape. With the Sun sinking in the west the lander's shadow masked ever more of the arm's operating area, so on 2 May the arm ended its operations by swinging over to the right to scrape a short fourth trench alongside the broad third trench.

The results of the arm operations indicated that the material was fine-grained and had sufficient cohesion to create loose aggregations up to several centimetres in size, although such 'clods' readily fell apart. When the scoop was pressed on the surface for a bearing-strength test, it left a smooth imprint which had a raised ridge of lumpy

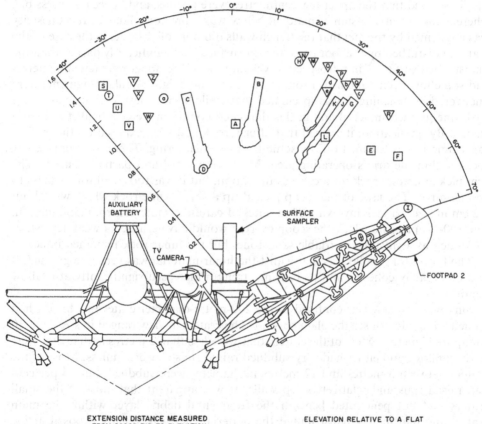

▽ IMPACT
☐ BEARING
◯ CONTACT

A	BEARING TEST 1	L	BEARING TEST 4	V	BEARING TEST 7
B	TRENCH 1	M˙	IMPACT TEST 1	W	IMPACT TEST 7
C	TRENCH 2	N	IMPACT TEST 2	X	IMPACT TEST 8
D	CONTACT 1	O	IMPACT TEST 3	Y	IMPACT TEST 9
E	BEARING TEST 2	P	IMPACT TEST 4	Z	IMPACT TEST 10
F	BEARING TEST 3	Q	IMPACT TEST 5	a	CONTACT 4
G	TRENCH 3	R	IMPACT TEST 6	b	IMPACT TEST 11
H	CONTACT 2	S	BEARING TEST 5	c	IMPACT TEST 12
I	FOOTPAD CONTACT	T	CONTACT 3	d	TRENCH 4
J	TRENCH 3 WIDEN	U	BEARING TEST 6	e	IMPACT TEST 13
K	TRENCH 3 WIDEN				

EXTENSION DISTANCE MEASURED
FROM SCOOP TIP TO ELEVATION
AXIS OF SURFACE SAMPLER

ELEVATION RELATIVE TO A FLAT
PLANE THROUGH THE BOTTOM
OF THREE FOOTPADS

The actions of Surveyor 3's soil mechanics surface sampler.

material around the edge. This implied that although the material was compressible, it was only moderately so, and after a certain compression the vertical force tended to displace material sideways. Impact tests were performed, but the 'spring constant' of the torsional spring proved insufficient to determine the density in this manner. In general, the first scrape of a trench excavated to a depth of about 7.5 cm, and each successive scrape on the same line gained an additional 5 cm – with the arm having to work harder to achieve this. Bearing tests on the floor of a trench showed that the strength of the material increased significantly at a depth of several centimetres. The deepest excavation achieved was about 18 cm, which was less than half of that for which the arm's range of operation had been designed. Nevertheless, it provided a valuable insight into the third dimension of the enigmatic fragmental debris layer. There was no indication of textural layering in the walls of the trenches. If there was any change in the grain size, this was on a scale finer than the camera's resolution. It simply seemed that the upper few centimetres were porous, and hence compressible, whereas the essentially similar material below was more consolidated. Its cohesivity was confirmed by the fact that the trench walls did not collapse. As in the case of the material disturbed by the foot pads, the subsurface was significantly darker than the undisturbed surface – in retrospect, it was more as if the uppermost few millimetres had somehow been lightened. From the fact that no bright angular fragments were uncovered in trenching, it was speculated that while buried they became coated with dark fine-grained material and in this darkened condition were difficult to see in a trench. By implication, it seemed that after a rock had been exposed on the surface for a time it was 'cleaned off' by some form of weathering. There were only a few rocks within the arm's operating area. Most were small and partially buried. The arm picked up one rock for a close examination, but it was too small for its mass to be measured. The jaws of the scoop picked up a small white rock which was about 1.2 cm in size – a task involving 90 minutes of careful remote-control manipulation. The 100-psi pressure which the scoop exerted would have crushed a weak terrestrial rock such as a siltstone or friable sandstone, but the lunar rock remained intact.

The lesson for Apollo was that whilst the lunar material was very fine-grained, it was moderately cohesive and its bearing strength increased significantly at shallow depth.

Surveyor 3's view was confined to the 200-metre-diameter crater in which it had landed – it could not see the plain beyond. The craters in view ranged in size from 10 cm up to 25 metres. Most of the craters that were less than 3 metres in diameter were fairly shallow, and either had very subdued raised rims or were rimless. Most of the craters between 3 metres and 12 metres in diameter were subdued, but 25 per cent had raised rims and relatively steep walls. It was apparent that most of the small craters had not penetrated beneath the fragmental debris layer within the main crater, and had merely redistributed the material that was already exposed at the surface. The size-frequency distribution was similar to that for this size range seen on the maria by Rangers 7 and 8.

The angular-to-rounded fragments ranged in size from tiny grains up to blocks of about 1.5 metres. The albedo of the undisturbed surface was 8.5 ($\pm$2) per cent, and in some cases the albedo of the blocks was one-third brighter. Although the camera operated most effectively when the Sun was high in the sky, in such illumination the absence of shadows made subtle terrain relief almost impossible to discern – but on the other hand in such illumination it was straightforward to chart the distribution of blocks. Most blocks were relatively angular, with many wedge-shaped and some even tabular. Some of the angular rocks were partially buried, but most of the well-rounded fragments were fairly deeply buried.

In addition to the sparse and random litter of blocks, there were two prominent 'strewn fields' of coarse blocks. One was clearly associated with a sharp raised-rim crater about 13 metres in diameter that was embedded in the northeastern rim of the main crater, some 80 metres from the lander. The other was associated with a pair of subdued craters that were located high on the southern wall. The line of sight provided a view inside the northeastern crater, revealing its interior to be full of similar blocks. Exterior to the rim, there were radial lines of blocks. The blocks associated with this crater were the largest, coarsest and most angular in the lander's field of view. It was evident that they were ejected by the impact that created the crater, and derived from material at a depth of 2 or 3 metres beneath the rim of the main crater. Some of the blocks had almost planar faces, as though they had broken along pre-existing joints. The tabular ones displayed grooves and ridges on their narrowest sides suggestive of lamination parallel to their longest dimension, such as would be produced in flow-banded lavas. The blocks associated with the southern craters were of similar size, but were more rounded and tended to be more deeply buried. Their source was probably the larger of the two craters there, which was 15 metres in diameter. These observations suggested that large blocks associated with subdued craters tended to be more rounded than those associated with sharp raised-rim craters, and those around subdued craters were more buried than those of sharp raised-rim craters. This suggested that freshly exposed blocks were not only eroded by the rain of meteoritic material, but also tended to be reburied as material accumulated – either by the arrival of further ejecta or as a result of downslope motion of loose debris. The fact that the rounded blocks had a pitted texture whereas the angular ones did not, implied that the pitting was caused by the same process that rounded off the angular blocks.

It was also apparent that the surface on the interior of a sizeable shallow crater on a mare plain was similar to that on a relatively level area between such craters. The 200-metre crater in which Surveyor 3 landed had probably been partially filled in by the downslope motion of material on its interior walls, thickening the debris towards the centre. Loose material piled up against the upslope sides of the larger blocks was interpreted as evidence of this process. The strewn fields of coarse blocks associated with 13–15-metre-diameter craters on or near the rim of the main crater implied that the fragmental debris layer was about 2 metres thick there. In contrast, the fact that a 20-metre crater near the centre of the main crater had *not* excavated blocks served to confirm that the layer there had been thickened. When the main crater was created, its floor would have been several tens of metres deeper and its rim several metres higher and sharper than it is today. By the 'hinge-flap' effect of an impact, the debris that formed the rim would have been excavated from the deepest point. In effect, a blocky crater on the rim of a larger crater serves as a 'drill hole'. The crater on the

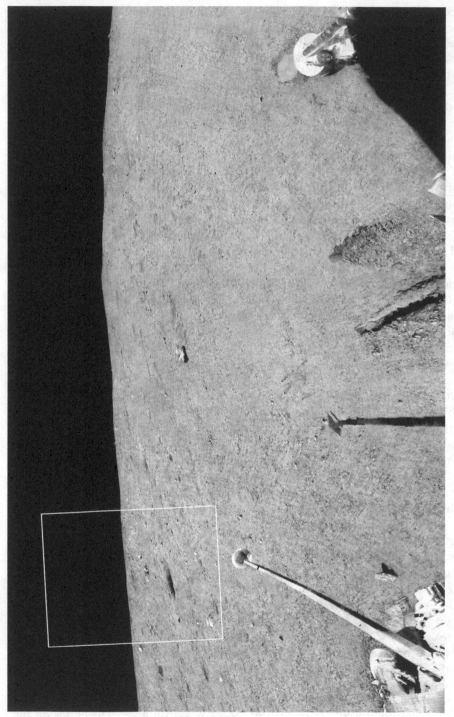

A northward-looking section of a panorama taken by Surveyor 3. The outline shows the area covered by the next illustration. (Courtesy of Philip J. Stooke, adapted from *International Atlas of Lunar Exploration*, 2007)

A portion of the previous illustration featuring a strewn field of boulders around a small crater on the northeastern rim of the crater in which Surveyor 3 landed. (Courtesy of Philip J. Stooke, adapted from *International Atlas of Lunar Exploration*, 2007)

northeastern rim would offer visiting astronauts an opportunity to recover material excavated from beneath the fragmental debris layer.[2]

After a detailed analysis of the Surveyor 3 imagery, Gene Shoemaker introduced the term 'regolith' to lunar science. This was familiar to terrestrial geologists as the collective name for the rock wastes of whatever origin and however transported that rest on bedrock and nearly everywhere form the surface of the land. On Earth, there are many erosional processes and the regolith includes volcanic ash, glacial drift, alluvial deposits, eolian deposits and soils rich in humus. On the Moon, the primary erosional process was meteoritic impact. Harold Urey had introduced 'gardening' for the manner in which the poorly sorted fragmental debris layer was turned over by impacts. The pictures from the two Oceanus Procellarum landing sites hinted that the thickness of the layer increased with age. To start with, the surface would have been bare rock. Any significant impact would have been capable of breaking up and scattering the material. This ejecta would have been progressively eroded by the rain of smaller projectiles. Over time, the layer of debris would have thickened, requiring ever larger impacts to reach the substrate. An important aspect of this process was that it would yield a continuous distribution of fragment sizes, which was expressed by saying that the lunar regolith was seriate.

One scientific task for Surveyor 3 was to monitor the Moon's passage through the Earth's shadow. With the Moon at 'full' phase and the Earth masking the Sun, this was a lunar eclipse to a terrestrial observer and a solar eclipse to the lander. It was the first opportunity to observe the thermal effects of such an event from the lunar surface and assess inferences drawn from telescopic studies. In particular, Surveyor science team member John M. Saari had been involved in infrared scanning of the Moon's disk during the lunar eclipse of 19 December 1964, the data from which was processed into isothermal contours that indicated the presence of many 'anomalies', mostly associated with craters, where the heat that had been absorbed while the Sun was shining at lunar noon was radiated again at the onset of the eclipse. In addition to monitoring the temperature, Surveyor 3 gave the scientists a bonus: the mirror of its TV camera had the same 35-degree elevation limit as previously, but because the lander was west of the meridian and inclined due west at an angle of 12 degrees by virtue of having settled on a slope, the field of view of the wide-angle frame was just able to include Earth, east of the zenith. Optical observations of the eclipse would yield the first direct measurement of the distribution of the refracted sunlight which weakly illuminates the lunar disk at such times.

Fortunately, the eclipse on 24 April occurred during a period when the Moon was still just above Goldstone's horizon. A total of 20 images were taken in two sets: the first at 11:24 and the second 37 minutes later. They were taken at two iris apertures, and with several exposures for each of the three colour filters. In the first set, an arc along the northwestern limb of Earth refracted light that varied greatly in brightness, and with a fainter glow at each end containing bright 'beads'. In the second set, the

---

[2]   And in fact Apollo 12 would do so at precisely this spot.

This picture taken on 20 November 1969 shows a similar view to the previous illustration, but includes the Surveyor 3 lander.

☆ TRANSIENT EVENTS

● INFRARED ANOMALIES

The locations of lunar transient events reported over the years by various observers and infrared 'hot spot' anomalies measured during a lunar eclipse on 19 December 1964. (Courtesy of John M. Saari and R.W. Shorthill, *Isothermal and Isophotic Atlas of the Moon*, NASA, 1967)

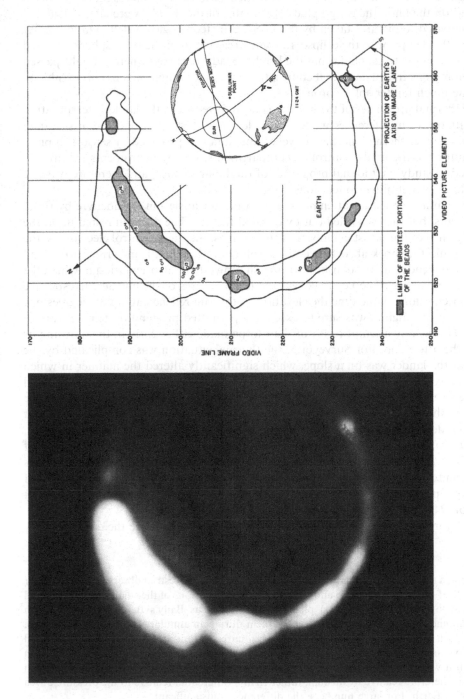

A picture taken by Surveyor 3 on 24 April 1967 showing Earth eclipsing the Sun, with associated exposition of how the Earth's atmosphere refracted sunlight.

brightest refraction had migrated to the northeastern limb. To assess the distribution of cloud on the limb, the geographic coordinates of the 'beads' were later calculated and compared to pictures taken by the ESSA 3 meteorological satellite in low polar orbit on the day prior to the eclipse. In some areas cloud in the troposphere occulted some of the refracted sunlight, but the bright 'beads' occurred where sunlight passed through regions free of cloud. Of course, the refracted sunlight made it impossible to view the much fainter solar corona.[3]

The thermal properties of the lunar surface inferred from the lander's temperature data differed from inferences made from telescopic studies, in that the in-situ data showed a higher thermal inertia. However, the data was provided by sensors in place to monitor the thermally controlled compartments, not by instruments specifically designed to study the thermal properties of the lunar surface, and therefore was too crude to draw definitive conclusions.

The plan had called for attempting a 'liftoff and translation' manoeuvre by firing the verniers, but this was ruled out by thermal factors. The decision rested upon the temperatures of the thrust chambers of the engines, the flight control electronics, the helium tank, the shock absorbers and the roll control actuator on vernier no. 1 – all of which depended upon solar heating and shadowing in the orientation in which the vehicle came to rest. The key issue was the temperatures of the engines. Irrespective of the orientation, by the time the elevation of the Sun reached about 35 degrees one or other of the engines was sure to exceed its permitted pre-ignition temperature of 105°C. Hence, any attempt to lift off had to be made either early in the morning or late in the afternoon. For Surveyor 3, the thermal situation was complicated by the fact that the lander was on a slope, which significantly altered the manner in which shadows were cast. To preserve the option of this experiment, the helium tank had not been vented. Also, given that at no time during the protracted landing had the forces on the shock absorbers imparted even half the load endured by Surveyor 1, it had been decided not to lock the legs. Whilst the helium was significantly depleted, as the tank absorbed solar heat its pressure increased to 2,735 psia.[4] By the time of the eclipse, however, it had been decided not to attempt to fire the verniers. The legs were monitored to determine whether the rapid decline in temperature at the onset of the eclipse prompted the shock absorbers to leak, but they retained their integrity. At 20:36 on 24 April, the helium tank was finally vented.

Surveyor 3 had two flat beryllium mirrors situated to enhance the camera's view

[3]   During a solar eclipse, when the Moon occults the Sun to terrestrial observers, the irregular profile of the lunar limb often allows light from small sections of the solar disk to be viewed during totality, giving rise to a phenomenon known as Baily's Beads after the British astronomer Francis Baily who first noted them during an annular eclipse on 15 May 1836.

[4]   The term 'psia' means pounds of force per square inch on an 'absolute' scale measured relative to zero. If a pressure gauge is calibrated to read zero in space, then at sea level on Earth it would read 14.7 psi, which is sea-level atmospheric pressure. A value specified in psia is therefore relative to vacuum, rather than a differential relative to the pressure at sea level on Earth. For large numbers, the difference is insignificant.

of the underside of the vehicle. That is, the camera was between legs no. 2 and 3 and the mirrors were affixed to leg no. 1. One mirror was 35 × 22 cm and gave a view of the lower portion of crushable block no. 3 and the area beneath vernier no. 3. The other mirror was 9 × 33 cm and viewed the area beneath vernier no. 2. The fact that the verniers were cut off at a height of only 3 feet suggested there might be signs of surface erosion, but since the vertical velocity at the third contact was only 1.5 ft/sec the crushable blocks probably did not strike the surface. Unfortunately, the hopes of viewing beneath the lander were foiled by the fact that when the Sun was low in the sky the pictures were 'washed out' by the glare from the coating on the main mirror of the camera, and when the Sun was high in the sky the area of interest was in the lander's shadow! It had been hoped to fire the downward-pointing cold-gas thruster on leg no. 2 to follow up on Surveyor 1's surface erosion experiment, but this time obtaining good 'before' and 'after' pictures of the test area. However, pictures taken when the Sun was low in the east were washed out by glare, pictures taken when the Sun was high lacked the shadows required to highlight the minute changes likely to result from such a weak thrust, and in the late afternoon the area was in the lander's shadow – so the test had to be cancelled.

Between 10:29 and 11:06 on 30 April, Surveyor 3 snapped wide-angle pictures of Earth illuminated as a crescent with the dawn terminator running the length of South America. These were the first colour pictures of Earth taken from deep space.[5] The filters had been revised to provide an improved spectral match to standard colorimetry functions.

Following sunset at 18:38 on 3 May Surveyor 3 monitored the rate at which the temperature fell, and at 00:02 on 4 May was commanded to hibernate.

In total, the camera took 6,326 pictures. Owing to the glare from contamination of the main mirror, usable images could be obtained only over a limited azimuth range during the early morning and late afternoon. This glare, combined with the difficulties in moving the mirror, made it impossible to obtain all of the systematic surveys of the landscape which had been planned. About 8 per cent of the pictures were taken at wide-angle to provide panoramas at specific illumination phases. The glare impaired detailed photometry, but the colorimetry confirmed that the surface was essentially grey. The glare precluded photographing the stars for use as celestial references to precisely determine the orientation of the camera, and hence the true orientation of the lander, but on one occasion it did manage to photograph Venus, which helped to some extent.

Although Surveyor 3 was unable to view the plain surrounding the crater in which it landed, scientists were delighted to have the opportunity to survey the interior of a medium-sized crater on a mare! In effect, therefore, its observations complemented those by Surveyor 1 of the open plain. In general, the character of the lunar surface material appeared to be similar at the two sites. The soil mechanics surface sampler

---

[5] The pictures taken by Lunar Orbiter 1 showing Earth against the lunar limb were in black-and-white.

was active for a total of 18.3 hours. It executed 5,879 commands, during which it made seven bearing-strength tests, thirteen impact tests and four trenches to provide data on the strength, texture and structure of the lunar material to a depth of 18 cm. In particular, it found that the bearing strength of the material increased with depth, even although there was no discernible change in the grain size – it was just that the uppermost few centimetres were more porous. Although sequences of commands to enable the sampler to perform complex operations had been stored on magnetic tape for step-by-step uplinking, for much of the time it was actually operated in real-time and monitored by TV.

Surveyor 3 evidently succumbed to the chill of the lunar night, because attempts to reactivate it after sunrise were unsuccessful.

OUTCOME UNKNOWN

Surveyor 4 was similar to Surveyor 3, with a soil mechanics surface sampler, but it also had a magnet on foot pad no. 2 to investigate whether there were magnetic particles in the surface material. It was to employ the last of the single-burn Centaur stages and fly essentially the same direct ascent trajectory as Surveyor 2 to aim for Sinus Medii. It lifted off from Pad 36A at 11:53:29 GMT on 14 July 1967. Both the Atlas and the Centaur performed satisfactorily, with translunar injection at 12:04:57. The spacecraft deployed its legs and omni-directional antenna booms, and, on being released, cancelled the inherited rates, acquired the Sun and deployed its solar panel. When commanded to acquire Canopus some 6 hours later, it did so without incident. It was decided to postpone the midcourse manoeuvre from the nominal 15 hours into the flight, and make it 24 hours later. The 10.5-second burn at 02:30:04 on 16 July imparted a change in velocity of 33.78 ft/sec to trim the initial divergence of 175 km from the centre of the 60-km-diameter target circle to a mere 8.5 km.

The pre-retro manoeuvre in which the spacecraft departed from its cruise attitude involved starting a roll of +80.4 degrees at 01:24:44 on 17 July and a yaw of +92.7 degrees at 01:29:34. This aligned the thrust axis with the velocity vector as that would be at retro ignition. The roll of –25.3 degrees at 01:35:05 was to optimise the illumination for post-landing imaging of crushable block no. 3. A landing on the prime meridian involved making an approach at 31.5 degrees to local vertical, as opposed to 23.6 degrees for Surveyor 3 at 23°W and 6.1 degrees for Surveyor 1 at 43°W. This would require a greater gravity turn in the vernier phase to force the trajectory to vertical. If successful, this mission would 'open the door' to sending future landers to targets in the eastern portion of the Apollo zone.

The altitude marking radar was enabled at 02:00:17, and issued its 100-km slant-range mark at 02:01:56.080. The programmed delay to the initiation of the braking manoeuvre was 2.725 seconds. The verniers ignited precisely on time, and the retro-rocket 1.1 seconds later – at which time the vehicle was travelling at 8,606 ft/sec. With everything apparently normal, the downlink fell silent at 02:02:41.018, when 40.9 seconds into the predicted 42.5-second duration of the retro-rocket's burn. The vehicle was at an altitude of 49,420 feet, travelling at 1,092 ft/sec, and nominally 2 minutes

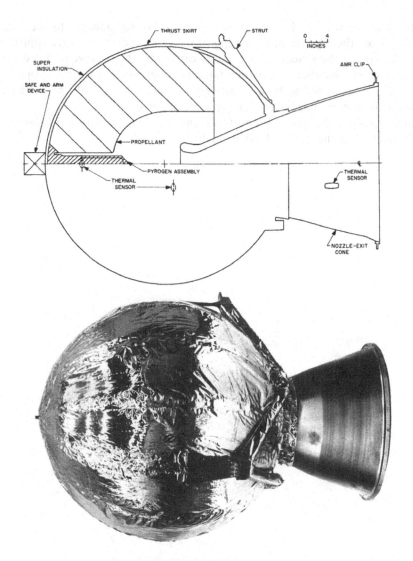

THRUST SKIRT
STRUT
0 4
INCHES
SUPER INSULATION
AMR CLIP
SAFE AND ARM DEVICE
PROPELLANT
PYROGEN ASSEMBLY
THERMAL SENSOR
THERMAL SENSOR
NOZZLE-EXIT CONE

Details of the Surveyor spacecraft's solid-fuel retro-rocket.

20 seconds from landing. The Deep Space Network was unable to re-establish contact with it.

The engineering team that studied the telemetry realised that whatever the fault was, it had cut the downlink within an interval of 0.25 millisecond without showing any indication in the preceding telemetry. The cause of the failure was not apparent. The only noteworthy unusual development was a slight modulation in the thrust of verniers no. 1 and 2, but it was not evident how this could have been relevant. The investigation listed four possible causes, without rating them in order of likelihood: (1) the breakage of a critical power lead in a wiring harness, or the failure of an electrical connector, or the failure of a solder joint; (2) damage to the spacecraft's circuitry from the rupture of the casing of the retro-rocket; (3) a transmitter failure; or (4) damage to the spacecraft's circuitry caused by the rupture of a pressure vessel such as a shock absorber or a helium tank, nitrogen tank or vernier propellant tank. Since there was judged to be a "relatively low probability" of any of these failure modes recurring, no hardware changes were ordered.

Interestingly, if Surveyor 4's problem was simply a transmitter failure, then it is highly likely that the vehicle landed safely.

13

Orbiters for science

GLOBAL MAPPING

In March 1967 the Surveyor/Orbiter Utilisation Committee agreed that since the first three Lunar Orbiter missions had achieved that project's commitment in support of Apollo, the next should "perform a broad systematic photographic survey of lunar surface features in order to increase scientific knowledge of their nature, origin and processes, and to serve as a basis for selecting sites for more detailed scientific study by subsequent orbital and landing missions". This plan had been conceived at the Summer Study on Lunar Exploration and Science held in Falmouth, Massachusetts, between 19 and 31 July 1965, in the hope that the opportunity to undertake it would arise. The primary objective was to obtain contiguous coverage of at least 80 per cent of the near-side of the Moon at a resolution better than 100 metres. In fact, if the project's priority had not been to reconnoitre specific areas in support of Apollo, the scientists would have *started* by mapping on a global basis.

To map in this way, the spacecraft would require to fly in a near-polar orbit with a perilune altitude fifty times greater than its predecessors, and as it would spend most of its time in sunlight the heat-rejection capacity of its protective base was enhanced by the installation of several hundred small quartz mirrors.

Lunar Orbiter 4 lifted off at 22:25:01 GMT on 4 May 1967. A midcourse burn of 60.8 m/s was required to deflect the trajectory away from the equatorial zone for a polar trajectory. This 53-second manoeuvre was made at 16:45 on 5 May. A further refinement was cancelled.

At 15:09 on 8 May the engine was reignited for 502 seconds to slow by 660 m/s and enter an orbit of 2,706 × 6,114 km with a period of 12 hours. The orbital plane was inclined at 85.5 degrees to the lunar equator, and oriented to enable the ground track to follow the migrating terminator to highlight topographic relief. The phase of the Moon was 'new' on 9 May; 'first quarter' would occur on 17 May and 'full' on 23 May. The photographic mission began at 15:46 on 11 May, while passing south to north on the eastern limb, and viewed Mare Australe and Mare Smythii. Given the high perilune, there was no requirement for motion compensation. The schedule was repetitive. During the perilune passage of each revolution, the spacecraft was to take sequences of four exposures centred at 42.5 and 14 degrees each side of the equator. In addition, on alternate orbits, it was to take pictures at 72 degrees from altitudes ranging between 3,340 and 3,600 km, with each such picture ranging poleward from 50 degrees latitude. On every fourth orbit, it was to take a picture of the far-side at apolune, with the flight programmer managing operations whilst the vehicle was out of contact with Earth. Given the long period of the orbit, it had been decided to scan and transmit the film as it advanced through the scanner to the take-up reel. This was a precaution against a repeat of the failure of the film transport system that had curtailed the rewind readout by Lunar Orbiter 3. As this was a mapping mission, the maintenance exposures to prevent the Bimat from sticking were to be taken with the thermal door closed.

A serious issue developed on the first day of photography. When the thermal door failed to close automatically after the second set of four exposures, a command was sent from Earth to make it do so. Next, when the third set of exposures were taken, the telemetry did not confirm that the door had opened. It was then manually commanded to open for the fourth set of exposures. It was an inauspicious start: the concern was that if the door were to close, it might never open again! But to operate with the door permanently open would risk light entering the photographic system and ruining the exposed film in the loopers between the camera and the Bimat

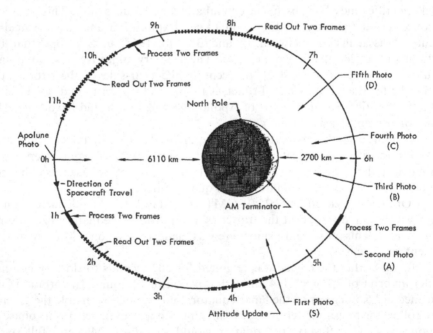

The sequence of events for Lunar Orbiter 4 on an orbit with imaging coverage of the south pole.

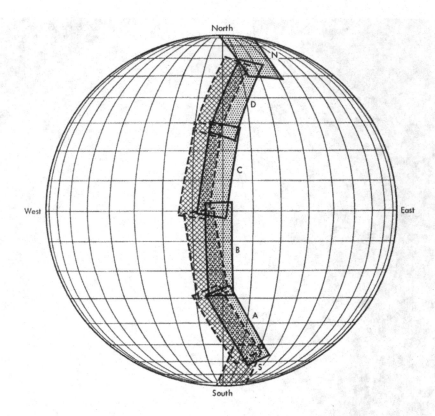

The Lunar Orbiter 4 imaging sequence was designed to provide comprehensive overlap in the high-resolution coverage.

processor. It would also risk moisture in the hermetically sealed compartment condensing on the lenses. It soon became evident that the longer the exposed film spent in the loopers before being processed, the greater was the light pollution. Tests by Boeing indicated that it should be safe to repeatedly partially close and fully open the door. When this was done, the light leakage was reduced to an acceptable level. To overcome the loss of image contrast arising from dew on the lenses, the vehicle was briefly oriented at the start of each orbit to let the heat of the Sun clear the condensation. By the time that the difficulties were completely overcome, the plane of the orbit had migrated about 60 degrees in longitude. However, it proved possible to rephotograph much of this area again from apolune later in the mission.

On 20 May the drive mechanism of the film scanner began to misbehave. Clifford Nelson, the Project Manager at Langley, debated the irrevocable step of cutting the Bimat strip immediately versus continuing in the hope that all would be well. Jack McCauley argued for extending the contiguous coverage beyond the western limb to document the Orientale basin. Nelson agreed. When the scanner problem worsened on 25 May, it was decided to cut the Bimat. Although the photography had reached 100°W, the readout was at only 70°W and the challenge was to coax the remaining

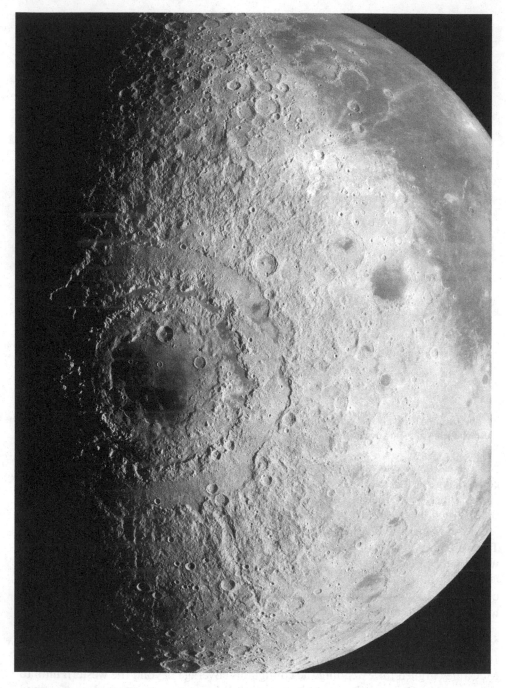

Lunar Orbiter 4 frame M-187 documented the Orientale basin in unprecedented detail.

processed exposures through the scanner in a manner which fooled the faulty logic unit. This task was successfully completed on 1 June.

The resolution of the mapping varied with altitude, but at perilune it was as fine as 60 metres, which was considerably better than was attainable from Earth. The results revealed hitherto unknown geological detail of the near-side polar and limb regions, and also increased to about 80 per cent the project's coverage of the far-side. Frame M-187, taken from an altitude of 2,723 km, showed the Orientale basin in startling detail. Secondary exposures included westward-looking oblique pictures of Apollo sites. The micrometeoroid experiment had reported two hits. Manoeuvres on 5 and 8 June lowered the orbit to 77 × 3,943 km to approximate that intended for Lunar Orbiter 5 and to obtain selenodesy to assist in the planning of that mission. (Meanwhile, tracking of Lunar Orbiters 2 and 3 was showing that a low perilune would decay unless maintained by engine firings.) Contact with Lunar Orbiter 4 was lost on 17 July, and calculations indicated that its diminishing perilune would have caused it to crash at the end of October 1967. There was no 'screening' after this mission, as the images were for scientific research rather than Apollo landing site certification.

SCIENTIFIC TARGETS

On 7 March 1967, several days after Lunar Orbiter 3's readout was curtailed, the project established a working group to develop the strategy for the final mission of the series. It was decided that if Lunar Orbiter 4 was successful in its mapping, then Lunar Orbiter 5 should undertake a scientific mission involving multiple targets. The photographic objectives were: (1) to obtain additional near-vertical, stereoscopic and westward-looking oblique frames of the eastern candidate sites for the early Apollo landings; (2) to accomplish broad survey coverage of those portions of the far-side which had been in darkness during previous missions; (3) to obtain pictures of sites of interest to the Surveyor project; (4) to reconnoitre potential targets for advanced Apollo landings – i.e. sites outside the equatorial zone; and (5) to take a close look at as many scientifically interesting sites as possible. The main criterion for a site being considered to be interesting was its perceived 'freshness'. The pictures from Ranger and the Lunar Orbiters to date had revealed most lunar terrain to appear subdued, so it had been decided to seek terrains which had not been exposed for long enough to have been significantly weathered by the incessant rain of material from space.

The preliminary plan was put to Boeing on 21 March, and a meeting on 26 May designed an orbit that would enable the spacecraft to address the widely distributed targets without violating any of its operating constraints. It would have to fly in a near-polar orbit to gain the latitude coverage, on a track with the Sun at an elevation of between 8 and 24 degrees to highlight the topography, and the perilune at about 100 km to provide the requisite 2-metre resolution.

Lawrence Rowan had led the US Geological Survey's participation in the Apollo site selection process, but he stood down after Lunar Orbiter 3. Donald E. Wilhelms took over this role for Lunar Orbiter 5, with its focus on advanced Apollo missions.

On 15 March Bellcomm had hired Farouk El-Baz, an Egyptian geologist with a PhD from the University of Missouri, and he undertook much of the organisational work. Whereas the sites short-listed for the early Apollo landings were on open plains with as few craters and rocks as possible, it was evident that advanced landings would be best made at sites where craters had excavated boulders. And, of course, there were mountains, rilles and features which appeared to be of volcanic origin. However, to be viable a target required a clear line of approach from the east, and this favoured interesting sites adjacent to smooth plains over which an Apollo lander could make its approach. Of course, if the landing site was several kilometres from the 'feature' that attracted the interest of the selectors, some form of surface transportation would require to be provided in order to enable the astronauts to reach their true objective.

On 14 June the Surveyor/Orbiter Utilisation Committee approved the overall plan. The initial target list was compiled largely from telescopic studies, but almost half of the items were revised following a review of the Lunar Orbiter 4 results. The agreed objectives were (1) to inspect 36 sites of scientific interest on the near-side, (2) to obtain additional views of five potential Apollo sites and a number of Surveyor sites, and (3) to map most of the far-side that had not previously been covered.

Lunar Orbiter 5 lifted off at 22:23:01 GMT on 1 August 1967. Its Canopus sensor had difficulty finding its target star, but locked on in time for the 26-second, 30-m/s midcourse manoeuvre at 06:00 on 3 August. A 508-second, 644-m/s insertion burn initiated at 16:48 on 5 August attained a 195 × 6,028-km orbit inclined at 85 degrees with a period of 8 hours 27 minutes. The phase of the Moon was 'new' on 6 August, and would be 'full' on 20 August. Most of the far-side pictures were to be taken in this initial orbit. The first picture was taken at 23:22 on 6 August, near apolune. An 11.4-second burn at 08:44 on 7 August lowered the perilune to 100 km. At 09:05 an impromptu picture was taken of Earth. A 153-second burn at 05:08 on 9 August lowered the apolune to 1,500 km and reduced the period to 3 hours 11 minutes. The Bimat was cut at 03:30 on 19 August, and the readout was concluded on 27 August.

The Apollo sites had been assigned 44 frame-pairs, which was about 20 per cent of the total. The sites of scientific interest on the near-side had included the rilles in Mare Serenitatis near the crater Littrow and near Sulpicius Gallus; some lava flow features in Mare Imbrium; the craters Copernicus, Dionysus, Alphonsus, Dawes and Fra Mauro; secondary craters associated with Copernicus; the Aristarchus plateau; and small domes near Gruithuisen, Tobias Mayer and Marius. All of these sites were regarded as possible targets for advanced Apollo missions.[1]

On 21 January 1968, during the extended mission, the 61-inch telescope of the University of Arizona successfully photographed the orbiter against the stars when it was at apolune. Ten days later, the spacecraft was deliberately crashed in Oceanus Procellarum.

---

[1] Both Apollo 15 and Apollo 17 were sent to sites imaged by Lunar Orbiter 5; although in the case of Apollo 17 the observations by Apollo 15 also contributed to the selection.

Astronomers had inferred that either the Moon's shape or its density distribution (or perhaps both) were irregular. The radio tracking of the Lunar Orbiters did *not* settle the question of the Moon's shape, but *did* yield a major discovery. The early missions in near-equatorial orbits had revealed the Moon's gravitational field on the near-side to be irregular – after allowing for the variation of velocity with altitude in an elliptical orbit, the vehicles kept speeding up and slowing down. The tracking of Lunar Orbiter 5 in its low polar orbit enabled a gravimetric map to be compiled with sufficient resolution for the 'anomalies' to be correlated with surface features. It was found that a vehicle was accelerated as it approached one of the 'circular maria' and decelerated afterwards.

Such sites included Imbrium, Crisium, Smythii, Serenitatis, Humorum, Nectaris, Humboldtianum, Orientale and Grimaldi. As the maria were low-lying rather than elevated landforms, it was apparent that they must be of a greater density than their surroundings. One early suggestion was that the 'attractor' was the buried body of the impactor which excavated the basin that was later filled in by mare material, but this hypothesis was rejected when it was realised that there were negative anomalies associated with basins that had not been filled in by maria. John O'Keefe argued that the attractor was the infill itself. That is, the anomalies were the due to magma from deep in the interior having erupted onto the surface – gravitational attraction falls off with the inverse square of range, and dense material on the surface would produce a significant *local* attraction. Negative anomalies included craters such as Copernicus, which had essentially excavated 'holes' in the maria. Since the positive anomalies represented concentrations of mass, they were dubbed 'mascons'. The discovery was reported in a paper in *Nature* in August 1968 by Paul M. Muller and William L. Sjogren.

If the lunar crust had been able to adjust isostatically to the eruption of dense lava onto the surface there would be no anomalies today; the fact that this had not occurred was evidence that the crust was sufficiently rigid at the time the lava was erupted to support its weight. In 1968 Ralph Baldwin provided an explanation. A basin formed and was 'dry' for a while, during which its floor began to adjust isostatically to the removal of the crustal material. But before it could achieve equilibrium, fractures in the floor allowed lava to well up and fill in the low-lying areas. This process of infill occurred in many pulses over an extended time. Being dense, the mare pool tended to sink, thereby forming compression wrinkles in its centre and opening rilles at its periphery. When it was unable to achieve isostatic equilibrium the result was a local mascon. A further realisation (obvious in retrospect) was that the sudden removal of crustal material in the excavation of a basin would relieve the pressure on the mantle below and induce deep melting, which would in turn cause a plume to rise, lift and fracture the floor of the basin, and drive enormous volumes of low-viscosity magma to the surface. A question for further research was why most of the basins on the far-side were 'dry' – was the crust thicker on that side of the Moon?

WRAPPING UP

The five Lunar Orbiter missions were launched within a 12-month interval. They suffered various technical problems, but the primary objective of providing pictures in support of Apollo was achieved. Only 78 per cent of the frames were classified as 'useful', but a large batch of useless ones were the H frames from the first mission. Although solar flares occurred during missions, photography continued. The worst radiation dose was on 2 September 1966, but the flood of energetic protons did not fog Lunar Orbiter 2's film. This confirmed the wisdom of using a very 'slow' film. The radiation detector data confirmed that the Apollo vehicles and spacesuits would protect astronauts from an average exposure to solar plasma, and indeed from short-term greater-than-average exposure. In all, the five Lunar Orbiters reported a total of 22 micrometeoroid strikes over their entire time in space. The hazard in lunar orbit proved to be about half of that in low Earth orbit. An additional benefit to Apollo from the extended missions of the Lunar Orbiters, was the experience gained by the Manned Space Flight Network in tracking vehicles in lunar orbit. If the existence of mascons had not been discovered prior to the first Apollo mission venturing out to the Moon, the astronauts would have found their orbit varying in an unpredictable (and alarming) manner. This was the value of making a thorough reconnaissance!

Boeing had sufficient parts to assemble a sixth spacecraft, and even before Lunar Orbiter 5 was launched the Office of Space Sciences and Applications considered an additional mission. On 5 July 1967 Lee Scherer explained that this could perform a survey of the far-side at a resolution similar to that provided by Lunar Orbiter 4 of the near-side. One suggestion was that it should carry the gamma-ray spectrometer built for Ranger, in order to make a preliminary map the composition of the surface. On 14 July Homer Newell wrote to Robert Seamans putting the case for launching a sixth mission in November. Seamans refused, in part because it would not directly contribute to Apollo – which was not capable of landing on the far-side.

The scientists had hoped to develop Lunar Orbiter Block II to conduct a series of missions utilising a variety of sensors. In late 1964 the Office of Space Sciences and Applications compiled an experiment list: a gamma-ray spectrometer to survey the abundances of radioactive isotopes on the lunar surface; and infrared experiment to map the surface temperatures; a photometry/colorimetry experiment to determine the variation in the photometric function and colour of the surface material; a radiometer to measure surface thermal gradients; an X-ray fluorescence spectrometer to survey the relative abundances of nickel and iron on the surface; a solar plasma experiment to measure the spatial and temporal variation in flux and energy distribution of low-energy protons and electrons; a magnetometer to determine whether the Moon had a magnetic field; an instrument to test for a low-density ionospheric plasma; and using the transmitter for a bi-static radar experiment to study the roughness and dielectric properties of the surface. But without the leverage of the Office of Manned Space Flight this proposal failed to attract funding.

14

Tasting the Moon

A RISKY DESCENT

Surveyor 5 was the first of the project to be assigned to an eastern portion of the Apollo zone. The target was a 60-km-diameter circle in the southwestern region of Mare Tranquillitatis, centred 80 km east of Sabine. The highlands to the west were characterised by prominent southeast-tending ridges and valleys that formed part of the Imbrium sculpture. The target was about 70 km north of the southern boundary of the mare, and 38 km northwest of Moltke. It was also 60 km southwest of where Ranger 8 struck. There were no mare ridges in the immediate vicinity, but the area was crossed by faint rays from Theophilus 350 km to the south. In fact, owing to the magnitude of the gravity turn required to cancel the approach angle relative to local vertical, this site was about as far east as a Surveyor was capable of venturing.

Launch from Pad 36B was at 07:57:01 GMT on 8 September 1967. The Centaur fired for 320 seconds and achieved parking orbit at 08:06:48. It emerged from the Earth's shadow about 10 seconds later. After coasting for 6.7 minutes, it reignited for the 113-second translunar injection. The spacecraft was released at 08:16:27 and established its cruise mode without incident.

The fact that the initial trajectory would reach the Moon just 46 km from the aim point made this the best performance to date for an American launch vehicle sending a payload towards the Moon. At 01:32:57 on 9 September the spacecraft began to manoeuvre to the attitude for the minor midcourse burn, and at 01:42:28 the squib was fired to open the helium regulator's valve to increase the six propellant tanks of the vernier propulsion system from their initial pressure of 264 psi to their operating pressure of 720 psi. Immediately beforehand, the helium tank pressure was stable at 5,160 psi. The decrease in helium pressure due to pressurising the propellant tanks was 182 psi, as expected. The burn was initiated at 01:45:02. It lasted 14.25 seconds, and the 45.5-ft/sec change refined the trajectory as desired. As the spacecraft was in the process of re-establishing its cruise attitude, the engineers checking the telemetry were alarmed to observe that instead of holding at 720 psi, the propellant pressure was rising. It was conjectured that a particle of contaminant had lodged in the seat of the helium regulator and was preventing the valve from closing properly.

There was nominally a 1-hour time constraint between vernier firings, but in view of the need for haste this was waived. Additional firings were made in unsuccessful attempts to clear the valve. As there were bladders in the propellant tanks, there was no mixing of helium and propellant, and whenever the pressure reached 825 psia a relief valve opened to vent the excess pressurant to space. As a result, the helium tank was losing pressure at a rate of about 10 psi per minute. This promised disaster when the vehicle attempted its descent to the Moon, as the verniers would fizzle out from propellant starvation far above the lunar surface.

The first impromptu vernier firing was made with the vehicle's main axis pointing at the Sun, because the vehicle had already re-established that attitude. The burn was made at 02:12:03, some 27 minutes after the midcourse manoeuvre. It lasted for 10.1 seconds and the change in velocity of 32.1 ft/sec had the effect of driving the aim point 1,600 km along the equator to longitude 77.5°E. This was only a temporary digression, however, because the vehicle was then yawed through 180 degrees and a second burn was initiated at 02:39:51. This lasted 23.0 seconds and the change in velocity of 73.9 ft/sec drove the aim point 2,685 km westward, to longitude 12°W. For these burns the pre-ignition temperatures exceeded the engines' operating limit, but they fired satisfactorily. Unfortunately, the helium regulator continued to leak.

After a period of deliberation, it was decided to make a third impromptu burn. In addition to a 'critical component' of 27.2 ft/sec to re-establish the aim point, there was a non-critical component of 36.1 ft/sec designed both to increase the flight time and to burn off propellant in order to reduce mass and thereby the velocity at retro burnout. To further improve the chances of clearing the regulator valve, the required 13-second duration would be built up by firing the engines for 12 seconds and then twice pausing for 1 second and 'blipping' them for half a second. This sequence was started at 04:18:48 and imparted a change in velocity of 45.3 ft/sec, but did not clear the valve.

At 05:50 one team of engineers was assigned to work out a manoeuvre that would enable Surveyor 5 to avoid the Moon and remain in an Earth orbit possessing a high apogee, in case it was decided to abandon the Moon as the target. Meanwhile, other engineers calculated that if sufficient propellant could be burned off to lighten the vehicle and thus reduce the velocity at retro burnout, and if retro ignition were to be postponed to a much lower altitude, then it might be possible to achieve a landing despite the helium problem. At 07:43 it was decided to attempt to accomplish the lunar mission.

The fourth impromptu vernier firing was to burn off propellant in order to further reduce the retro burnout velocity and increase the volume available in the propellant tanks for the helium leaking from the regulator. The latter would both minimise the rate of venting and maximise the impulse available for operating the engines in the vernier descent. The burn was initiated at 08:24:03, lasted 33.0 seconds, changed the velocity by 106.0 ft/sec, and drove the aim point into the central highlands. For the next 13 hours the spacecraft was tracked to precisely define its trajectory. Given the uncertainty about whether a soft-landing would be achieved, it was decided to test the alpha-scattering instrument to check its calibration and assess whether there was a significant cosmic background. Power was applied to the instrument at 10:36. The calibration rates transmitted by an omni-directional antenna were monitored for two 10-minute periods by the Deep Space Network station at Canberra in Australia, then the instrument was switched off at 11:25. The fact that the background was very low was good news; if this should prove to be the only scientific data from the mission, it would assist with the next one. The fifth and final impromptu burn had a critical component of 12.7 ft/sec to draw the trajectory 267 km back towards the target, and a non-critical component of 11.8 ft/sec to further reduce the retro burnout velocity and increase the gas volume in the propellant tanks. It was made at 23:30:58, lasted 5.45 seconds and imparted a change in velocity of 17.3 ft/sec that returned the aim point to within 30 km of the target. By 23:46:37 the spacecraft had re-established its cruise attitude. The midcourse manoeuvre itself had used 11.3 pounds of propellant, and the impromptu firings had consumed another 67 pounds.

A new descent profile had been devised to minimise the total impulse requirement for the verniers, to enable them to accomplish the landing by operating solely on the residual helium pressurant. On a normal descent the altitude of retro burnout was selected to allow ample time for the verniers to initiate the gravity turn in advance of reaching the 'descent contour', but now this margin had to be minimised to reduce the duration of the vernier-only phase of the descent.

The altitude marking radar would still issue its mark at a slant range of 100 km, but the programmed delay would be increased to 12.325 seconds in order to reduce the altitude at which to initiate braking. In the case of Surveyor 1, whose target was at 43°W, the approach angle had been at 6.1 degrees to the local vertical; for Surveyor 3, at 23°W, the angle was 23.6 degrees; for Surveyor 4, on the meridian, it was 31.5 degrees; for Surveyor 5, aiming for 23°E, the angle would be 46.5 degrees – even on the original plan, this was to have been the most demanding descent to date. Instead of aligning the thrust axis precisely with the velocity vector, it had been decided to offset it by 0.78 degree in order to enable a component of the powerful retro-rocket's thrust to contribute to reducing the approach angle, and thereby reduce the gravity turn the verniers would have to perform. And to ensure the most accurate alignment of the retro axis the pre-retro roll was timed for when the divergence in the Canopus sensor would be zero, and the yaw for when the Sun would be precisely aligned in that sensor – as calculated by monitoring the oscillations in the alignment. Whilst the pre-retro sequence in which the vehicle departed from its cruise attitude would usually include a roll to optimise post-landing operations, for this descent the only factor considered was the RADVS, and because the first roll would satisfy this requirement the second roll was deleted. In essence, therefore, the powered descent would be initiated later than usual in order that instead of retro burnout occurring at an altitude of 35,000 feet this would take place at 4,260 feet, and instead of the total velocity being 400–500 ft/sec it would be just 100 ft/sec. In addition, the release of the spent retro-rocket casing was revised to advance the onset of RADVS-controlled flight by 4 seconds – since this time every second would count! These changes were designed to make the most efficient use of the remaining helium pressurant, but the descent would be much more risky than the usual profile.

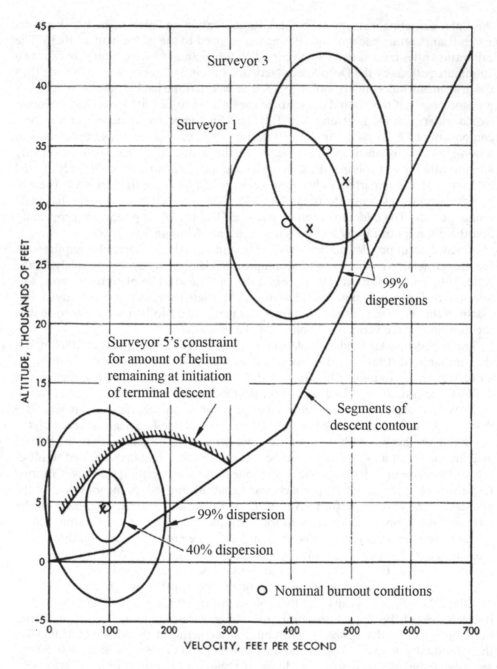

The helium pressurant problem suffered by Surveyor 5 required the burnout of its retro-rocket to occur at an unprecedented low altitude. Note that the 99% dispersion ellipse for retro burnout extends below ground level!

The pre-retro roll of + 73.8 degrees was started at 00:12:15 on 11 September, and the yaw of + 119.6 degrees at 00:16:20. The altitude marking radar was enabled at 00:43:01 and issued its 100-km slant-range mark at 00:44:39.118, at which time the vehicle was travelling at 8,441 ft/sec. During the specified delay of 12.325 seconds it would travel 104,000 feet, and, in view of the angle of the trajectory, the altitude would reduce by some 74,000 feet prior to retro ignition – this delay being to fly the revised powered descent profile.

The verniers ignited at 00:44:51.443, and the retro-rocket at 00:44:52.533. The helium pressure at vernier ignition was 836 psia, which was sufficient to provide the desired total vernier thrust of 152 pounds for this phase of the descent. The RADVS was switched on at 00:44:53.456, and the radar altimeter locked on at a slant range of 16,000 feet. The acceleration switch noted the peak thrust of 9,740 pounds fall to 3,500 pounds at 00:45:31.355, indicating a burn duration of 39 seconds. The spent casing was jettisoned at 00:45:40.395. The flight control system throttled the three verniers to full thrust at 00:45:40.955 to ensure that the casing fell clear. At burnout, the angle between the thrust vector and velocity vector was 45 degrees.

At 00:45:42.395 control was passed to the RADVS. The altitude was 4,139 feet (a slant range of 6,300 feet owing to the approach angle) and the velocity was 79 ft/sec (the vertical component of which was 46 ft/sec). Attitude control was switched from inertial to radar, and the vehicle immediately resumed the gravity turn by holding a constant deceleration of 0.9 lunar gravity whilst maintaining the thrust in line with the instantaneous velocity vector. At 00:46:19.697 the radar indicated the 1,000-foot mark. On a normal descent, the vehicle would intercept the 'descent contour' at a slant range of 18,000 to 25,000 feet, travelling at 400 to 500 ft/sec with 100 seconds remaining to landing. On this revised profile it was to have reached the contour at a slant range of 1,500 feet, but it did so at 806 feet with just 22 seconds remaining for the closed-loop throttling phase. The 10-ft/sec mark was issued at 00:46:37.097, at a height of 50 feet. Some 5.6 seconds later, at 00:46:42.697, the verniers were cut off with a sink rate of 5.2 m/s, and Surveyor 5 fell freely for 1.7 seconds.

After leg no. 1 hit the surface at a vertical rate of 13.5 ft/sec, the vehicle pitched over at an angular rate which was faster than the gyroscopes were able to measure, and 190 milliseconds after first contact the other legs slammed down at 13.8 ft/sec and it came to rest tilted at 19.5 degrees. It was the harshest landing of the project to date, but still within the design specification. Fortunately, the lateral velocity component was negligible and the widely splayed legs prevented the lander from toppling over.

The time spent under RADVS control was a mere 62 seconds. The descent had burned 60 pounds of propellant – a little less than was consumed by the impromptu firings after the midcourse manoeuvre. Although about 45 pounds remained, the issue had not been a shortage of propellant; it was a shortage of helium pressurant to force the propellants into the engines, and by the end of the descent the helium, fuel and oxidiser pressures were identical at 560 psia – in effect, the vernier propulsion system had been in 'blow down' mode. Some engineers had estimated the likelihood of achieving a soft-landing at no better than 40 per cent, so this success came as a great relief! The engineering recommendation was to improve quality control in the

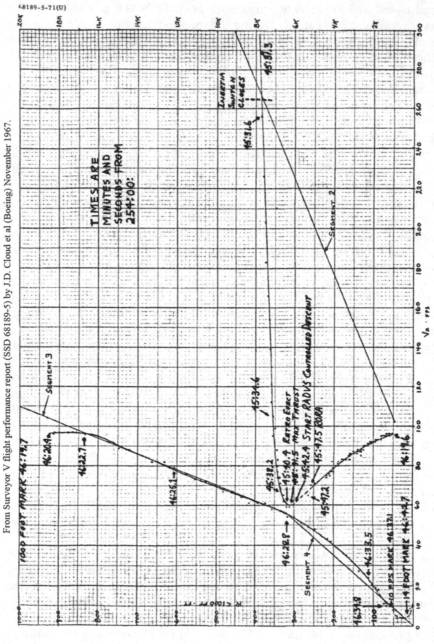

From Surveyor V flight performance report (SSD 68189-5) by J.D. Cloud et al (Boeing) November 1967.

The descent of the Surveyor 5 spacecraft depicted in two sections, one for slant ranges above 1,000 feet and the other below 1,000 feet. The delayed retro-rocket braking meant the vehicle did not join the contour until the third segment of the linear approximation to the ideal parabolic trajectory.

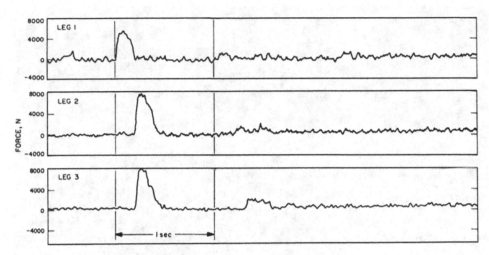

The dynamics of Surveyor 5's landing as indicated by the axial strain gauges on the shock absorbers of the legs.

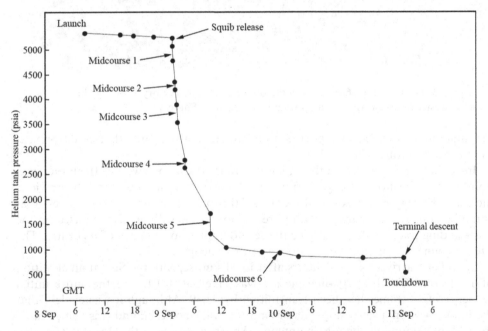

The history of Surveyor 5's helium pressurant from launch to lunar touchdown, showing the leak that developed when the valve of the regulator was activated and the stabilisation of the depleted level by the impromptu midcourse manoeuvres.

A picture taken by the 36-inch refractor of the Lick Observatory showing the southwestern part of Mare Tranquillitatis assigned to Surveyor 5.

manufacture of the helium regulator to eliminate contaminants that could prevent the proper reseating of the valve.

In-flight tracking indicated that it landed some 30 km northwest of the target, at a point that was just off the edge of frame H-78 taken by Lunar Orbiter 5 the previous month – the only high-resolution picture of this particular area – and the best that could be done was to mark a medium-resolution frame with an ellipse based on the post-landing tracking data and give the selenographic coordinates of its centre. The site was either in, or close to, a faint ray from Theophilus.

Surveyor 5 arrived some 35 hours after local sunrise, with the Sun at an elevation of 17 degrees. The first 200-line picture was taken at 02:01, and the entire initial sequence of 18 frames had been transmitted by 02:39. Although it was evident that the lander was on a steep slope, its orientation was uncertain, and this complicated the task of aligning the high-gain antenna. This began to scan the sky at 02:58, with Goldstone monitoring the strength of the received signal and directing the search. Meanwhile, the solar panel locked onto the Sun at 04:10. The antenna finally locked on at 05:21, and the first 600-line picture was transmitted at 05:30.

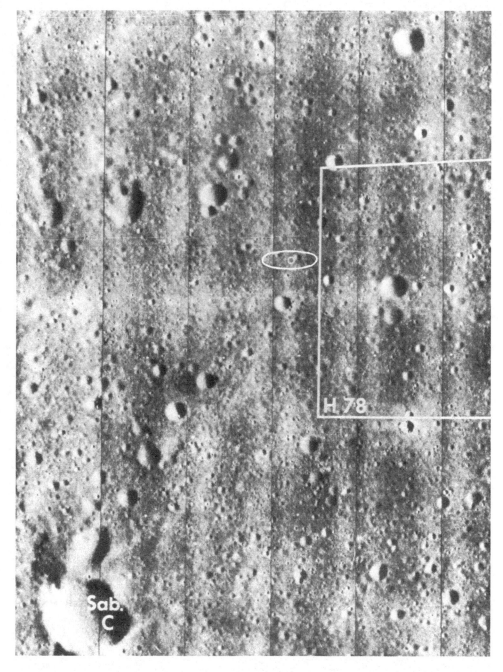

A portion of frame M-74 by Lunar Orbiter 5 showing where Surveyor 5 landed. The precise point could not be identified from lander imagery. Unfortunately the ellipse is just outside the orbiter's frame H-78.

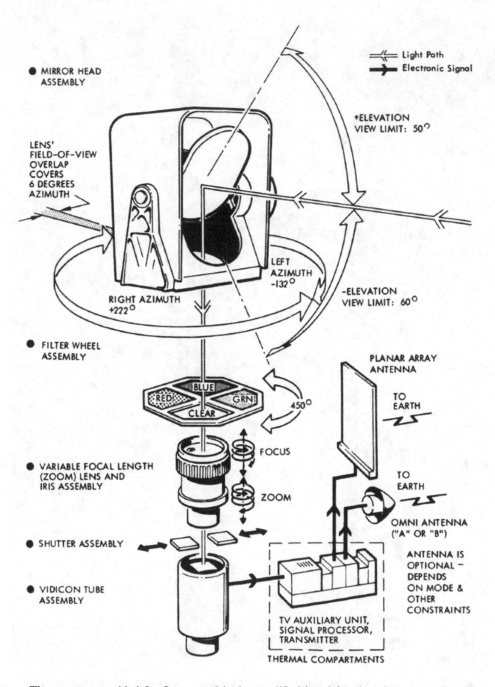

The camera provided for Surveyor 5 had a modified hood for the mirror assembly.

It was soon realised that foot pad no. 1 was resting on the southwestern margin of a small irregularly shaped crater 12 metres long by 9 metres wide and a fraction over 1 metre deep, and the other legs were inside the cavity. The crater was rimless, with the slope increasing towards the centre, where there was a distinctly concave floor spanning about 2.4 metres. In fact, the crater was of a type which had been classified in the Ranger imagery as a 'dimple', and the lack of a rim had led some people to argue that such pits marked where the loose surface material had drained into a subterranean space – with the implication that such features were *not* of impact origin.[1] The major axis of the crater was northwest-southeast, and the vehicle was on the southwestern interior wall, tilting northeast. It seemed to be the largest member of a chain of small elongated craters, all oriented with their major axes in line with the chain, and there were other chains in the area sharing this directionality.

Pad no. 1 had struck just outside the crater, and the others on its interior slope. As the vehicle rebounded, it slid downslope, causing pads no. 2 and 3 to scrape furrows about 1 metre long before they came to rest near the base of the wall. So, although this lander did not possess a soil mechanics surface sampler with which to scrape a trench, its legs served this role. As the vehicle slid down into the crater, pad no. 1 was dragged closer to the edge and the fragmental material that it displaced came to rest directly beneath the camera. Furthermore, as pads no. 2 and 3 ploughed their furrows they hinged and drove their outer edges into the surface, causing material to spill onto their upper surfaces where it was conveniently situated for examination by the camera. The furrows varied in depth between a few centimetres and 12 cm, most likely because the slide coincided with the rebounding of the legs following the initial impact. Furthermore, in addition to piling up material downslope, pads no. 2 and 3 splattered ejecta for a distance of about 1 metre across the concave floor – in fact, this was the greatest displacement of lunar material of any landing to date. The cohesion of the aggregates was evident from the fact that one clod remained intact on settling on the floor of the crater. The frequency-size distribution of the lumpy fragmental material pushed downslope was much coarser than that on the undisturbed surface.

The walls of the crater in which Surveyor 5 landed provided a view into the upper half-metre of the fragmental debris layer. There were bright angular fragments that seemed to be rocks, rounded aggregates of fine-grained material with bright angular chips bound into a dark matrix, and dark lumpy aggregates of aggregates. Beneath a depth of 10 cm it appeared to be a uniform mass of fine-grained material containing small rocks and shock-compressed aggregates several centimetres in size. Although the albedo of the disturbed material was comparable to that at the previous sites, the undisturbed surface was less bright, making the difference less pronounced. The flat mirrors which had been installed on leg no. 1 of Surveyor 3 had been superseded by convex mirrors to improve the view beneath the lander. Material around the bottom

---

[1] The 'dimple' craters are now known to be secondaries made by the fall of ejecta issued by larger impacts.

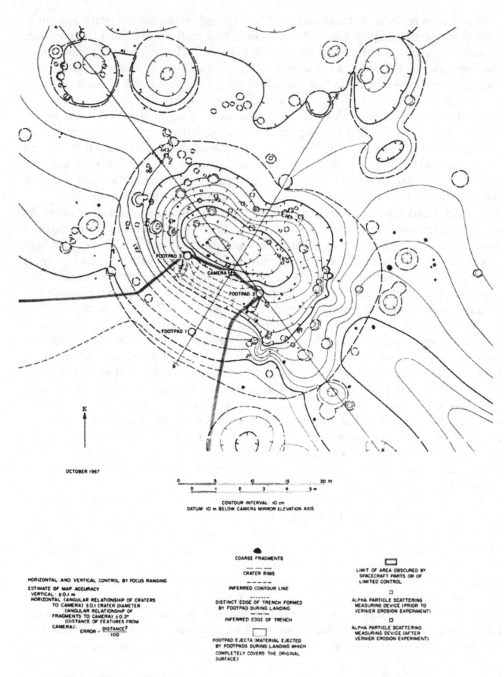

A detailed map of the small crater in which Surveyor 5 landed and the immediately surrounding plain, produced by R.M. Batson, R. Jordan and K.B. Larson of the US Geological Survey.

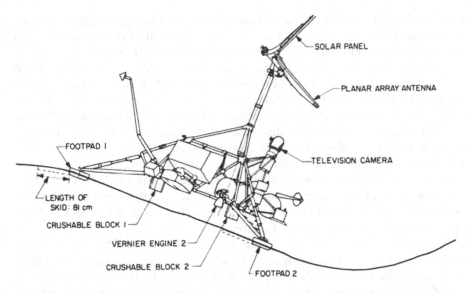

SOLAR PANEL

PLANAR ARRAY ANTENNA

FOOTPAD I

TELEVISION CAMERA

LENGTH OF
SKID: 8I cm

CRUSHABLE BLOCK I

VERNIER ENGINE 2

CRUSHABLE BLOCK 2

FOOTPAD 2

Surveyor 5 came to rest with two legs inside a small crater and the third on the lip.

As Surveyor 5 slid down into the small crater on landing, foot pad no. 2 scraped a shallow furrow.

edge of crushable block no. 3 implied that it struck the surface at the time of landing, but any imprint had been masked by the material thrown forward by pad no. 1 when this was dragged towards the crater as the other legs slid down the interior slope.

As the mirror of the camera was only 1.2 metres above the plane of the foot pads and was between legs no. 2 and 3, it too was 'inside' the crater, with the result that more than 80 per cent of the field of view was within a range of 6 metres. The view of the peripheral terrain was highly foreshortened. In fact, although the mirror was about 80 cm above the northern rim, the fact that the rim was higher to the southeast meant that the mirror was only 30 cm above the southern rim. For a lander on open ground the horizon would have been 2 km away. In this case, the camera was able to see to a distance of about 1 km to the north and west, but less far to the south and east. In one sector to the south the horizon was barely 100 metres away because it was the raised rim of a 20-metre-diameter crater. There was a strewn field of blocks associated with this crater that extended almost to the crater in which the lander was situated. When the closest blocks of this field were examined at high resolution they proved to be angular to sub-rounded and less than 50 cm in their longest dimension. Some had a mottled appearance reminiscent of a rock at the Surveyor 1 site. There was another strewn field associated with a 15-metre crater situated 200 metres to the north. On the basis of these strewn fields and other raised-rim craters in the middle distance, it was possible to estimate the fragmental debris layer in this part of Mare Tranquillitatis to be no greater than 5 metres thick.

On Surveyor 5, the soil mechanics surface sampler was replaced by an instrument to study the composition of the lunar surface material. This was developed by a team headed by Anthony L. Turkevich, a nuclear chemist at the University of Chicago. It comprised an electronics package in a thermally controlled compartment which was mounted high on the vehicle's frame, and a box-shaped sensor head that was stowed midway between legs no. 2 and 3, immediately to the left of where the electronics package for the surface sampler would have been if this were present. Including the ancillary hardware, the mass of the experiment was 13 kg. The head was held by its deployment mechanism until needed, then lowered on a nylon cable to the surface with a ribbon cable linking it to its electronics. The head was about 13 cm tall with a 17×16-cm cross section. To prevent the box from sinking into soft material, it had a D-shaped 15-cm-radius plate on its base. Six curium-242 alpha-particle sources inside the head were collimated to irradiate the surface through a 10.8-cm hole at the centre of the base plate. This isotope was chosen for its short (163-day) half-life, to obtain a high emission rate and a narrow energy distribution in the irradiation. A pair of sensors were positioned to detect alpha particles which were scattered back from the surface. There were also four sensors for any protons issued by nuclear reactions resulting from the irradiation of the surface. The instrument would be calibrated by alpha particles emitted by a 'standard sample' of einsteinium-254. The instrument was completed in September 1966 and, after acceptance tests, was delivered to Hughes on 18 January 1967 for use on this mission. Although the alpha-scattering technique could measure the abundances of elements with masses ranging from carbon up to iron, its ability to identify atomic weights at the heavier end of this range relied on achieving a high signal-to-noise ratio in the energy spectrum. The data was to be transmitted to Earth as a sequence of 10-bit words in real-time (although not on a continuous basis since the instrument and the camera could not use the high-gain antenna simultaneously) and stored on magnetic tape for later analysis. The elemental abundances would yield no direct information of how the elements were combined as chemical compounds, nor of how these compounds were combined as minerals – such insight would have to be inferred from assumptions about the nature of the sample.

A view to the north beyond the crater in which Surveyor 5 landed. (Courtesy of Philip J. Stooke, adapted from *International Atlas of Lunar Exploration*, 2007)

A strewn field of blocks nearby Surveyor 5 to the south.

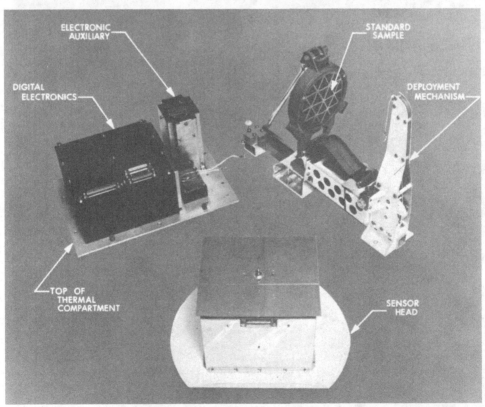

Alpha-scattering instrument and auxiliary hardware. The sensor head is the part of the instrument which was lowered to the lunar surface. In its stowed position, the head was held on the deployment mechanism in contact with the standard sample. The digital electronics and auxiliary electronics were housed in thermal compartment 'C'.

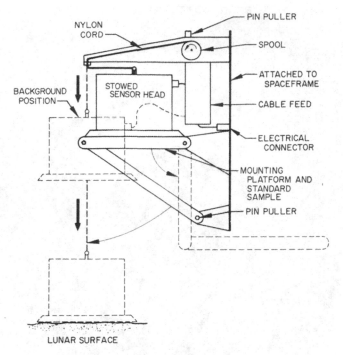

The operation of the alpha-scattering instrument's deployment mechanism.

The operating procedure was designed to obtain, in turn: data on the performance of the sensor with the head in its stowed position; the background radiation in the lunar environment with the head partially deployed; and the composition of the lunar surface with the head fully deployed. The first stage would be done with the head in its stowed configuration, in which it could view the standard sample. About 2 hours after landing, the instrument was checked out. Power was applied at 02:50 and the system interrogated, then it was switched off at 04:40. Following the handover from Goldstone to Canberra, the camera surveyed the area where the head would deploy, both directly and by using an auxiliary mirror affixed to leg no. 1, and revealed that the head would drop onto material which had been dislodged from the wall of the crater by the vehicle's arrival. Although mainly aggregates of fine-grained material, the sample area was fairly smooth with the largest aggregate about 3 cm in size. The instrument's beam of alpha particles was capable of penetrating only a few microns, but the aggregates were probably typical of the subsurface, and hence representative of the mare. Ideally, the calibration would be given 6 hours, as would measuring the background. But a complication now arose.

At about 08:00 it was decided that during the second Goldstone session a static firing of the verniers should be performed, before the engines overheated as the Sun continued to rise in the sky. As this left only about 12 hours in which to deploy the alpha-scattering instrument, this had to be abbreviated. Three 20-minute readings of the standard sample proved the detector to be working satisfactorily. It was therefore

A view of the underside of the alpha-scattering instrument's head showing the relative positions of the sources and sensors.

decided to remove the standard sample. To save time, the scheduled pictures were cancelled. Instead, Canberra monitored the data in real-time using an oscilloscope, and when the command was sent at 12:14 the rate fell by a factor of ten, confirming that the standard sample had swung clear to expose the aperture in the base plate of the head about 60 cm above the ground. After measuring the background of cosmic rays, solar protons and possible surface radioactivity for 170 minutes, it was decided to proceed with the deployment to obtain some sample data before the vernier firing. The head was to be lowered to the surface by a nylon cord wrapped around a geared cylinder. A pyrotechnic charge released the locking pin, and the cord unreeled under the weight of the head, lowering it in several controlled steps. As previously, the TV coverage was cancelled and real-time monitoring of the data by Madrid confirmed that the head touched down at 15:36. On the surface, the count rate sharply increased – almost to the level of the standard sample. A surface sample would require a total integration time of at least 24 hours – the longer the better, to improve the signal-to-noise ratio. About 5 hours of data had been obtained when Goldstone took over. By then it had been decided to postpone the vernier test by 24 hours, and so Goldstone spent its session photographing the surface in the immediately vicinity

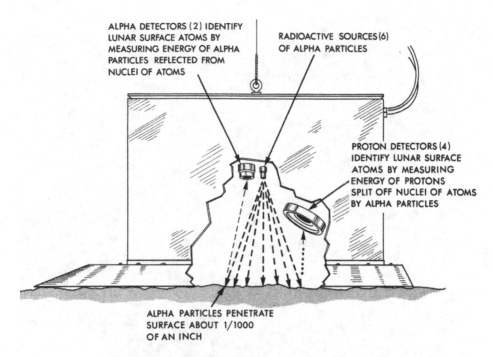

ALPHA DETECTORS (2) IDENTIFY
LUNAR SURFACE ATOMS BY
MEASURING ENERGY OF ALPHA
PARTICLES REFLECTED FROM
NUCLEI OF ATOMS

RADIOACTIVE SOURCES (6)
OF ALPHA PARTICLES

PROTON DETECTORS (4)
IDENTIFY LUNAR SURFACE
ATOMS BY MEASURING
ENERGY OF PROTONS
SPLIT OFF NUCLEI OF ATOMS
BY ALPHA PARTICLES

ALPHA PARTICLES PENETRATE
SURFACE ABOUT 1/1000
OF AN INCH

The operation of the alpha-scattering instrument's head.

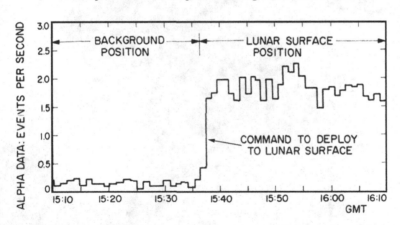

Real-time monitoring of the counts provided by the alpha-scattering instrument showed
that the head had deployed to the surface.

of the lander as a point of reference for the study of erosional effects. The pictures
showed that in settling on the slope, the base plate of the head had partially
embedded itself on the downslope side. The instrument renewed integrating at 06:18
on 12 September, via Canberra. Twelve hours of data was accumulated over the next
17 hours, and then the instrument was switched off. A total of 18.3 hours of data had
been obtained for the surface sample.

A view of the head of Surveyor 5's alpha-scattering instrument on the lunar surface.

After the verniers were fired at 05:38 on 13 September, Goldstone took a second set of pictures. These showed that the engine blast had further embedded the base plate of the head and tilted it sufficiently to raise the near side of the plate 2 cm off the ground. About 3 hours later, the instrument was restarted and established to be still functional. Because the aperture of the plate had been displaced by more than one diameter's distance, the new data was treated as a second sample. About 9 hours of data had been obtained by the end of that day. The maximum permitted operating temperature of the head was 55°C, and that for the electronics box was 50°C. Only 1 hour of data was obtained on 14 September and the instrument was not used at all on 15 September. But on 16 September the shadow of the high-gain antenna fell on the head and sampling was able to resume. Noon was at 01:47 on 17 September. By the time the final data was taken on 23 September the temperature of the head had fallen to −56°C, which was 16°C below its operating minimum. The next day, the Sun set. In all, 66 hours of data had been obtained for the second sample.

The three most abundant elements found by Surveyor 5 proved to be the same as those which are most prevalent in the Earth's crust – namely (in decreasing order): oxygen, silicon and aluminium. By the process of gardening, the uppermost layer of material could be expected to be primarily fragments of what lay beneath (although there would be some material tossed in from elsewhere) and hence could reasonably be presumed to be characteristic of the bedrock at that locality. Although the retro-rocket burn was made at a much lower altitude than usual, there was little chance of solid particles of aluminium oxide in the exhaust contaminating the analysis that was later performed by the lander, especially as the sample was not undisturbed surface material.

As a chemist, Harold Urey believed the Moon to be 'pristine' material condensed from the solar nebula, and hence of an ultrabasic composition. The alpha-scattering data showed that the analysed material was too poor in magnesium to be ultrabasic, and too rich in iron and calcium to be an acidic rock such as granite. A good match was a basalt created by chemical fractionation of ultrabasic silicates. The similarity in the general character of the three mare sites visited by Surveyors implied that this analysis was probably representative. Superposition relationships of geological units indicated that the 'circular maria' were not created at the same time as the basins in which they reside. This ruled out the maria forming by fractionation of a splash of impact melt – they simply had to be volcanic flows. This was clear evidence that the Moon was thermally differentiated and argued in favour of the 'hot Moon' theory expounded by Gerard Kuiper.[2]

On Surveyor 5, foot pad no. 2 had two bars mounted on its vertical side in view of the TV camera, one of which was magnetic to assess the presence of material with a

---

[2]   In the late 1950s J.J. Gilvarry argued that the maria were once water oceans, and hosted life. He said the now-dry plains were sedimentary rock, and dark owing to the presence of organic material. He claimed the elemental abundance data from the alpha-scattering instrument matched mudstone even better than it did basalt.

high magnetic susceptibility, and the other was non-magnetic to act as the 'control' by indicating the extent to which the lunar material would adhere to a metal bar. The magnet was strong enough to attract the ferrimagnetic mineral magnetite – which is common in igneous rocks on Earth, with different types of rock possessing differing percentages of this mineral – and the ferromagnetic metals iron, nickel and cobalt. To expose the magnet to contact, the pad needed to penetrate the surface to a depth of at least 6 cm. Serendipitously, this pad was one of those that scraped a furrow as the vehicle slid into the crater, and the magnet was in direct contact with material for most, if not all, of this time. However, in the orientation in which the lander came to rest, the magnet was in the shadow of the foot pad and could not be inspected until later in the lunar day, when it was revealed to be stained with a small quantity of magnetic material of a grain size finer than the camera's resolution – which at that range was between 0.5 and 1 mm. After making comparative laboratory tests, it was concluded that the observations were consistent with pulverised basalt in the 40–50-micron size range, with a 10–12 per cent fraction of magnetite and no more than 1 per cent of admixed metallic iron. Because it had been believed there would be an accumulation of meteoritic iron on the surface in the form of dust, this was much less metallic iron than most people had expected.

To follow up the observation that Surveyor 3's verniers seemed to have disturbed the surface during its 'hot' touchdowns, it was decided that Surveyor 5 should fire its verniers for a duration of 0.55 second at a total thrust less than its weight in lunar gravity, in order not to cause it to move. One motivation was that, like Surveyor, an Apollo lander was to cut off its engine a few feet above the surface, and a grounded Surveyor firing its verniers at low thrust would impart a comparable loading on the lunar surface to an Apollo lander in the final part of its descent – so this test would provide information that would enable the Apollo planners to estimate the effects of their lander's arrival. The specific objective was to determine the type and degree of erosion caused by the interaction of rocket efflux with the surface material, thereby estimating its cohesivity and permeability. Three processes of erosion were possible. Viscous erosion was entrainment of particles as the gas flowed across the surface. Gas diffusion erosion was due to gas penetrating the pores in the surface material while the engine was firing, with an upward force during and after the firing causing the movement of material. Explosive cratering would occur if the pressure of the gas impinging on the surface exceeded the material's bearing capacity, compressing and excavating it. In the lunar environment, where full expansion of the plume produced by the verniers at low thrust would be possible, explosive cratering was not expected to play a significant role. At its second bounce, Surveyor 3 appeared to have caused some viscous erosion, but because it had lifted off again there was no sudden engine shutdown and hence no opportunity to study diffused gas eruption. The aim of the Surveyor 5 experiment was therefore to investigate the effects of engine shutdown. Prior to the test, the surface in the immediate vicinity of the lander was documented in detail.

At 09:30 on 12 September the flight control system was powered up to verify its status. The temperatures of the components of the vernier propulsion system, and in particular the pre-ignition temperatures of the solenoid-operated propellant valves of the engines, were the dominant factors in deciding when to perform the test. The fact that the vehicle was inside a crater having steep interior slopes affected the heat-rejection capability of the thermally controlled compartments, because the slopes re-radiated heat onto the vehicle in a manner that would not have occurred if the lander had been standing on a level surface. Shortly after landing, vernier no. 2 significantly exceeded its pre-ignition temperature limit. It cooled down, but did so only slowly. Meanwhile, with leg no. 1 'exposed' on the rim of the crater, vernier no. 1 progressively rose in temperature. Early on 13 September it was decided to make the test without further delay. At 05:38:05, some 53 hours after arrival, the verniers were ignited at minimum thrust. Despite their high temperatures, the solenoid-operated propellant valves functioned correctly. The engines fired for about 0.57 ($\pm$0.15) seconds, with vernier no. 1 delivering 22 pounds of thrust, no. 2 delivering 17 pounds and no. 3 delivering 26 pounds. Despite the slope, the vehicle did not slide any further down into the crater. When the shock absorbers had been commanded to lock following landing, the legs on the downslope side had not done so, and their flexure during the vernier firing increased the tilt by 0.2 degree.

Because no pictures could be taken whilst firing the engines, the analysis utilised before-and-after views of the surface. A key source of data was the surface beneath vernier no. 3, both by direct viewing and by the auxiliary mirror on leg no. 1. This suffered two types of erosion. As the engine cut off, the sudden removal of pressure on the surface evidently allowed the gas which had diffused into the surface to erupt and displace material to erode an arcuate depression some 20 cm in diameter and up to 1.3 cm deep directly under the engine, whose nozzle was 13 cm in diameter with its aperture 37 cm off the ground. There was also viscous erosion which removed a 1-cm-thick layer of material out to a radius of 60 cm and had lesser effects out to about 2 metres. Most, if not all, of the small objects in the immediate vicinity were displaced. The largest fragment known to have been displaced was 4.4 cm in size. The permeability suggested that most of the particles in the surface material were in the 2–60-micron size range – which was consistent with an analysis of the imprint left by one of Surveyor 3's foot pads. The efflux blew away the dust which had adhered to the control strip and brackets of the magnet experiment on pad no. 2, but did not significantly alter the coating on the magnet itself. It also deposited material on the radiator on top of one of the equipment compartments. A clod must have been ejected by the diffused gas erupting from the surface at engine cutoff and followed an almost vertical trajectory, breaking apart on striking the mirrored surface. When the blast hit the vertical sides of the 2.2-kg head of the alpha-scattering instrument, this was driven about 10 cm downslope and rotated 15 degrees. The ability of the fine-grained material to adhere to a smooth vertical metal surface was indicated by a stain on the previously highly reflective head.

Sunset was at 10:56 on 24 September. Between 11:02 and 14:28 Surveyor 5 took 37 images to observe the solar corona with the solar disk just below the horizon. The longest exposures of 10 minutes showed coronal streamers out to 6 solar radii from the centre of the disk, and provided the first measurements of the brightness of the corona to a distance of 30 radii. After reporting the temperature for about 115 hours, the lander was ordered into hibernation at 06:37 on 29 September. Since Surveyor 1

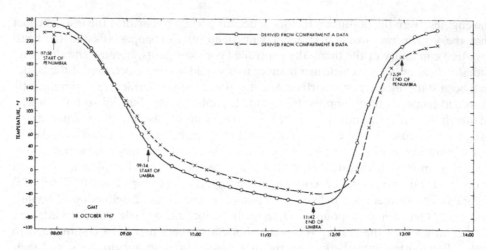

Lunar surface brightness-temperature profiles from sensors in Surveyor 5's thermal compartments as the Moon passed through the Earth's shadow on 18 October 1967.

had reported for 48 hours and Surveyor 3 for a mere 2 hours, this monitoring greatly increased the post-sunset data.

Surveyor 5's camera had suffered none of the problems that impaired Surveyor 3. There was no evidence of dust or vernier efflux on the mirror, and the azimuth and elevation actions worked perfectly. It returned 18,006 pictures of excellent quality in all illuminations, and the 180 mosaics that it produced made the crater in which this vehicle landed the best documented feature on the Moon! To improve the accuracy of its pointing system for celestial observations, a new procedure was used in which the rotational matrix of the camera relative to the frame of the lander was calibrated by taking a set of pictures of fixed reference points. The alpha-scattering instrument also functioned perfectly, and provided just over 83 hours of data for a total of two samples of surface material.

After being allowed to 'warm up' for 147 hours after sunrise, Surveyor 5 replied promptly to a command sent at 08:07 on 15 October. As the temperature declined about 12 hours before sunset of the first lunar day, the unlocked shock absorbers had compressed, deflecting leg no. 2 by 4.4 degrees and leg no. 3 by 6.9 degrees, further increasing the tilt of the vehicle. The fact that the shock absorbers re-extended at the start of the second lunar day indicated that their relaxation at sunset was not due to a pressure leak but to a decrease in fluid volume owing to the declining temperature. The camera's electronics had suffered from the cold during the night, but it was able to provide another 1,048 pictures. The alpha-scattering instrument was reactivated for 22 hours, but the low signal-to-noise ratio indicated that it had deteriorated and this data was rejected. On 18 October the Moon passed through the Earth's shadow. Surveyor 5 reported thermal data but, in spite of its tilt, was unable to view Earth. After sunset on 24 October, the lander monitored the temperatures for 215 hours until it was ordered into hibernation again at 12:15 on 1 November 1967.

To date, Surveyor 1 had provided a view of an open mare plain, Surveyor 3 had inspected the interior of a medium-sized mare crater, and Surveyor 5 had landed in a very small crater on a mare plain. The main conclusion was that Mare Tranquillitatis was generally similar to the sites in Oceanus Procellarum. In fact, perhaps the most significant observation was that the three sites were so similar that it was difficult to tell them apart!

ON THE MERIDIAN

In June 1967 the Surveyor Scientific Evaluation Advisory Team considered sending Surveyor 6 to a 'scientific' target, with one option being the hummocky Fra Mauro Formation, but NASA headquarters specified Sinus Medii, which would be the 'first backup' for an Apollo primary target in the eastern hemisphere. Surveyor 6 would be the project's third attempt at the meridian – Surveyor 2 had been lost attempting its midcourse manoeuvre, and contact had been lost with Surveyor 4 towards the end of its retro-rocket burn.

Sinus Medii was a relatively small mare plain about 170 km across, bounded to the north and south by highlands. The fact that the northwest–southeast structural trends of the adjacent terrain were radial to Imbrium indicated its origin as sculpture from the creation of that basin. The shapes and trends of the wrinkle ridges, crater chains and small shallow trenches on the plain reflected this structural pattern. The fact that the mare had a larger number of craters with diameters exceeding several hundred metres indicated its surface to be older than most maria. Telescopic studies showed it to have a higher average albedo than most maria. The largest crater on the plain was Bruce, at 7 km in diameter. The centre of the 60-km-diameter target circle was 55 km southwest of Bruce. It was hoped that the lander would set down within sight of a wrinkle ridge.

The launch window for Surveyor 6 was 7–12 November 1967. Although the first unmanned test of the Saturn V launch vehicle was due on 7 November, preparations for the lunar mission went ahead because if it were to become evident that the other mission would not meet its schedule, Surveyor 6 would attempt the first day of its window; otherwise it would be slipped – the Cape's tracking system required at least 24 hours to reconfigure for different types of vehicle. In the event, the Saturn V was postponed.

Surveyor 6 lifted off from Pad 36B at 07:39:01 GMT on 7 November. Since this was a predawn launch, the Centaur achieved parking orbit in darkness. It flew into sunlight at 07:53:22, initiated the 115-second translunar injection at 08:01:35 and released the spacecraft at 08:04:30. At 02:15:59 on 8 November, once the spacecraft had adopted the attitude for the midcourse manoeuvre, the helium valve was opened to pressurise the vernier propellant tanks. In raising the propellant to 764 psi, the helium fell by 180 psi from its initial 5,182 psia. The burn at 02:20:02 lasted 10.3 seconds, and the 33.1 ft/sec change in velocity moved the aim point 90 km closer to the centre of the target circle. After declining by 208 psi during the burn, the helium regulator maintained the propellant pressures constant throughout the remainder of the cruise.

The pre-retro manoeuvre in which the spacecraft departed from its cruise attitude involved initiating a roll of +82.0 degrees at 00:25:20 on 10 November, a yaw of +111.8 degrees at 00:29:38 and a final roll of +120.5 degrees at 00:34:56. The initial approach was at 24.3 degrees to the local vertical. The altitude marking radar was enabled at 00:56:16, and issued its 100-km slant-range mark at 00:57:57.038. The delay to the initiation of the braking manoeuvre was specified as 5.875 seconds.

The verniers ignited precisely on time, and the retro-rocket 1.1 seconds later. At that time the vehicle was travelling at 8,460 ft/sec. The RADVS was switched on at 00:58:05.798. The acceleration switch noted the peak thrust of 9,700 pounds fall to 3,500 pounds at 00:58:43.397, indicating a burn duration of 39.4 seconds. After allowing time for the solid rocket thrust to tail off, the verniers were throttled up to their maximum thrust at 00:58:53.297 for a duration of 2 seconds, during which the motor was jettisoned. At burnout, the angle between the vehicle's thrust vector and velocity vector was 26 degrees. The RADVS-controlled phase of the flight began at 00:58:57.737, when the slant range was 40,574 feet (and because the velocity vector at burnout was offset to vertical, the altitude was 36,625 feet) and the total velocity was 515 ft/sec (and since the vehicle had maintained its thrust along the velocity vector extant at the time of retro ignition, the longitudinal rate was 463 ft/sec). The vehicle immediately aligned the thrust axis along the velocity vector extant at retro burnout and flew with the verniers at 0.9 lunar gravity, very slowly accelerating as it descended. When the altimeter locked on at 00:58:59.892, at a slant range of 35,924 feet, attitude control was switched from inertial to radar and the thrust axis was swung in line with the instantaneous velocity vector to initiate the gravity turn. On intercepting the 'descent contour' at 00:59:21.276, the slant range was 24,730 feet and the speed was 552 ft/sec. By the 1,000-foot mark at 01:00:40.534, the vehicle was descending very nearly vertically at 106 ft/sec. The 10-ft/sec mark was issued at 01:00:57.634 at a height of 50 feet.

On receiving the 14-foot mark at 01:01:04.133, the flight control system cut the verniers. At that time the rate of descent was 4.6 ft/sec. After falling freely for 1.3 seconds, the vehicle touched down at 01:01:05.467 with a vertical rate of 11.2 ft/sec. Foot pad no. 1 made contact first, then legs no. 2 and 3 some 25 and 40 milliseconds later, respectively. It rebounded slightly, then settled, with the lateral rate of 1.0 ft/sec in the direction of leg no. 1 causing each pad to produce a pair of overlapping imprints. The gyroscopes indicated that it was within 1 degree of local vertical. It was a perfect landing!

The verniers had consumed 8.4 pounds of propellant in the midcourse manoeuvre, 41.1 pounds in the retro phase of the descent and 96.8 pounds in the vernier phase – a total of 146.3 pounds of the initial propellant load of 182.6 pounds. The total time spent under RADVS control was 2 minutes 6 seconds, with 1 minute 43 seconds of that flying the descent contour. In contrast, in its improvised descent Surveyor 5 had spent just 62 seconds under RADVS control. In order to have their full functionality available in the event of attempting a 'lift off and translation' experiment, it had been decided not to lock the legs as part of the post-landing sequence.

The first 200-line picture was sent at 01:50, and this 24-frame survey of the arc between foot pads no. 2 and 3 continued to 02:35. At an elevation of 3 degrees, the

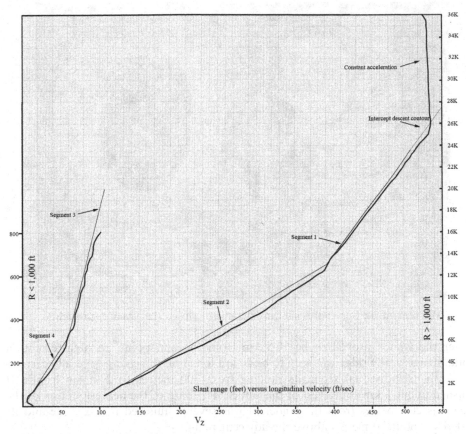

The descent of the Surveyor 6 spacecraft depicted in two sections, one for slant ranges above 1,000 feet and the other below 1,000 feet. (Compare this nominal descent with the improvised profile of Surveyor 5).

Sun was barely above the horizon. At 02:55 the solar panel began to scan in azimuth for the Sun, and located it at 03:19. With the landing site at the centre of the Moon's disk, Earth near the zenith and the vehicle upright, the alignment of the high-gain antenna was simple. It locked on at 03:40. The first 600-line picture was transmitted at 04:02. This camera was the first to have the new box-shaped hood, the mirror of which could seal the aperture to prevent dust or efflux from penetrating the optical system during landing.

The first 360-degree wide-angle panorama was completed by 05:00 and showed a relatively smooth, heavily cratered plain, but there was a feature on the southeastern horizon which, in the low-angle illumination, looked as if it *might* be a ridge. When this was examined again on Goldstone's second pass, with the Sun about 13 degrees higher, this identification was confirmed and a series of narrow-angle pictures were taken to record it in detail. The ridge was identified in Lunar Orbiter 2 frame M-113, and when the individual features visible to the lander were located on H-121 by that

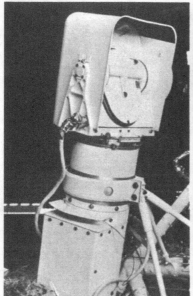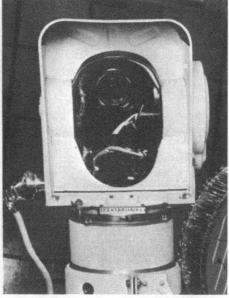

The camera for Surveyor 6 showing the mirror closed (left) and open (right).

orbiter the lander proved to be 10.5 km from the aim point. In high-resolution orbital imagery, the ridge was seen to be 40 km long and to zig-zag generally east-to-west with its individual segments ranging from 300 metres to 2 km in length. It vanished about 1 km southwest of the lander. The base of the nearest section of the ridge was 200 metres from the lander, it was several hundred metres wide and its crest rose about 30 metres above the adjacent plain.

To improve visibility of the surface beneath the vernier engines, Surveyor 6 was provided with three convex mirrors instead of two. Its orientation on the surface was determined by star sightings. Like its predecessor, it had a magnet on foot pad no. 2 to study the concentration of magnetic particles in the surface material. The colour filters had been superseded by polarising filters, and pictures were taken of selected areas during successive Goldstone sessions to build up a dataset in which the Sun's elevation changed at intervals of about 13 degrees, and thus measure the variation of the polarised component of surface reflection as a function of solar phase angle; the results proved to be insignificant, however.

The fragments displaced and ejected by the foot pads were composed primarily of aggregates of fine-grained material, and in many cases included small bright rock chips. In the immediate vicinity of the lander there were fewer fragments exceeding 2 cm in size than at the other sites, but a greater number smaller than this size. There was also a relative paucity of blocks within 50 metres of the lander – there were only six larger than 20 cm, and the largest was about 50 cm in size. Some were tabular, resembling the layered rocks seen by Surveyor 3 in its medium-sized crater. Most of the fragments within this range were subangular to subrounded, and although many were resting on the surface others were partially buried.

On the plain, craters up to about 150 metres in diameter generally possessed low subdued rims, but some were rimless. The fact that the *smallest* craters observed by Surveyor 6 on the plain to possess blocky rims exceeded this size indicated that the fragmental debris layer was up to 20 metres thick. For one crater the rim was not actually visible to the lander, just the associated field of blocks. This was visible in high-resolution Lunar Orbiter pictures, which also showed a bench in the wall of the crater at a depth of about 20 metres that could have marked the contact between the fragmental debris layer and the substrate.

In contrast, the lander observed a crater on the flank of the ridge about 30 metres in diameter and one on the crest of 20 metres diameter with blocky rims, indicating that the fragmental debris layer on the ridge was only 8 to 10 metres in thickness. On the crest to the south of the lander there was a crater 180 metres in diameter that was littered with blocks ranging up to 3 metres in size. In the high-resolution Lunar Orbiter pictures, it was possible to see blocks up to 6 metres in size elsewhere on the ridge. The coarse blocks within strewn fields were angular and faceted, and mostly appeared to be exposed on the surface. In terms of small craters, the size-frequency distribution on the ridge was comparable to that of the plain at the landing site. But a close inspection of the lander's pictures and the high-resolution orbital imagery indicated there to be many more coarse blocks on the ridge than on the adjacent plain – in this respect the ridge was similar to other examples of wrinkles, suggesting that it was representative. The origin of the ridge was disputed. One idea was that it marked where lava had extruded from a fracture (a dyke) and solidified in place. If

A mosaic of narrow-angle frames of a ridge to the south of Surveyor 6. The crest at this point was about 650 metres from the lander. There was a crater 180 metres in diameter on the crest strewn with blocks up to 3 metres in size. (See also page 357)

A picture of Sinus Medii taken by the 100-inch reflector of the Mount Wilson
Observatory. The outline shows the area covered by the next illustration.

this were the case, then the ridge could have formed at any time since the lava flow
that made the plain. But there was no evidence in the cratering to suggest that the
ridge was significantly younger than the plain. Another theory was that such ridges
were produced when a mare plain was deformed by compressional stress. In this
case, the manner in which the ridge zig-zagged suggested that its formation was
controlled by regional structures. Such stresses could have been imposed at any time
after the formation of the mare. The fact that the fragmental debris layer on the ridge
was thinner than on the plain was explicable by the slow but progressive flow of
loose material downslope. The profusion of large blocks on the crest was certainly
consistent with such 'mass wastage'. The effect was to smooth the transition between
the plain and the ridge. Indeed, in frame H-121 provided by Lunar Orbiter 2 it was
difficult to precisely identify the outline of the ridge.

The experiment in which Surveyor 1 pulsed a cold-gas attitude control thruster to
study surface erosion had been inconclusive, so Surveyor 6 was to repeat this test by
firing a thruster continuously. Since any disturbance of the surface would be subtle,
the test was made on 11 November, while the Sun was still low in the east to
maximise shadow detail in the impingement area. The downward-aimed thruster on
leg no. 2 was fired at 03:23 for 4 seconds, and again at 03:47 for 60 seconds. The
ground beneath it was surveyed by the camera prior to, between and after the firings.
The nozzle was 10.4 cm above the surface and inclined at 24 degrees to the lander's
vertical axis. Both firings displaced fine grains and individual clumps, and produced

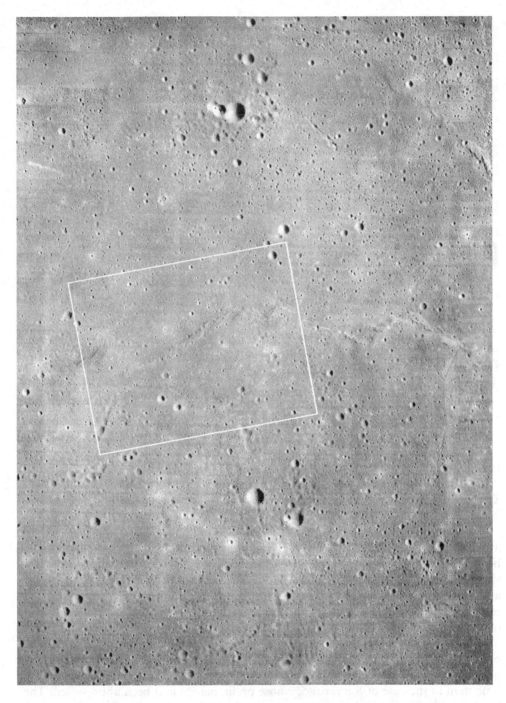

A portion of the II-P-8 mosaic showing the ridge on Sinus Medii in the target area assigned to Surveyor 6. The outline shows the area covered by the next illustration.

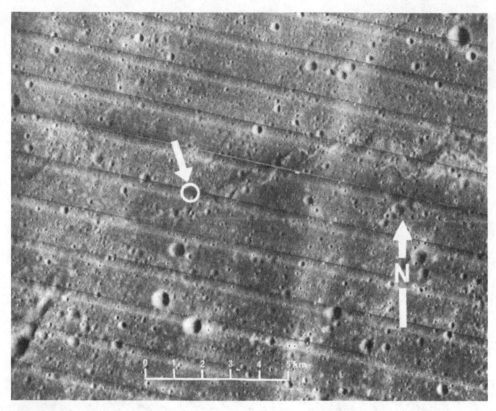

A portion of frame M-113 taken by Lunar Orbiter 2 on 22 November 1966 which shows the relative positions of the Surveyor 6 landing site and the nearby ridge.

partial erosion of some of the clumps that were too large to be moved. The radius of disturbance was 15 cm for the 4-second firing, and 25 cm for the 60-second firing. The fact that no crater was formed implied that the dimple beneath the mildly pulsed jet on Surveyor 1 had been coincidental.

Surveyor 6 was equipped with an alpha-particle instrument. This was powered up at 05:38 on 10 November. After two 10-minute calibrations of the standard sample between 05:41 and 06:21, activity was suspended for 3.5 hours in order to allow TV surveys to be conducted before the Moon set for Goldstone. Calibration of the alpha-scattering instrument resumed when Canberra took over. A total of 318 minutes had been obtained by 21:00, and at 21:18 Madrid commanded the head to deploy ready to measure the background. The first session began at 21:37, and lasted 33 minutes. Then operations reverted to Goldstone, which undertook TV work. The background measurements resumed at 05:00 on 11 November, and a total of 367 minutes of data had been obtained by 12:07. These operations were allowed more time than in the case of Surveyor 5, whose preliminaries had been abbreviated. The head was finally lowered to the surface at 12:08, some 35 hours into the surface mission. The sample was undisturbed surface. There were few fragments exceeding

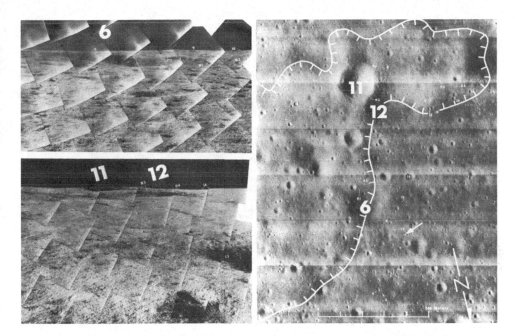

Many of the features visible to the Surveyor 6 lander could be identified in frame H-121 taken by Lunar Orbiter 2. The position of the lander in the overhead view is indicated by the arrow (although it was not present when the picture was taken).

several millimetres in size, and the largest *in* the sampled area was about 1.5 cm in size. Some 7.2 hours of data had been obtained by 23:00, when the instrument was switched off in order to resume TV work. Data collection resumed at 07:48 on 12 November, and a total of 15.7 hours of data had been obtained by 19:55. Activity had to cease at 23:39, when the head exceeded its maximum operating temperature of 50°C. The instrument was off through local noon, but was able to resume sampling at 16:50 on 16 November, when the Sun's elevation had decreased to 79 degrees and the shadow cast by the solar panel allowed the head to cool. By the time the instrument was switched off at 03:30 on 17 November a total of 30.5 hours had been obtained.

Surveyor 6 was to investigate further how the lunar surface was affected by rocket exhaust. The static vernier firing by Surveyor 5 had produced both viscous erosion and gas diffusion erosion – the latter resulting from the fact that the pressure on the surface was relieved suddenly as the engines were cut off whilst the vehicle was still on the surface. In the case of Surveyor 6, the engines were to deliver a greater thrust and for longer to emphasise viscous erosion, and because such a burn would lift the vehicle off the ground the pressure on the surface would be relieved slowly and thus minimise the disruptive effects of gas diffusion. And since the vehicle was to lift off, it had been decided to impart a horizontal displacement so that upon touchdown the camera would be able to view the original imprints made by the foot pads and the erosional effects of firing the verniers.

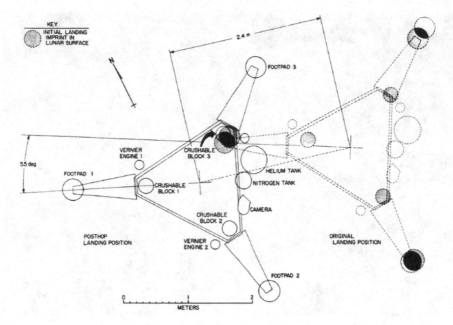

The positions of Surveyor 6's foot pads prior to and after it performed its liftoff and translation 'hop' manoeuvre.

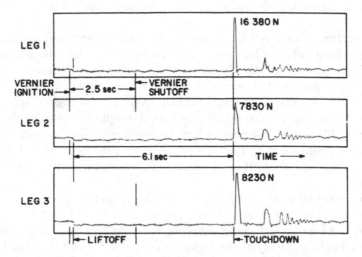

The dynamics of Surveyor 6's 'hop' as indicated by the axial strain gauges on the shock absorbers of the legs.

This 'liftoff and translation' was scheduled for 17 November. As a preliminary, high-resolution pictures were taken to document the state of the area immediately in front of the camera. As the Sun was high in the sky, the solar panel and high-gain antenna were temporarily repositioned to shade and cool the engines to a permissible

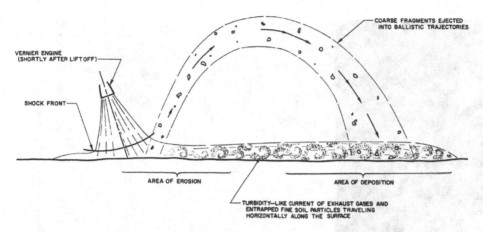

VERNIER ENGINE
(SHORTLY AFTER LIFTOFF)

SHOCK FRONT

COARSE FRAGMENTS EJECTED
INTO BALLISTIC TRAJECTORIES

AREA OF EROSION

AREA OF DEPOSITION

TURBIDITY—LIKE CURRENT OF EXHAUST GASES AND
ENTRAPPED FINE SOIL PARTICLES TRAVELING
HORIZONTALLY ALONG THE SURFACE

Detail of the viscous erosion effects of the efflux from vernier no. 2 as Surveyor 6 performed its 'hop'.

pre-ignition temperature. At 08:00 the flight control system was powered up for 35 minutes to verify its status. When the solar panel was stowed in order to prevent its being damaged by the stresses of the manoeuvre, this placed the vehicle on battery power. As the camera installed between legs no. 2 and 3 was on the east side of the vehicle, the flight control system was to fire vernier no. 1 at a lower thrust than the other two engines to make the vehicle lift off inclined at an angle of 7 degrees in the direction of foot pad no. 1, thereby displacing the vehicle to the west whilst causing the material eroded from the surface to be displaced preferentially in the opposite direction. Afterwards, the camera should have a good view of the erosional effects. At 09:46 the flight control system was reactivated, and at 10:32:02 the verniers were ignited and throttled to deliver a total thrust of 150 pounds. The intended period of firing was 2.0 seconds, but the cutoff failed and by the time the backup command took effect a total of 2.5 seconds had elapsed. The manoeuvre consumed 1.5 pounds of propellant. Once the telemetry had been examined to verify the systems, the solar panel and high-gain antenna were redeployed, and within 35 minutes photography had resumed.

The 'hop' lasted about 6 seconds, peaked at a height of 12.5 feet, and ended about 8 feet from the initial position in a direction slightly north of west. The vertical rate on making contact with the surface was 12.3 ft/sec, and the horizontal rate was 1.8 ft/sec – which was greater than that of the original landing and caused the foot pads to displace material as ejecta. The post-hop pictures showed the double imprints of pads no. 2 and 3 and the single imprints of the crushable blocks on those legs made at the time of the lander's arrival. But because the vehicle rolled 5.5 degrees in an anticlockwise direction around its main axis during the hop the imprints of pad no. 1 ended up beneath crushable block no. 3 and thus were not visible for inspection. The imprint of the alpha-scattering head in between legs no. 2 and 3 was obliterated by the blast. At the initial landing, the verniers had been cut off at a height of 12 feet to minimise disturbing the surface, but for the hop they had been fired at even greater

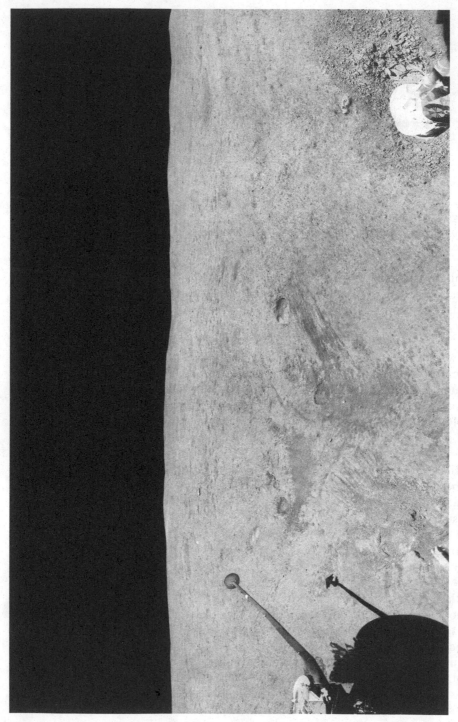

A section of a panorama taken by Surveyor 6 after its 'hop', showing the original imprints and the erosional effects of firing the verniers. (Courtesy of Philip J. Stooke, adapted from *International Atlas of Lunar Exploration*, 2007)

thrust within a foot of the ground. Nevertheless, there was no evidence of explosive cratering – the surface was sufficiently cohesive to resist bearing capacity failure at the imparted gas pressure. Furthermore, although in the case of Surveyor 6 the pressure of the gas on the surface from firing the verniers was *thrice* that of the static test by Surveyor 5 and the higher pressure would have increased the diffusion into the surface, the rate at which the gas pressure on the surface declined as the vehicle rose was sufficiently slow to inhibit the gas diffused into the surface from escaping violently, with the result that the gas diffusion erosion was no worse than the static test. However, viscous erosion blew dark subsurface material across the undisturbed surface, and there was a striking pattern of fine rays radiating from below where the verniers had been when they ignited. Most of the displaced material was from where the surface had been previously disturbed by the foot pads and crushable blocks. The fact that 1–2-cm fragments left dark trails as they rolled on the undisturbed surface was evidence that the lighter-toned surface material was at most several millimetres thick. Some larger fragments were ejected on ballistic trajectories. One clod of fine-grained material splattered the photometric target on omni-directional antenna boom 'B', almost obscuring its pattern.

When Surveyor 6 arrived, the magnet on foot pad no. 2 had made no contact with the surface material. No changes were observed after firing the cold-gas thruster on that leg. But when pad no. 2 came into contact with the surface following the hop it penetrated to a depth of 10 cm, bounced and came to rest about 12 cm away, thereby not only leaving an overlapping imprint for the soil mechanics team to study but also finally giving the magnet scientists a coating of material to examine. The horizontal displacement from the hop also provided the camera with a baseline for stereoscopic

Two views of the photometric calibration chart on omni-directional boom 'B' of Surveyor 6, showing the lunar surface material which coated it during the 'hop'.

imaging. Later photogrammetric analysis enabled an extremely detailed topographic map to be produced extending out about 50 metres from the lander.

After the hop, the sensor head was observed to have come to rest upside down! It was switched on at 12:48 on 17 November and found to be too hot, so was turned off again at 12:52 and allowed to cool before undergoing a test to determine whether it could provide any worthwhile data in this orientation – it could monitor solar wind protons bathing the lunar surface, and was operated in this manner for a total of 13 hours between 18 to 20 November and 22 to 24 November, with this experiment concluding at sunset. The alpha-scattering instrument operated for a total of 108.3 hours during which it provided 59 hours of science data, but only 30.5 hours of this was of the surface material and 10 per cent of the data was rejected because it had a low signal to noise ratio – which left 27 hours of surface data for integration.

In the case of Surveyor 5, whose ad hoc descent had required the retro-rocket to operate to within 4,200 feet of the ground, it was conceivable that the aluminium abundance measured by the alpha-scattering instrument was inflated by efflux from the solid-propellant rocket motor. But Surveyor 6 jettisoned its motor at a height of 35,000 feet and measured essentially the same abundances, and this implied that the Surveyor 5 data was valid. The analyses at the two sites suggested that the elements in the lunar surface material were in the form of oxides, and formed compounds and minerals that were familiar on Earth. It was not pristine material condensed from the solar nebula. As in the case of Earth, the Moon has undergone significant chemical differentiation. Although it was concluded that the maria were of a basaltic composition, the data was insufficient to identify the particular type of basalt. The observations of the magnets on these landers were consistent with the fine-grained material being pulverised basalt with little (if any) admixed meteoritic iron.

On 19 November the oxidiser part of the vernier propulsion system developed a leak, possibly owing to the degradation of a rubber o-ring seal. This automatically opened the helium regulator to top up the pressure, which was impossible – with the result that by 25 November both the oxidiser and helium had been completely lost. This leak pre-empted a tentative plan to perform a second hop.

With sunset imminent, the lander recharged its battery to sustain itself through the lunar night. At 16:08 on 22 November the shock absorbers of the legs were locked in order to preclude the deflections suffered by Surveyor 5 when its unlocked legs relaxed upon being chilled.

Sunset was at 13:53 on 24 November. Over the next 6 hours, pictures were taken using the polarising filters to study the solar corona. Between 16:23 and 16:50, and between 19:05 and 19:28, pictures were also taken of foot pad no. 2 illuminated by Earthshine. At 19:03, at the start of the final 10-minute corona exposure, the upper limb of the Sun was about 10 solar radii below the horizon. Camera activity ended at 20:04. It sent some 14,500 pictures prior to the liftoff and translation experiment and by the time it was switched off it had provided a total of 29,952 pictures. The final data from the alpha-scattering instrument on the protons impinging on the Moon was obtained 4 hours after sunset. Temperature monitoring was concluded at 06:41 on 26 November, after 41 hours – it had been hoped to obtain 130 hours of such

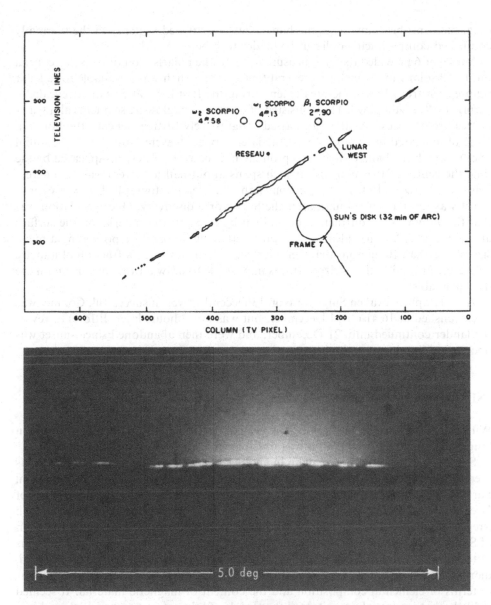

A picture of the 'horizon glow' phenomenon taken by Surveyor 6 at 14:25 GMT on 24 November 1967, about half an hour after sunset. The sketch shows the position of the solar disk in relation to the horizon and the 'beads' at that time. The position of the Sun was determined in relation to the marked stars, the magnitudes of which are indicated beside the circles. The grid coordinates are relative to the digital frame. The diffuse glow is the solar corona.

data, but a problem involving the bimetallically activated switches in the thermally controlled compartments obliged the lander to hibernate early.

Surveyor 6 provided the first measurements of the polarisation of the solar corona out to 30 solar radii, which was several times further than was attainable for a solar eclipse seen from Earth. Pictures taken during the first hour after sunset revealed a surprising 'horizon glow'. This consisted of a number of glowing segments along a 5-degree arc due west. As the Sun passed progressively further beneath the horizon, these disappeared in groups. Whilst the later exposures were longer than the initial ones in which the band of light was prominent, it had completely disappeared by the time the centre of the solar disk (which spans about half a degree) was 1.2 degrees below the horizon. In fact, this phenomenon had been photographed by Surveyor 1, but it was not recognised until after the Surveyor 6 discovery. One speculation was that the glow was the diffraction of sunlight by fine-grained material on the surface at the horizon. Another idea was forward scattering by particles possessing a mean size of less than 10 microns that were electrostatically levitated a fraction of a metre above the ground at the horizon. It was impossible to draw a firm conclusion on the data available.

An attempt to awaken Surveyor 6 on 13 December was unsuccessful. Contact was re-established at 16:41 on 14 December, but was lost 3 hours later. Efforts to revive the lander continued until 21 December, and were then abandoned since sunset was once again imminent.

INFERENCES ABOUT THE MARIA

With four mare sites in the Apollo zone visited, distributed more or less uniformly in longitude from 23°E to 43°W, it was possible to draw some generalisations.

Surveyor 1 inspected a level plain in an ancient 100-km-diameter crater known as the Flamsteed Ring that had been 'inundated' in some way by Oceanus Procellarum; Surveyor 3 landed in a subdued medium-sized crater situated on the open plain of Oceanus Procellarum; Surveyor 5 provided a detailed inspection of a very small irregularly shaped crater in Mare Tranquillitatis; and Surveyor 6 inspected the plain of Sinus Medii within sight of a mare ridge. All four sites were very similar in terms of topography, and in terms of the structure of the surface layer and its mechanical, thermal and electrical properties; and the surfaces at the latter two sites were similar in terms of elemental composition and the content of magnetic material. It seemed unlikely that terrestrial sites situated thousands of kilometres apart and selected in a manner similar to that by which the lunar targets were chosen would prove to be so similar.

At all sites, the undisturbed fine-grained surface material was lighter toned than the subsurface. This difference was as much as one-third for Surveyors 1, 3 and 6, but less for Surveyor 5. The fact that the albedo of the subsurface was the same at all sites meant the exceptional case of Surveyor 5 was due to the surface material being less bright. Observations of the erosion of the fine-grained surficial material by the vernier efflux during the 'hop' performed by Surveyor 6 and of the tracks left by the fragments that were rolled across the surface, indicated that the bright surficial layer was limited to the uppermost few millimetres. The existence of such a well-defined 'contact' in a nominally undisturbed surface at four widely spaced sites on the maria implied the action of a process (or combination of processes) which had the effect of increasing the albedo of the material at the surface, for otherwise such a fine layer would be destroyed by the gardening of meteoritic bombardment. Furthermore, the fact that the material at all depths below the surface was uniformly dark, as opposed to there being a gradation in albedo, indicated that whenever an impact mixed the lightened surficial material into the subsurface, it became dark. Perhaps the process which altered exposed material had not had long to act on the material in the small fresh-looking crater in which Surveyor 5 landed. At all sites, the bright rounded rock fragments visible on the surface had textures featuring knobs and pits, whereas these were absent on the highly angular faceted blocks. This hinted that the process which produced the rounding – undoubtedly the relentless meteoroid bombardment – also gave rise to the 'worn' texture.

At all sites the fine-grained material was cohesive, and whilst the surficial layer was mildly compressible, its bearing strength increased rapidly with depth. But there was no observable variation in grain size with depth – evidently it was simply a case of the porosity decreasing with depth. It was estimated that the bulk density of the upper centimetre of undisturbed material was in the range 0.7 to 1.2 g/cm^3, and that by a depth of several centimetres this had increased to 1.6 g/cm^3.

The size-frequency distribution of small craters at all sites matched that expected for a steady-state population resulting from the protracted bombardment of primary meteoroids and the fall of ejecta from such impacts. Furthermore, this distribution was independent of individual differences in the mare surfaces and of the population of craters larger than several hundred metres in size.

The thickness of the fragmental debris layer on the mare plains was clearly related to the abundance of craters with diameters ranging between 1 and 10 km. In the part of Mare Tranquillitatis where Surveyor 5 landed the size-frequency distribution of such craters was twice that of the Oceanus Procellarum inundation of the Flamsteed Ring where Surveyor 1 landed, and the minimum size of the blocky rimmed craters at those sites indicated that the fragmental debris layer was several times thicker in Mare Tranquillitatis than in Oceanus Procellarum. Of all the maria, Sinus Medii had one of the highest size-frequency distributions of craters with diameters larger than several hundred metres, and the fragmental debris layer on the plain near Surveyor 6 was thicker to match. The fact that the cratering indicated the surface of Sinus Medii to be older than the other maria was evidence that the older the surface the thicker its fragmental debris layer. The size distribution of the material on the surface was also related to the thickness of the fragmental debris layer. When the mare lava flow was fresh and its rocky surface was exposed, small impacts were able to excavate it. As a layer of fragmental debris accumulated, it took larger and larger impacts to reach the substrate. Over time, the loose fragments were both reduced in size and increased in number. The trend was therefore towards a thickening layer of ever finer fragments. That is, the regolith 'matured'.

The implication for Apollo was that an older surface would be a safer landing site.

When viewed from afar, an older surface might *look* rough by virtue of having large craters with blocky rims, but since only large craters would be able to excavate the substrate this meant that the plethora of small craters (which must be present) would not possess blocky rims. The task for the site selectors was therefore to measure the smallest craters with blocky rims on a mare surface to measure the thickness of the fragmental debris layer, and then seek a flat patch of open ground situated between such craters where it was likely to be relatively free of blocks.

THE SOUTHERN HIGHLANDS

The successful mission of Surveyor 6 completed *all* requirements established for the project in direct support of Apollo landing site selection. Nevertheless, the Office of Space Sciences and Applications decided to fly the final mission. The target was hotly contested. It was decided that the most important objective was to investigate a site as different as possible from the maria already visited, and preferably a site that offered the greatest likelihood of being different in terms of geology and chemistry. Harold Urey had proposed that the Moon was 'pristine' material condensed from the solar nebula and therefore must have an ultrabasic composition, but several lines of evidence implied that the maria were volcanic lava of a basaltic composition. H.H. Nininger proposed in 1936 that 'tektites' were ejected by impacts on the Moon and had acquired their aerodynamic shape and a glassy skin during hypersonic entry into the Earth's atmosphere. But their acidic composition posed a problem. The viscosity of an acidic magma such as granite is several orders of magnitude greater than that of basalt. If the tektites originated from the Moon, they must therefore represent the highlands.[3]

The 85-km-diameter crater Tycho in the southern highlands was widely believed to be the result of a hypervelocity impact but Jack Green thought it was a volcanic caldera, and there were a variety of intermediate theories speculating that an impact promoted volcanism. Because the prominent system of bright rays indicated it to be the youngest crater of its size on the near-side of the Moon, it ought to be relatively uncontaminated by ejecta from elsewhere. Infrared observations made during a lunar eclipse on 19 December 1964 had shown Tycho to be one of the most striking of the thermal 'anomalies', implying that there would be lots of rocks on the surface out to a distance of one crater's diameter beyond the rim crest. It was also bright at radar wavelengths, which also implied rockiness. Lunar Orbiters 4 and 5 had photographed the crater and its immediate environs from an overhead perspective at high resolution. It was decided to aim for a point 30 km north of the rim crest. But a

[3]   Although NASA was unaware of it, a gamma-ray spectrometer operated in lunar orbit by Luna 10 in 1966 had provided a rudimentary analysis of the composition of the lunar surface across a wide range of latitudes, and the results showed there to be no significant exposures of acidic rock in the highlands.

The crater Tycho in a portion of frame H-119 taken by Lunar Orbiter 4 during its mapping mission.

'target' was a circle within which the vehicle had a given probability of landing, and in such rough terrain it proved necessary to trim the diameter of the circle from 60 km to 20 km, which offered barely 10 per cent of the area and would require an extremely accurate trajectory.

It was decided to select a backup target at a similar longitude in order to have the same illumination for surface operations, and make the choice of which target to aim for in-flight by the accuracy of the translunar injection. If the trajectory was unlikely to yield the accuracy required for Tycho, the spacecraft would be diverted to a site in the Fra Mauro Formation. Although this blanket of ejecta lying peripheral to the Imbrium basin would be hummocky and heavily cratered, it ought to be much less demanding than Tycho, and a landing there offered the prospect of determining the composition of material excavated from many tens of kilometres beneath the surface – perhaps even subcrustal material whose chemistry would provide valuable insight into the interior of the Moon. But because the Fra Mauro Formation was 'ancient', it was probably contaminated by ejecta from elsewhere. The target was at 13°W, 5°S, just northeast of the 95-km-diameter crater Fra Mauro, and the target circle was the usual 60 km in diameter.

Surveyor 7 was launched at 06:30:01 GMT from Pad 36A on 7 January 1968. For the first time in the series, Gene Shoemaker went to watch the launch. The Centaur achieved orbit at 06:39:54. Owing to the predawn launch, the vehicle emerged from the Earth's shadow at 06:50:19. It reignited at 07:02:15, shut down at 07:04:15, and released the spacecraft at 07:05:16. Because the decision about the target was to be delayed until after the performance of the Centaur had been ascertained, the nominal aim point for the translunar injection was Hipparchus, near the centre of the disk. It was decided to aim for Tycho. The midcourse manoeuvre at 23:30:09 on 7 January lasted 11.4 seconds, and the change in velocity of 36.4 ft/sec placed the interception point within the target circle. The trajectory was so accurate that an optional second refinement was cancelled.

A study of Tycho and its immediate environs had been made using the medium-resolution pictures taken by Lunar Orbiter 5. On average the rim crest stood 2.5 km above the surrounding highlands, but this was difficult to specify since there was no level plain for reference. The floor of the crater was 4.5 km below the rim crest. The prominent central peak rose over 2 km above the floor, and had hills nestled close alongside it. The theory of impact crater formation implied the central peak complex was a mass of rock thrust up from a great depth by the 'rebound' in the final stage of the process. The interior wall was a series of terraces produced when large blocks of material slumped on steeply inclined concentric faults. The high-resolution pictures from Lunar Orbiter 5 revealed the presence of flow features in low-lying areas of the wall terracing, and on the crater's floor. The exterior was an annular belt 80–100 km wide that could readily be subdivided by albedo and texture into several geological facies.[4]

---

[4] The term 'facies' was introduced to geology in 1838 by the Swiss stratigrapher Amanz Gressly to specify a body of rock having given characteristics.

The innermost of the concentric rings extended from near the crest of the rim out to about 10–15 km, was asymmetric, widest on the northern side of the crater, and comprised irregular hills and intervening depressions which presented a hummocky texture. It contained many well-developed flow features, some as long as 8 km. The second ring, extending from 15 km out to 35–40 km, comprised subradial ridges and valleys, with the ridges typically 2–5 km in length and 0.5–1 km in width, etched on broad undulations 5–20 km across that were recognisable as craters swamped by the ejecta from Tycho. The inner ring had an albedo of 16–17 per cent, and the second was darker at 13–14 per cent.[5] Most parts of the rim and the inner two rings were broken by closely spaced radial, arcuate and circumferential faults. Displacements on the radial faults had produced many small radial ridges and troughs interpreted as horsts and grabens respectively. Next was a ring of closely spaced craters ranging from one to several kilometres in diameter that were made by the fall of individual blocks of ejecta from Tycho. Beyond, out about as far as 100 km – a little over one crater's diameter – the ejecta was discontinuous and transitioned into the system of rays composed of smaller craters which were much less closely spaced. The pictures from Ranger 7 showed a ray from Tycho crossing Mare Nubium to comprise craters ranging in size from 100 metres to 1 km. The thickness of the ejecta was expected to range from several hundred metres near the rim crest, where the 'hinge flap' placed the material excavated from the deepest point, to only a few metres in the peripheral zone. At Surveyor 7's target, on the second ring, the ejecta was expected to average several tens of metres in thickness.

The pre-retro manoeuvre in which the spacecraft departed from its cruise attitude involved starting a roll of +80.5 degrees at 00:27:17 on 10 January, a yaw of +96.1 degrees at 00:35:52 and a roll of –16.5 degrees at 00:41:09. The initial approach was at 34.8 degrees to the local vertical. The altitude marking radar was enabled at 01:00:33.7, and it issued its 100-km slant-range mark at 01:02:11.892. The delay to the initiation of the braking manoeuvre was specified as 2.775 seconds.

The verniers ignited precisely on time, and the retro-rocket 1.1 seconds later. At that time the vehicle was travelling at 8,580 ft/sec. The RADVS was activated at 01:02:15.752. The acceleration switch noted the peak thrust of 9,200 pounds fall to 3,500 pounds at 01:02:58.973, giving a burn duration of 42.9 seconds. The verniers were throttled up to their maximum thrust at 01:03:09.250 for 2 seconds while the motor was jettisoned. At burnout, the angle between the vehicle's thrust vector and velocity vector was 19 degrees. The RADVS-controlled phase of the flight began at 01:03:13.090, when the slant range was 51,259 feet (and because the velocity vector at burnout was offset to vertical, the altitude was 41,510 feet) and the total velocity was 452 ft/sec (and since the vehicle had maintained its thrust along the velocity vector extant at the time of retro ignition, the longitudinal rate was 428 ft/sec). The vehicle immediately aligned the thrust axis along the velocity vector extant at retro burnout

---

[5]   Although described here as 'darker', the outer ring of ejecta surrounding Tycho was still much brighter than the average lunar albedo of 8 per cent.

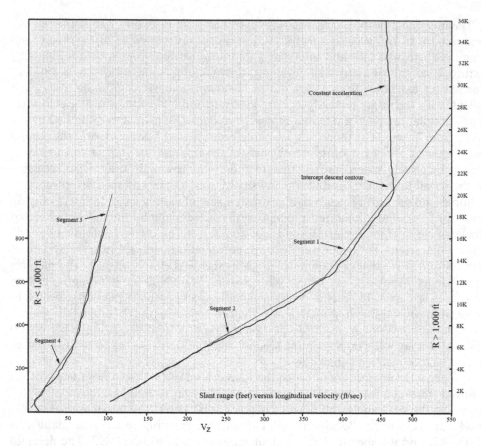

The descent of the Surveyor 7 spacecraft depicted in two sections, one for slant ranges above 1,000 feet and the other below 1,000 feet.

and flew with the verniers at 0.9 lunar gravity, very slowly accelerating as it descended. When the altimeter locked on at 01:03:17.649, at a slant range of 41,673 feet, attitude control was switched from inertial to radar and the thrust axis was swung in line with the instantaneous velocity vector to initiate the gravity turn. On intercepting the 'descent contour' at 01:04:03.018, the slant range was 20,246 feet and the speed was 464 ft/sec. By the 1,000-foot mark at 01:05:13.285, the vehicle was descending very nearly vertically at 102.5 ft/sec. The 10-ft/sec mark was issued at 01:05:30.184 at a height of 46 feet.

The verniers were cut off at 01:05:36.284, and after falling freely the vehicle touched down at 01:05:37.620 with a vertical rate of 12.5 ft/sec and a lateral rate of 0.3 ft/sec. Leg no. 1 was the first to make contact, followed rapidly by legs no. 2 and 3 in that order. There had been a fair chance that the vehicle would be disabled on trying to land in such rough terrain, so its survival gave rise to wild applause in the Space Flight Operations Facility.

The camera was of the type introduced by Surveyor 6 – the hood was of the boxy configuration, the elevation range of the mirror was 70 degrees, and it had polarising filters. The first 200-line picture was sent at 01:47. After 15 pictures had been taken for a preliminary study, the solar panel and high-gain antenna scanned for the Sun and Earth respectively, locking on by 03:21. The first 600-line picture was taken at 03:42. The foot pads had displaced as the legs rebounded on contact, but overlapped their original imprints. Pad no. 2 had nudged aside a rock that was about 18 cm long and at least 10 cm high. Pad no. 3 landed partially on top of a semi-buried rock, and in the process suffered localised deformation and tearing. The pads had penetrated to a depth of 4 cm and displaced clods to a radius of about 40 cm, but there was barely any lunar material on their upper surfaces. As on the maria, the disturbed material was darker than the undisturbed surface. The orientation of the lander put the camera on the north-facing side. To the east, south and west the horizon was less than 200 metres away, but because the local surface sloped down to the north the view in that direction was spectacular, with a succession of ridges on the horizon. Despite being the roughest-looking target to date, the landscape still bore little resemblance to the depictions of the lunar surface in contemporary popular fiction. The slope on which the lander stood was about 3 degrees. Most of the landscape on view was no steeper than 10 degrees. The steepest flank of a ridge on the horizon was 34 degrees, and the summit was rounded.

It turned out that Surveyor 7's trajectory was very accurate and it landed a mere 2.5 km from the aim point. The coordinates were difficult to determine because the selenographic grid presumed the Moon to be spherical, which was not the case. For points above the mean sphere and situated some distance from the centre of the lunar disk, the measured coordinates were greater than the actual coordinates. Also, at this location the latitudinal circles were significantly curved. This complicated the drawing of a local grid. Instead, features on frame M-128 from Lunar Orbiter 5 were identified on a picture taken in 1919 by the 100-inch telescope at the Mount Wilson Observatory, and the site pin-pointed on M-128 was transferred first to the telescopic picture and then to the coordinate system of the relevant sheet of the *Orthographic Atlas of the Moon* (based on that picture) which had been issued by D.W.G. Arthur and E.A. Whitaker in 1961 as a supplement to the *Photographic Lunar Atlas* produced by Gerard Kuiper.

A number of geological units were identified in the high-resolution pictures taken by Lunar Orbiter 5. The most widespread unit was described as 'patterned debris'. This was the major unit of the second ring of ejecta deposits. The size-frequency distribution of craters exceeding 8 metres in diameter was the highest of all the units in the ejecta debris.

To the lander, the patterned debris extended several tens of kilometres to the west, north and northeast, and the large craters had raised rims which, in many cases, were relatively smooth. This was well demonstrated by a crater about 650 metres to the north that was 60 metres in diameter. Although it was 10 metres deep, its rim was smooth. It was apparent that the patterned debris was a blanket of unconsolidated material. The crater contained a few large blocks, but they were no more numerous than were lying around between the craters. The absence of strewn fields associated with craters on the patterned debris indicated it to be at least 20 metres thick, which

The northward-looking portion of a panorama taken by Surveyor 7. The outline shows the area covered by the next illustration. (Courtesy of Philip J. Stooke, adapted from *International Atlas of Lunar Exploration*, 2007)

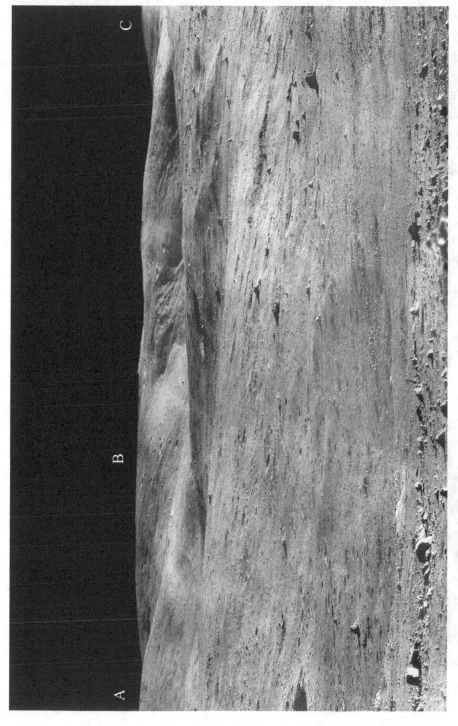

Three ridges at successively greater distances to the north of the Surveyor 7 lander. The letters are explained in a subsequent Lunar Orbiter illustration. (Courtesy of Philip J. Stooke, adapted from *International Atlas of Lunar Exploration*, 2007)

A mosaic derived from frames M-125 and M-128 taken by Lunar Orbiter 5 of the crater Tycho. The outline shows the area covered by the next illustration.

Frame M-126 by Lunar Orbiter 5 showing the ejecta to the north of Tycho's rim. The outline shows the area covered by the next illustration. The dot indicates where Surveyor 7 landed, and the letters identify the ridges observed from the surface.

Landing point

A portion of frame H-128 by Lunar Orbiter 5 showing where Surveyor 7 landed.

A preliminary geological map of the immediate vicinity of the Surveyor 7 landing site, produced by E.M. Shoemaker and E.C. Morris of the US Geological Survey.

was consistent with expectation. Surveyor 7 actually set down about 50 metres from the western margin of a 'patterned flow', beyond which was the dominant patterned debris. There was no relief at the contact between the two – the difference evident in overhead imagery was only a distinction in the surface texture. The patterned flow extended to the north and east, but most of it was to the south of the lander and hence beyond the near horizon, which was only a few hundred metres off. In the Lunar Orbiter pictures, the surface of the flow comprised low hills and depressions ranging up to several hundred metres across. But superimposed on these broad irregularities was a pattern of north-trending low ridges and grooves similar to those on the patterned debris (but less well defined) with swarms of fissures running along the ridges which suggested they had undergone slumping. There were a great variety of rock fragments in the vicinity of the lander, ranging up to 1 metre in size. Two craters to the southwest of the lander – one 20 metres in diameter and the other 30 metres – had rims littered by coarse blocks up to 75 cm in size. They were on the patterned flow, and had excavated these rocks. There was an irregular crater about 3 metres in diameter 5 metres north of the lander that contained coarse blocks up to 60 cm in size and issued a strewn field extending to the northwest, but this crater was undoubtedly made by the fall of ejecta from another event, and the blocks were the debris of the secondary projectile.

The lander could also observe an area of 'smooth patch material' to the northeast. Such units occurred in enclosed depressions several hundred metres across and – in addition to being smooth – were relatively dark. The fact that the smallest crater on this material to have a blocky rim was only 5 metres in diameter indicated a source of rocks at a depth of about 2 metres. As this particular smooth patch material was superimposed on the patterned flow, the craters had undoubtedly punched through to the patterned flow. This in turn indicated that this particular spot of smooth patch material was a thin veneer. On all three types of terrain, most of the craters with diameters in the range 8 to 16 metres were elongated with their major axes radial to Tycho.

The principal difference between the patterned debris and the patterned flow appeared to be that whereas the patterned debris material rapidly settled on being ballistically deposited, the patterned flow gained its distinctive texture by flowing for distances ranging between several tens and several hundreds of metres.

In the immediate vicinity of Surveyor 7, small craters were as abundant as on the maria, but the size-frequency distribution of those exceeding 10 metres across was significantly less. There was a greater variety of rock types than at any of the maria sites, and they varied in albedo up to 22 per cent. Some blocks were plain, but others were spotted. The spots were of various sizes, had irregular margins, and appeared to be surface protrusions. One particularly striking rock 2 metres away had spots that ranged in size from less than 1 mm to about 30 mm and covered about 30 per cent of the visible face. It was speculated that the spots were fragments of light-toned rock assimilated into a dark matrix – a mechanically assembled conglomerate known as a breccia. Some rocks had well-developed linear structures, and others appeared to be vesicular – both of which were suggestive of lava.

The maria were lava flows that solidified as coherent rock and were subsequently progressively pulverised by meteoroid bombardment to accumulate a regolith which matured over time into ever finer fragments. Although the excavation of Tycho laid down a blanket of ejecta, it did so essentially instantaneously. Such material would contain blocks of all sizes with a size-frequency distribution different to a regolith. Some blocks would have come to rest on the surface as the ejecta was laid down, but most would have been buried. Some would later be excavated and tossed around. If (as the superposition relationships indicated) Tycho was formed recently, then there could not have been much time for the rain of meteoroids to produce a true regolith on top of the ejecta blanket. A theoretical study which incorporated all that had been learned to date about the rates of small impactors, predicted that the regolith at this site ought to be about 10 cm thick on average.

In early 1967 the plan was to fly the soil mechanics surface sampler on Surveyors 3 and 4, and then the alpha-scattering instrument on the remaining missions. But it had been decided that Surveyor 7 should have both. The intention was to conduct an elemental analysis of undisturbed surface, then activate the sampler and use this to reposition the sensor head to analyse subsurface material excavated by the arm. The alpha-scattering instrument was powered up at 09:28 on 10 January. The

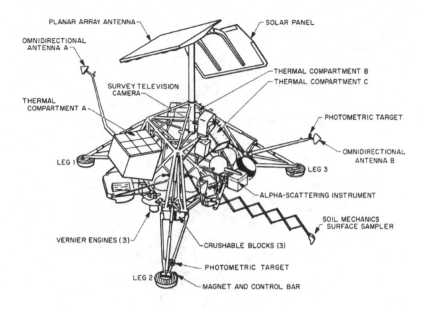

The configuration of the Surveyor 7 lander.

standard sample was measured between 09:28 and 15:29, yielding 5.2 hours of data. At 15:49 the standard sample was removed to enable the head to measure the background, and 4.8 hours of data was obtained between 16:13 and 21:59. The auxiliary mirror on leg no. 1 orientated to provide the camera with a line of sight beneath the vehicle in order to investigate where the head of the alpha-scattering instrument would take its first sample showed a uniform grey; evidently during the landing it had been completely coated with fine-grained material! There was a partial coating on the mirror for viewing crushable block no. 2, but it was still possible to establish that there was an imprint. The area beneath block no. 3 was in the lander's shadow for most of the time. The command to release the head was issued at 22:01 on 10 January. The instrument's counts were monitored in real-time for 6 minutes, awaiting the increase that would confirm that the head was on the surface; but there was no change. The command was reissued at 22:09 and the counts monitored for 10 minutes – again with no increase. A series of pictures taken between 22:41 and 23:44 of the head and the deployment mechanism showed that the squib had fired, the pin had been pulled and the door of the cable compartment was open. However, the head had not moved. It was concluded that the escapement mechanism for the nylon cord had either failed to function or the cord had become stuck in the mechanism. It was decided to activate the arm and try to use this to lower the sensor head.

The soil mechanics surface sampler was powered up at 01:00 on 11 January, then exercised to verify its functionality. To enable the arm to manipulate the head of the alpha-scattering instrument, its mounting structure had been redesigned to draw the arc of its azimuth range to the left; but this was at the expense of the other end of its range, with the result that it could no longer reach foot pad no. 2. On Surveyor 3, an

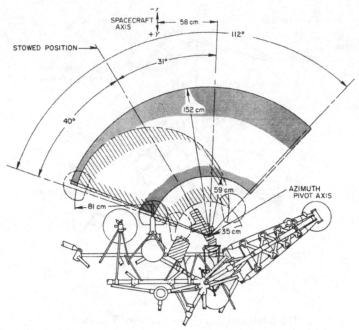

The operating area of the soil mechanics surface sampler carried by Surveyor 7, and the extent to which it could reposition the alpha-scattering instrument.

attempt had been made to measure the force that was applied by the arm but the time-resolution of the motor current telemetry had been insufficient to provide an accurate measure. Originally, the motor current readout had allowed a maximum of eight current samples for a 2-second motor actuation. The upgraded system provided samples at 50-millisecond intervals. In a static bearing test the arm would lower the flat face of the scoop's door over the selected spot using a series of either 0.5 or 2.0-second commands until the elevation motor stalled, indicating the force against the surface. When driving the scoop into the surface with its door open for trenching, it would be possible to measure bearing strength as a function of depth.

After the arm had made two bearing tests well clear of where the alpha-scattering instrument was to sample – in the process obtaining readings very similar to those at the Surveyor 3 site, where an identical test had been made – the arm began its efforts to deploy the balky instrument. Between 07:23 and 08:09 the arm was manipulated until it rested its scoop on the base plate of the head, and then from 08:15 to 08:50 it applied a series of light taps to the plate. It had been hoped that this would cause the head to unreel all the way down to the surface, but TV pictures showed that it swayed on its cable without unreeling. It had not, in fact, been possible to impart much downward force because with the head dangling freely on its cord a force applied to one side of the plate simply caused the head to tilt and swing away. Further pictures were taken between 23:14 and 23:41 to inspect the escapement mechanism.

On 12 January the arm began its day with another three bearing tests, and then manoeuvred to pick up a rounded rock (Rock 'A') that was about 5 cm in size, with

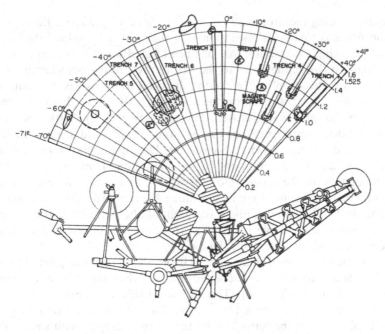

The actions of Surveyor 7's soil mechanics surface sampler, indicating the three points sampled by the alpha-scattering instrument.

Surveyor 7's soil mechanics surface sampler at work on 21 January 1968.

the motor current being monitored to estimate the mass of the rock. Between 05:48 and 07:33 the arm was first positioned alongside the right-hand side of the head and then manipulated using a procedure that had been tested on an engineering model to jam the head against the helium tank to the left, to prevent it from swinging when a downward force was then applied. In driving the head down by several centimetres, sufficient room was gained to position the scoop directly above the head, near the eye-bolt to which the cable was attached. This was achieved by 08:01, and between 09:11 and 10:05 a downward force was applied to drive the head against the tension of the fouled cord all the way down to the surface. During the hiatus, the instrument had obtained an additional 7.2 hours of background data. There were a number of fairly large fragments on the surface close to where the head deployed, and in fact not only was there a rock some 4 cm in size under the inboard side of the base plate that was tilting the head, there was also a fragment of about 1.5 cm in the aperture at the centre. The analysis of sample 1 started at 16:42 on 12 January, and 6.2 hours of data had been acquired by the end of the day.

On 13 January the arm made an attempt to lift Rock 'B', but this proved to be just out of reach. Instead, Rock 'C' was scooped up, but when the arm was elevated to weigh the rock it fell out of the scoop. Trench 1 was made at the extreme right of the arm's operating area. This was foiled by a subsurface obstacle at a depth of 2.5 cm which proved able to resist the retraction motor on two scrapes, with the result that the penetration was limited to 5 cm. This obstacle proved to be a buried rock with an irregular upper surface. The alpha-scattering instrument had to be switched off early on 13 January when the head exceeded its maximum operating temperature of 50°C, but data-taking resumed at 10:55 after the arm positioned its scoop to shade the head and allow it to cool off. There was a hiatus between 20:21 and 22:36 while the scoop again provided shade, and operations ended at 23:33. The arm devoted the whole of 14 January to Rock 'D'. At first sight this appeared as a rounded knob poking out of the surface and sufficiently small to fit into the scoop for weighing, but when it was excavated it proved to be too large. The angular shape of the buried portion of the rock was an indication of the efficiency of the erosional process that had rounded off its exposed face.

Some alpha-scattering data was obtained on 14 January, but for most of the time the head exceeded its maximum operating temperature. When the lander arrived, 30 hours after sunrise, the Sun was at an elevation of 13 degrees. At the equatorial sites visited by previous missions the Sun passed close to the zenith at noon, but at this site its maximum elevation was 48 degrees above the northern horizon. A pre-flight study indicated that the surface temperature at noon would be approximately 100°C, as opposed to 127°C on the equator, but the analysis presumed a smooth surface and did not take account of local influences such as slopes, craters and blocks that could not be predicted in detail. Although the incident heat would be less harsh at this site, the fact that the Sun did not pass near the zenith meant that the shadow of the mast-mounted panels never fell on the vehicle. As a result, the scientific payload, which faced north, grew hotter during the mid-morning, noon and mid-afternoon than at any previous site, and both the head of the alpha-scattering instrument and the electronics of the sampling arm exceeded their maximum operating temperatures.

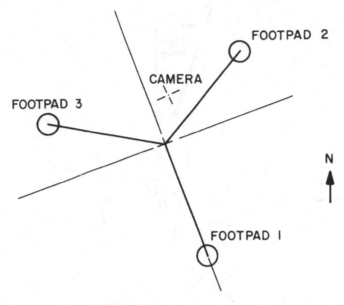

Owing to the orientation in which Surveyor 7 came to rest, the camera, the alpha-scattering instrument and the soil mechanics surface sampler were all exposed to the Sun as it tracked across the northern sky.

For the three days during the mid-day period, therefore, the arm only manoeuvred to maintain the shadow of its scoop on the head in an effort to keep this below its 'survival limit' of 75°C – which was only just achieved. The thermal stress on the camera was worse than on earlier missions, as at noon the Sun shone on the side of the cylinder rather than down upon its top. It was rotated in azimuth away from the Sun to minimise the solar absorption by the black interior of the hood. During the 5-day period near local noon its use was severely restricted – at worst to only 5 minutes per hour.

Arm work resumed on 19 January by re-weighing Rock 'A'. A flat 22×60-cm mirror just below the upper collar of the mast allowed the camera to view a roughly triangular area of about 0.25 square metres lying between 1.7 and 3 metres from the camera, a portion of which was accessible to the sampling arm. Carrying the rock, the arm swung into the field of view of the mirror and stereoscopic pictures were taken to enable the volume of the rock to be estimated and, in turn, the density. With a density in the range 2.4 to 3.1 g/cm^3 – the most likely value being in the range 2.8 to 2.9 g/cm^3 – the rock could not be very porous. The arm was then elevated to drop the rock from a height of about 60 cm – after making a small indent in the surface, it bounced or rolled about 12 cm. The strength of this rock was tested by having the scoop squeeze it, but it did not break. The arm then retrieved the rock and deposited it closer to the head of the alpha-scattering instrument – in preparation for possibly placing the head on top of the rock to determine its elemental abundance and further characterise it.

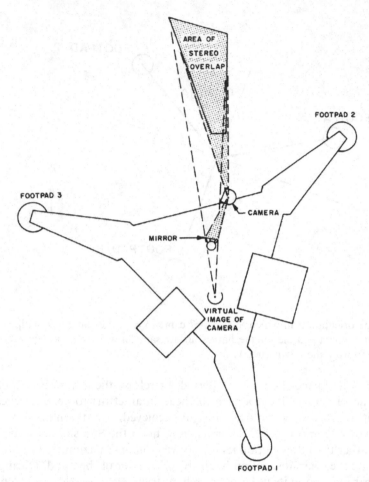

For Surveyor 7, the surface in the stereoscopic overlap area was viewed directly by the camera and indirectly from a position corresponding to the virtual image of the camera in the mirror mounted on the mast.

On 20 January the arm made trench 2 using four scrapes. A buried obstacle near the start of the trench deflected the scoop to the left. Once finished, the trench was 75 cm long and varied in depth between 15 and 18 centimetres. Trench 3 was placed close alongside as a single scrape. In scraping trench 4, the scoop's door was kept closed in order to measure the ability of the surface to resist lateral force. In addition to having magnets on both foot pads no. 2 and 3, Surveyor 7 had a pair of horseshoe magnets embedded in the door of the scoop. In fact, the pads penetrated only 6 cm, which was insufficient to bring the magnets on their sides into contact with the lunar material. However, the magnets on the arm were able to test selected spots. As these magnets repeatedly came into contact with the lunar surface they attracted material with a granularity finer than the resolution of the camera, which was about 1 mm at that distance. As an experiment, the scoop was nudged against a chip of rock about

1.2 cm in size that had a smooth shape, a low albedo and a lustre. When the scoop was raised, the rock clung to the magnets on the door. Prior to being disturbed, this rock was partially embedded, with no evidence on the surface of it having landed or rolled into position. The fact that its exposed surface was much darker than most of the rocks in the arm's operating area implied that it was atypical. Its strong magnetic susceptibility would have been consistent with a nickel-iron meteorite or a material rich in magnetite – it was not possible to say which (if either). Unfortunately, as the arm was manoeuvred to facilitate stereoscopic photography the acceleration caused the fragment to fall off.

The alpha-scattering instrument was reactivated at 21:14 on 20 January and found to have cooled down to 40°C, so data-taking was resumed. By 07:21 on 21 January it had accumulated a total 27.3 hours of data for its first sample. With two days remaining to sunset, it was decided to reposition the head. On previous missions the cord was attached to an eye-bolt installed directly on the head's upper surface, but this time it was attached to a knob that had been designed to enable the arm to grasp the head. On 21 January, the scoop lifted the head. The motor current was monitored in order to provide a calibration check of this method of weighing rocks. The plan to analyse Rock 'A' had been rejected in favour of an undisturbed rock – the one selected was adjacent to Rock 'D'. It was lighter in tone than its surroundings and about 5 × 7 cm in size. In manoeuvring the head the arm caused the rock to shake, which indicated that it was sitting on the surface rather than embedded in it. The arm engineers were delighted with the precision of its movements. At 12:30 the head was left perched on the rock. In fact, part of the rock protruded through the aperture into the cavity of the head. The instrument analysed the 'exposed' upper surface of the rock for 10 hours. Meanwhile, the arm scraped first trench 5 with the door open and then trench 6 with the door closed. On 22 January the arm excavated trench 7 in between the trenches scraped the previous day – again with the door closed. With three trenches in close proximity, there was a significant amount of debris piled up at their inner ends. The alpha-scattering instrument was moved onto this disturbed material and 6.7 hours of data was obtained between 12:06 on 22 January and 14:40 on 23 January, which was some 8.5 hours after sunset, and by the time it was switched off at 15:36 the head of the instrument had chilled down to –20°C and the associated electronics to –50°C.

While the alpha-scattering instrument was performing its third analysis, the arm made a series of bearing tests – one of which was adjacent to trench 2, with pictures being taken to look for any evidence of the wall of the trench collapsing in response to the pressure. Then it was decided to attempt to crack open a rock to obtain a fresh surface which could be photographed using the polarising filters to seek insight into its structure. The arm positioned the scoop directly above Rock 'E', raised the scoop to a height of about 35 centimetres and let it fall with its door open on the rock. The rock was about 5 cm in size, and the impact broke off a small piece – either because it was intrinsically weak or because it contained a fracture. While scraping trench 1, the arm had been fouled by a buried rock, so as a finale the scoop was lowered back into this trench and the arm retracted to re-engage the obstacle. Then after sunset on

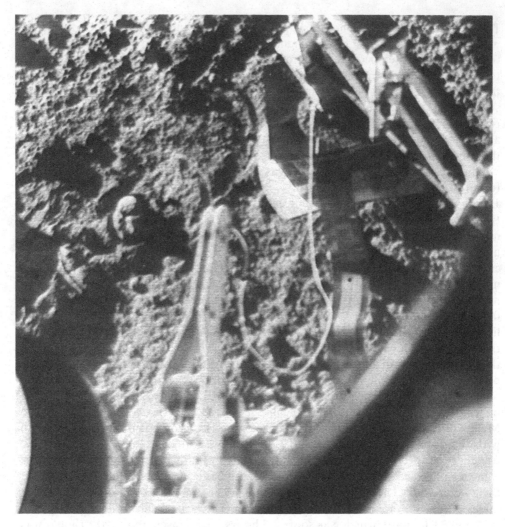

A picture taken at 11:21 GMT on 22 January 1968 showing Surveyor 7's sampler repositioning the alpha-scattering instrument in preparation for its third sample.

23 January, exploiting the increased torque of the retraction motor as a result of it having chilled to –110°C, the arm tried to dislodge the rock. However, the fact that the applied force partially compressed the shock absorber on leg no. 2 indicated the rock to be firmly embedded! This final experiment over, the arm was switched off at 08:41 that same day.

Over a total of 36.3 hours of operation the arm made 4,397 mechanical motions. It made two impact tests, seven trenches, two of which used more than one scrape, and sixteen bearing tests, five of which were with the scoop door open. One of the trenches was directed by a command tape which took a picture between each action, and the frames were later compiled into a movie. There were fragments up to 10 cm in size within the arm's reach, most of which seemed to be dense coherent rock. The subsurface was predominantly fine-grained granular material but, as was revealed by the trenching, there were fairly large rocks at a shallow depth. As on the maria, the fine-grained material was slightly cohesive, partially compressible and consolidated with depth. The fact that the undisturbed surface was brighter than the subsurface on both the maria and in the highlands indicated that this was not simply a property of the maria but a universal weathering process. Whereas on the maria the albedo of the undisturbed surface was 7.2 to 8.2 per cent and the subsurface was 5.5 to 6.1 per cent, in this case the undisturbed surface was 13.4 per cent and the subsurface was 9.6 per cent. The fact that at this site the *subsurface* was slightly brighter than the *undisturbed* surface of the maria was probably a consequence of the highlands being generally brighter than the maria.

From the point of view of the alpha-scattering investigation, Surveyor 7 was the most productive mission, with the increase in yield being derived from collaboration with the sampler. In fact, fully 8.75 hours of the arm's operating time was devoted to deploying and subsequently tending to the alpha-scattering instrument.

The high latitude and the increased elevation range of the camera's mirror enabled narrow-angle pictures to be taken of Earth. On its first day of operation, the lander photographed Earth through polarising filters – the first time this had been done, and the highly polarised component was inferred to be specular reflection of sunlight by the ocean.

An experiment assigned to the first Apollo lunar landing mission was to emplace on the lunar surface an instrument that comprised an array of corner-cube reflectors designed to reflect a laser beam fired from Earth back towards its source, so that the intervening distance could be precisely measured to analyse the secular components of the Moon's motion.[6] To test the first stage in this process, Surveyor 7 was to take pictures of Earth while laser beams were being fired. Six sites were established, each with a laser directed through a telescope to reduce the divergence of the beam during its transit to the Moon. The first test on 14 January was a failure. After a pause while Earth and the Sun were close together in the lunar sky, the experiment resumed on 19, 20 and 21 January. The first clear detection was on 20 January, with the pictures simultaneously capturing the beams from the 60-inch McMath solar telescope of the Kitt Peak National Observatory, Tucson, Arizona, and the 24-inch telescope of the Table Mountain Observatory – the latter operated by JPL in Wrightwood, near Los Angeles, California. Although the power of the lasers was just 1 to 2 watts and they illuminated a footprint on the lunar surface that was about 3 km in diameter, which had the effect of diluting the energy across a wide area, to the lander's camera they appeared against the dark part of the crescent Earth as points of light rivalling Sirius, the brightest star in the sky.

Because the Moon rotates once in the same time as it takes to pursue one orbit of

---

[6]   The secular components (as distinct from the periodic components) of the Moon's motion are cumulative over time.

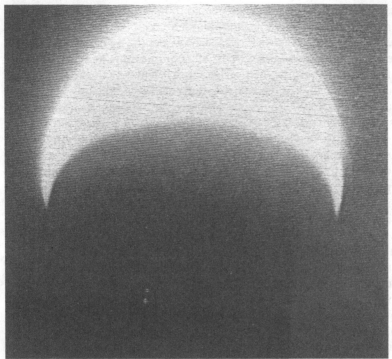

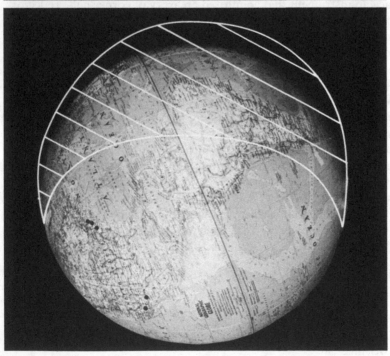

A 3-second exposure taken by Surveyor 7's camera at 09:06 GMT on 20 January 1968 recorded laser beams fired at it by two telescopes on Earth. The lasers were fired by the Kitt Peak National Observatory near Tucson, Arizona, and the Table Mountain Observatory near Los Angeles, California. They are indicated by black circles on the globe (left). Owing to the long-exposure required to detect the laser beams, the illuminated part of Earth is washed out.

Earth, our own globe remains more or less stationary in the lunar sky, rotating on its axis on a 24 hourly basis and waxing and waning in illumination over the period of a month. Starting at 17:11 on 22 January, the lander took sets of pictures of Earth through polarising filters at 2–3-hour intervals over a period of 26 hours. The goal was simply to determine what such observations could discern, starting with how the reflectance of the atmosphere varied as a function of the changing cloud distribution over one diurnal period.

It had been hoped to make at least one 'hop', and the predicted windows for this were when the elevation of the Sun was climbing between 23 and 31 degrees in the morning and falling between 23 and 16 degrees in the afternoon, but even at these times parts of the vernier propulsion system were too hot. Because leg no. 2 was on the north side of the vehicle, it suffered the longest period of direct illumination. The thrust chamber assembly of vernier no. 2 exceeded its pre-ignition temperature limit in the morning – in fact, it grew hotter than any engine on any previous vehicle. And before this engine could cool in the afternoon, vernier no. 3 had grown too hot. Only vernier no. 1, on the south side, remained usable in this respect. Furthermore, the temperature of the shock absorber on leg no. 1 plummetted to –53°C, well below the minimum temperature at which it could be expected to work properly during a hop. With leg no. 2 cooling in the afternoon, at 23:55 on 20 January the temperature of vernier no. 2 sharply dropped from 31°C to –18°C. At the same time there was an increase in the temperature of the fuel line indicating a flow from the tank, which was at 70°C. The engine was being chilled by the vaporisation of fuel leaking from the shutoff valve poppet. At 17:21 on 22 January the helium regulator automatically opened in a fruitless effort to re-establish the fuel pressure. The shock absorbers of the legs had not been locked following landing, but at 15:32 on 21 January, with no prospect of attempting a hop, they were commanded to lock. The squibs fired, but at sunset leg no. 2 deflected 2.4 degrees, indicating that it had failed to lock.

Sunset occurred at 06:06 on 23 January. Over the next 15 hours the camera took further pictures of the Earth, stars and the solar corona. Exposures of 20–30 minutes detected the corona out as far as 50 solar radii, which was about five times further out than was feasible for a solar eclipse viewed from Earth – sufficiently far, in fact, to study the hitherto unobservable transition zone between the solar corona and the inner zodiacal light zone. The pictures taken during the first 90 minutes after sunset provided further evidence of the 'horizon glow' discovered by Surveyor 6. When viewed from the equator the glow had remained due west, but for Surveyor 7, some 40 degrees south, the axis of the glow tended to migrate northward along the horizon with time. As for the cause of this phenomenon, it was concluded that electrostatic levitation of fine dust, if it occurred at all, would be minimal with insolation at a grazing angle. This left diffraction of the last rays from the upper limb of the solar disk approximating a point source on the horizon – as this geometry migrated beyond the local horizon the glow would persist for a time, with the pattern of the gaps in the line changing according to how the lunar surface features lying beyond the horizon cast their shadows onto the local horizon. Camera operations ceased at 21:10, having taken a total of 20,993 pictures.

Less post-sunset temperature data was obtained than hoped, once again owing to problems involving the bimetallically activated switches in the thermally controlled compartments, and the hibernation command was enacted at 14:12 on 26 January, 80 hours after sunset.

Surveyor 7 was reactivated at 19:01 on 12 February, some 120 hours after sunrise on its second lunar day. The shock absorber on leg no. 1 had compressed during the night, causing a deflection of 23.5 degrees – in effect completely collapsing the leg. Leg no. 2 had also compressed at sunset, but the fact that it recovered meant that its deflection had merely been the result of fluid contraction. In the case of leg no. 1, however, it appeared that high-pressure gas had leaked from the shock absorber. The alpha-scattering instrument resumed taking data on 13 February. The following day the arm was sent a single-step extension command, simply to verify that it was still operational. When the camera was activated, it proved unable to scan pictures in the 600-line mode due to an electrical fault. It was still functional in the 200-line mode, but a problem with the rechargeable battery limited its use. Indeed, after 45 frames the rate at which the camera could take pictures had decreased to one per hour and it was decided to cancel further operations. The arm test was successfully repeated on 20 February. By that time the power supply was so critical that the alpha-scattering instrument had to be switched off. The final communication from Surveyor 7 was at 00:24 on 21 February 1968 – there would not be another NASA transmission from the lunar surface until the first landing by an Apollo crew.

During the first lunar day the alpha-scattering instrument was operated for a total of 136.5 hours. This included 310 minutes of calibration using the standard sample, 727 minutes measuring the background (longer than planned, owing to the difficulty in deploying the sensor head) and 64 hours of science data – of which 44 hours was of sample material: 27.3 hours on the undisturbed surface, 10 hours on the rock and 6.7 hours on the subsurface that had been exposed by the arm. On the second lunar day the instrument provided another 34 hours of data for the third sample area over a 35-hour period, but only 20 hours of this were deemed to be usable owing to a low signal-to-noise ratio in the transmission. The total usable surface data was therefore 64 hours.

The main results of the alpha-scattering instrument were that for the fine-grained material the aluminium abundance was higher than measured at the maria sites, and the 'iron group' with atomic masses ranging from titanium to nickel were a factor of two less abundant. In the case of the relatively light-toned rock that was analysed the iron content was lower still. On seeing the elemental abundances, some people inferred that the lunar highlands must be an alumina-rich form of basalt, but Gene Shoemaker countered that the dominant rock in the Tycho ejecta – which was drawn from deep within the crust – was anorthositic gabbro. Such a feldspathic rock was a further indication that the Moon had undergone thermal differentiation. The lower iron content suggested that the highlands had a significantly lower density than the mare material. Later, John A. Wood would pick up on this and argue that the Moon had been so hot in the final stage of its accretion as to be molten to a considerable depth. In this 'magma ocean', the heavier elements sank to create a magnesium and iron silicate mantle whilst the lighter elements floated to the surface and cooled to create a crust. There was no evidence of the acidic rocks that would be required to account for the tektites. Overall, the elemental analyses performed by the Surveyors strongly indicated that the Moon was not a pristine body of ultrabasic composition, and this, in turn, ruled against the hypothesis that the most abundant meteorites on Earth – known as chondrites – originated from the Moon.

The Surveyor project achieved its primary objective of yielding sufficient insight into the nature of the lunar surface to allow Apollo to proceed in confidence, free of concern that the lander might sink into a sea of dust or fall through a brittle surface into a subterranean cavity.

Table 14.1: Surveyor sites – selenographic coordinates

Spacecraft	Longitude	Latitude	Description
Surveyor 1	43.22°W	2.45°S	Level mare floor of the Flamsteed Ring
Surveyor 3	23.34°W	2.97°S	Inner wall of 200-metre-diameter crater
Surveyor 5	23.20°E	1.42°N	Near top of a 9 × 12-metre crater
Surveyor 6	1.40°W	0.53°N	Level mare area, near a ridge
Surveyor 7	11.47°W	40.86°S	In the hilly ejecta blanket of Tycho

As derived from *Orthographic Atlas of the Moon* issued by D.W.G. Arthur and E.A. Whitaker in 1961 as Supplement 1 to the *Photographic Lunar Atlas*.

Table 14.2: Surveyor landing times and lighting

Spacecraft	Landing time (GMT)		Sun angle (degrees)
Surveyor 1	06:17:36	2 June 1966	28
Surveyor 3	00:04:17	20 April 1967	11
Surveyor 5	00:46:42	11 September 1967	17
Surveyor 6	01:01:04	10 November 1967	3
Surveyor 7	01:05:36	10 January 1968	13

Table 14.3: Surveyor photography

Spacecraft	1st day	2nd day	4th day	Total
Surveyor 1	10,341	899	–	11,240
Surveyor 3	6,326	–	–	6,326
Surveyor 5	18,006	1,048	64	19,118
Surveyor 6	29,952	–	–	29,952
Surveyor 7	20,993	45	–	21,038
				87,674

Note – these are finalised figures from the *Surveyor Program Results*, SP-184, 1969, which states that it corrects figures in the individual mission reports.

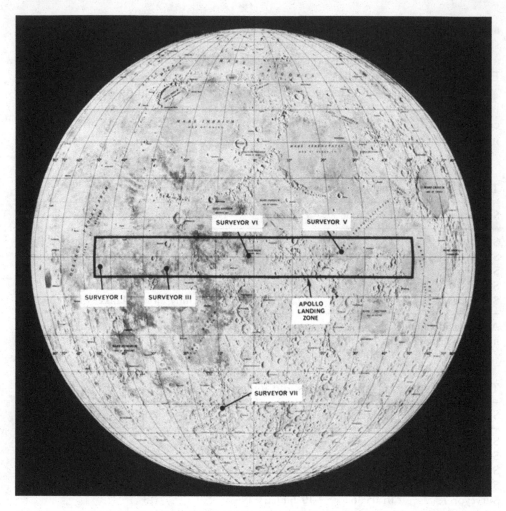

All but one of the successful Surveyor landers examined sites in the Apollo zone.

ROVING PLANS

It had been hoped that Surveyor would advance beyond the Block I engineering model. Planning for the Block II was terminated on 13 December 1966. Each would have weighed about 100 kg more than the original model, and have carried a greater scientific payload. In April 1964 Bendix submitted to JPL the outcome of a 6-month study to assess the feasibility of having a Block III deliver a 45-kg Surveyor Lunar Roving Vehicle. The plan was for the rover to be remotely controlled from Earth as it conducted a systematic study of a site which had been short-listed on the basis of Lunar Orbiter imagery as an Apollo target, to provide the 'ground truth' required to inform a final decision. It would be equipped with a scanning and digitising camera

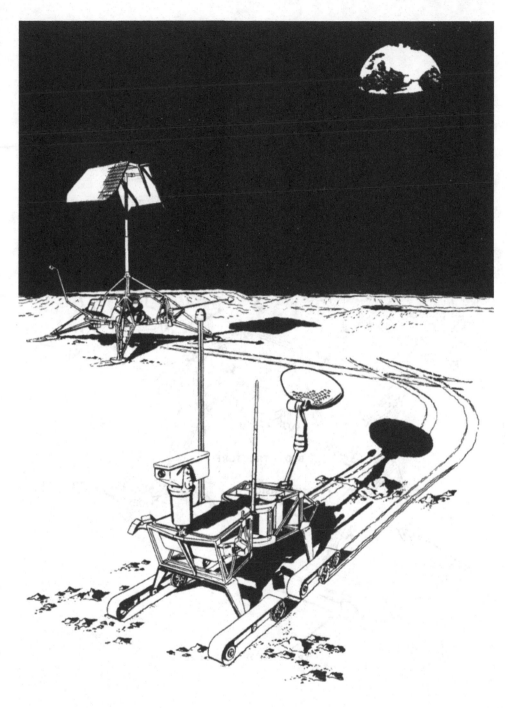

A depiction of the Surveyor Lunar Roving Vehicle proposed in 1964 by the Bendix Corporation.

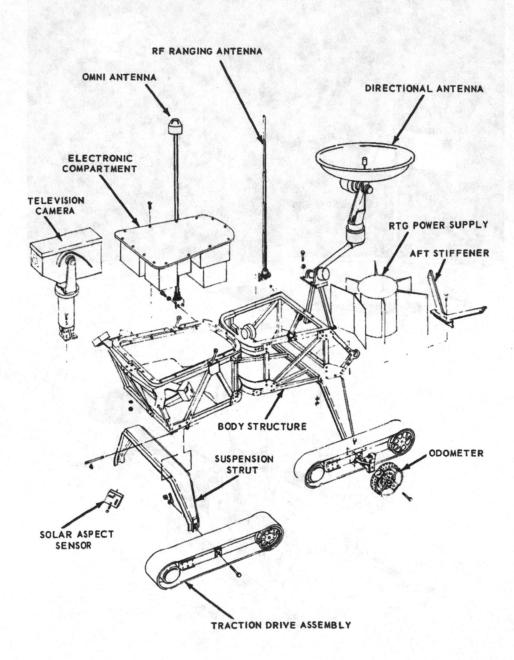

RF RANGING ANTENNA

OMNI ANTENNA

DIRECTIONAL ANTENNA

ELECTRONIC
COMPARTMENT

TELEVISION
CAMERA

RTG POWER SUPPLY

AFT STIFFENER

BODY STRUCTURE

SUSPENSION
STRUT

ODOMETER

SOLAR ASPECT
SENSOR

TRACTION DRIVE ASSEMBLY

Detail of the Surveyor Lunar Roving Vehicle proposed by the Bendix Corporation.

for stereoscopic imagery from which the local relief could be mapped on the scale of interest to the site selectors seeking 'clear' areas large enough to accommodate an Apollo lander. The rover would also have a penetrometer with which to measure the roughness and bearing strength of the surface along its route – something that could not be done from orbit. One survey method would involve an ever-widening spiral. It might make several such spirals, driving some distance cross-country in between, in total driving up to 25 km over an interval of several months – working during the lunar day and hibernating at night. There would be a trade-off between conducting a wide-area survey and certifying a given site for an advanced Apollo landing.[7] It had been hoped to launch five such missions in the 1970s, but the development funding was never forthcoming.

[7] Note that the plan presumed that an Apollo spacecraft would be able to set down precisely at a preselected point.

15

Apollo development

ORGANISATION

On 15 January 1962 the Manned Spacecraft Center reorganised the Apollo Project Office as the Apollo Spacecraft Project Office (ASPO) and appointed Charles W. Frick as Manager, with Robert O. Piland, head of the extinct Apollo Project Office, as Deputy Manager. In recognition of the scale of the project, on 13 February 1963 Piland's responsibility was narrowed to the LEM and James L. Decker was assigned as Deputy Manager for the CSM.

James Webb again reorganised NASA's top management on 30 October 1962. In addition to being Director of the Office of Manned Space Flight, Brainerd Holmes became a Deputy Associate Administrator and as such took direct responsibility for the field centres primarily engaged in manned space projects (i.e. the Marshall Space Flight Center, Manned Spacecraft Center and Launch Operations Center) which had previously reported to Robert Seamans. On 20 February 1963, Holmes made George M. Low his Deputy Director for Programs and Joseph F. Shea his Deputy Director for Systems in order to increase their authority over the Directorates of the Office of Manned Space Flight. On 10 May 1963 the Manned Spacecraft Center separated development from operations – as Deputy Director for Development and Programs, James C. Elms was to manage manned space flight projects and plan, organise and direct all administrative and technical support; and as Deputy Director for Mission Requirements and Flight Operations, Walter C. Williams was to manage the writing of mission plans and rules, the training of crews, and the provision of all ground support and mission control facilities. On 9 October 1963 James Webb announced a reorganisation of headquarters to become effective on 1 November. This introduced three Associate Administrators under Robert Seamans. Thus, George Mueller, who on 1 September had replaced Holmes by taking the post of Director of the Office of Manned Space Flight, now also became the Associate Administrator for Manned Space Flight with responsibility for the three field centres most directly involved in manned programs. The Goddard Space Flight Center, JPL and related facilities were assigned to the newly merged Office of Space Sciences and Applications headed by

Associate Administrator Homer Newell. Other facilities were assigned to the Office of Advanced Research and Technology under Associate Administrator Raymond L. Bisplinghoff.

On 22 October 1963 Joseph Shea was reassigned to Houston as Apollo Spacecraft Program Manager. George Low expanded his duties in the Office of Manned Space Flight to include Shea's post. On 27 August the Manned Space Flight Management Council at headquarters had decided to create a Deputy Associate Administrator for Manned Space Flight Operations so that the Director of the Office of Manned Space Flight could divest himself of this subsidiary role, and on 22 October Mueller drew Walter Williams from Houston for the job.

PLANS AND SCHEDULES

In view of the reason for his predecessor's resignation, George Mueller ordered a review of Apollo, and this confirmed the project to be in trouble. On 29 October 1963 Mueller informed the Manned Space Flight Management Council that the only way to recover time would be to reduce the number of development flights. The plan drawn up by the Marshall Space Flight Center in March 1962 envisaged a series of launches of the Saturn V in which the stages were tested in sequence – with only the first stage being 'live' on the first test scheduled for late 1965. The aim was to 'man rate' this vehicle by the summer of 1967, then use it to launch at least six manned missions in Earth and lunar orbit prior to attempting a lunar landing in late 1968 or early 1969. Mueller proposed to reduce this research and development phase by 'all up' testing in which each launch would use only 'live' stages, modules, systems and spacecraft. Wernher von Braun and Robert Gilruth objected, but Mueller had the support of James Webb.

In addition, a recent study by Bellcomm had recommended reassigning the early tests of the Apollo Block I spacecraft from the Saturn I to the Saturn IB, and so on 30 October Mueller cancelled the four manned test flights with the Saturn I that had been set for 1965. The development of the Saturn IB for manned missions would be accelerated and the 'all up' testing strategy employed in this case too. After coming to terms with this, Gilruth asked von Braun whether the Saturn IB could lift both the CSM and LEM, and was advised that it would be feasible only if their weights were controlled. At the White Sands Missile Range in New Mexico on 7 November the Apollo launch escape system successfully performed its first 'pad abort' test. On 18 November 1963 Mueller directed that if the LEM was not ready in time, the early Saturn IB flights would fly without it. But it must be phased into the test program as quickly as possible. Furthermore, Mueller directed that two successful development flights for each of the Saturn IB and Saturn V would serve to 'man rate' them. The schedule that he issued on 31 December 1963 listed the first Saturn IB test in early 1966 and the first manned mission later that year. The first Saturn V test was to be in the first quarter of 1967, with the first manned flight (hopefully on the third launch) later that year. Mueller then established the Apollo Program Office with himself as Director, and hired Samuel C. Phillips, who had managed the development of the Air Force's Minuteman missile, as Deputy Director.

On 17 September 1962 NASA had announced the nine men of its second intake of astronauts.[1] At the same time, Deke Slayton was appointed Coordinator of Astronaut Activities, reporting to Robert Gilruth.[2] In addition to the administrative tasks of the Astronaut Office, which Slayton managed in the manner of a military unit, he was responsible for making flight crew assignments. On 18 October 1963 the fourteen men of the third astronaut group were announced.[3] By now James Elms was Deputy Director of the Manned Spacecraft Center, and on 5 November 1963 Gilruth inserted Assistant Directors under Elms in order to strengthen the local management of flight operations: Chris Kraft was redesignated as Assistant Director for Flight Operations, Deke Slayton as Assistant Director for Flight Crew Operations and Maxime Faget as Assistant Director for Engineering and Development. In addition, Merritt Preston was assigned to manage Manned Spacecraft Center operations in Florida. However, on 17 January 1964 Elms resigned, and two days later George Low was reassigned from headquarters to replace him.

President Kennedy flew to Cape Canaveral on 16 November 1963 to inspect the 'moonport' which NASA was beginning to construct on nearby Merritt Island. He was shown models to illustrate the enormous size of the Saturn V. On 22 November he was assassinated in Dallas, Texas, and later that day Lyndon Johnson was sworn in as his successor. In a TV address on 28 November Johnson directed that Cape Canaveral be renamed Cape Kennedy, and the next day he signed an executive order in which the Launch Operations Center was renamed the John F. Kennedy Space Center.[4]

On 15 January 1964 the Manned Spacecraft Center proposed to Apollo Spacecraft Program Manager Joseph Shea that two of the Saturn IB 'all up'

---

[1]   They were: Lieutenant Charles 'Pete' Conrad Jr, Lieutenant Commander James Arthur Lovell Jr, and Lieutenant Commander John Watts Young from the Navy; Major Frank Frederick Borman II, Captain James Alton McDivitt, Captain Thomas Patten Stafford, and Captain Edward Higgins White II from the Air Force; Neil Alden Armstrong, a former naval aviator, now a civilian test pilot for NASA; and Elliot McKay See Jr, a civilian test pilot for the General Electric Company.

[2]   Slayton had been grounded in 1962 owing to a heart irregularity while training for a Mercury mission.

[3]   They were: Major Edwin Eugene 'Buzz' Aldrin Jr, Captain William Alison Anders, Captain Charles Arthur Bassett II, Captain Michael Collins, Captain Donn Fulton Eisele, Captain Theodore Cordy Freeman, and Captain David Randolph Scott from the Air Force; Lieutenant Alan LaVern Bean, Lieutenant Eugene Andrew Cernan, Lieutenant Roger Bruce Chaffee, and Lieutenant Commander Richard Francis Gordon Jr from the Navy; Captain Clifton Curtis Williams from the Marines; Ronnie Walter Cunningham, a research scientists at the RAND Corporation; and Russell Louis 'Rusty' Schweickart, a research scientist at the Massachusetts Institute of Technology.

[4]   This name change officially took effect on 20 December 1963.

Group 1, seated (left to right): Captain Leroy Gordon Cooper Jr, Captain Virgil Ivan 'Gus' Grissom, Lieutenant Malcolm Scott Carpenter, Lieutenant Commander Walter Marty Schirra Jr, Lieutenant Colonel John Herschel Glenn Jr, Lieutenant Commander Alan Bartlett Shepard Jr and Captain Donald Kent 'Deke' Slayton. Group 2, standing (left to right): Captain Edward Higgins White II, Captain James Alton McDivitt, Lieutenant Commander John Watts Young, Elliot McKay See Jr, Lieutenant Charles 'Pete' Conrad Jr, Major Frank Frederick Borman II, Neil Alden Armstrong, Captain Thomas Patten Stafford and Lieutenant Commander James Arthur Lovell Jr.

Group 3, seated (left to right): Major Edwin Eugene 'Buzz' Aldrin Jr, Captain William Alison Anders, Captain Charles Arthur Bassett II, Lieutenant Alan LaVern Bean, Lieutenant Eugene Andrew Cernan and Lieutenant Roger Bruce Chaffee; standing (left to right): Captain Michael Collins, Ronnie Walter Cunningham, Captain Donn Fulton Eisele, Captain Theodore Cordy Freeman, Lieutenant Commander Richard Francis Gordon Jr, Russell Louis 'Rusty' Schweickart, Captain David Randolph Scott and Captain Clifton Curtis Williams.

During a visit to Cape Canaveral on 16 November 1963 John F. Kennedy is briefed by
George E. Mueller on the 'mobile launcher' concept for Apollo. To Kennedy's right are
(in turn) James E. Webb, Robert C. Seamans, Kurt H. Debus and George M. Low. To
his left are Hugh L. Dryden, Wernher von Braun, General Leighton I. Davis and
Florida Senator George A. Smathers.

development flights be used to test the heat shield of the Apollo command module,
because this would enable the early tests of the Saturn V to be classified as
'demonstration' rather than 'development' for the spacecraft.[5] On 7 February
Grumman was directed to provide two LEM test articles (LTA) and eleven
flightworthy LEMs, the first three of which were to be capable of either manned or
unmanned operation. On 23 March George Mueller ordered that if the first two
unmanned CSM test flights were successful, the next mission would be a long-
duration manned flight, after which there would be two tests of the LEM, the first

---

[5]   On 26 October 1962 a nomenclature was introduced by which the pad abort tests were to
run in sequence from PA-1; the Little Joe II flights were to start at A-001; missions using
the Saturn I were to start at A-101; missions using the Saturn IB were to start at A-201; and
missions using the Saturn V were to start at A-501, with the 'A' standing for 'Apollo'. The
'SA' prefix was employed by the Marshall Space Flight Center (giving precedence to the
launch vehicle) and the 'AS' prefix was used by the Manned Spacecraft Center (giving
precedence to the spacecraft). In addition, the term 'space vehicle' was introduced to
describe the integrated 'launch vehicle' and 'spacecraft', with the latter comprising the
CSM, the LM (if present) and the SLA structure.

Table 15.1: Outline schedule for Apollo drawn up in November 1964

Mission	Payload	Launch Date
AS-201	CSM-009 (unmanned)	1965
AS-202	CSM-011 (unmanned)	1966
AS-203	No spacecraft (S-IVB development flight)	Jul 1966
AS-204	CSM-012 (manned)	Oct 1966
AS-205	CSM-014 (manned)	Jan 1967
AS-206	LEM-1 (unmanned)	Apr 1967
AS-207	CSM-101 (manned) and LEM-2	Jul 1967

one unmanned and the second together with a manned CSM – so long as the Saturn IB proved capable of lifting both vehicles together. In November 1964 Joseph Shea, George Mueller and Sam Phillips drew up an outline schedule for testing Apollo hardware in advance of the introduction of the Saturn V, but it remained uncertain whether the weights of the two spacecraft were sufficiently constrained for them to be lifted together by a Saturn IB for the joint mission. On 16 December Shea directed that the Block I manned missions must use low orbits from which the spacecraft could use its reaction control system thrusters to de-orbit itself in the event of the failure of the service propulsion system; and in the event of these too failing, the orbit must decay naturally and result in re-entry within an acceptable duration.[6]

On 31 August 1964 Lead Flight Director Chris Kraft appointed John D. Hodge, Eugene F. Kranz and Glynn S. Lunney to alternate in round-the-clock flight operations. On 24 December Everett E. Christensen was made Director of Mission Operations, a position which effectively superseded Deputy Associate Administrator for Manned Space Flight Operations – vacant since the resignation of Walter Williams in April. At the same time, two posts of Mission Director were also created, with the intention that the appointees would run alternate missions. In addition, activities at the Cape were consolidated, with Kurt H. Debus being made Director of Launch Operations and Merritt Preston, who had been managing the Manned Spacecraft Center's activities at the Cape, becoming his Deputy.

Joseph Shea, Chris Kraft and Deke Slayton were briefed on 18 January 1965 by the Mission Planning and Analysis Division of the Manned Spacecraft Center about the Saturn IB and early Saturn V flights. On 21 January, in response to a question by Sam Phillips, Shea said the current estimate was that the Saturn IB would be able to insert 35,500 pounds into a circular orbit at 105 nautical miles. This, however, was less than the combined 'control weights' of the CSM and LEM by 870 pounds, and both vehicles were currently *above* their control weights. Shea argued that in view of the difficulty in constraining the weights, the best solution would be to find a way of increasing the launcher's capacity by 1,000 pounds. In fact, the Saturn IB had a

---

[6] This precaution was reputedly a headquarters response to the situation depicted by Martin Caidin in his recent novel *Marooned*.

'control payload' which was the specified minimum mass that it was to be capable of placing into the reference orbit, and a 'design goal' which exceeded this. On 23 February Phillips told Shea that the Marshall Space Flight Center would endeavour to increase the payload by 1,000 pounds. The development version of the cluster of eight H-1 engines had yielded 1.3 million pounds of thrust, but the fifth flight of the Saturn I had introduced an upgraded cluster that finally achieved its specification of 1.5 million pounds of thrust. In August 1963 Rocketdyne had proposed an upgrade for 1.6 million pounds of thrust, and on 8 November of that year NASA had ordered this be done. By 23 April 1965 the improved engine had completed its qualification testing. On 12 May Huntsville reported that it would be possible to uprate the engine by an additional 5,000 pounds of thrust, to raise the total to 1.64 million pounds. But the rocket engineers were fighting a losing battle, as by then both spacecraft had put on even more weight.

On 13 January 1965 Shea had established the Configuration Control Board, with himself in the chair. This was to rule on all proposals for engineering changes to the spacecraft. On 10 May he faced a dilemma: the 'all up' testing regime required that all spacecraft incorporate a full set of subsystems, but it had been proposed that the landing radars be omitted from LEM-1 and LEM-2 on the basis that a radar would serve no function on an Earth orbital mission. Omitting the radar on these early test flights would save money and assist in the effort to trim the weight of the vehicle at this critical juncture, but doing so would establish the precedent for a series of one-of-a-kind spacecraft, each tailored to achieving specific development objectives and with none demonstrating all of the systems in conjunction. Furthermore, by relieving the pressure on the effort to trim weight in the short term, such compromises might jeopardise it in the long term. On 27 May the Manned Spacecraft Center reaffirmed that LEM-1 must test the radar. But on 25 June ASPO Assistant Manager Harry L. Reynolds warned Owen E. Maynard, Chief of the Systems Engineering Division at the Manned Spacecraft Center, that it was "becoming increasingly clear that we are going to have a difficult job keeping the LEM weight below the control weight". On 6 July Grumman requested to be allowed to deliver the early LEMs without some subsystems installed, but Shea insisted they must all leave the factory in a fully functional condition. At that time, LEM-1 was to be delivered to the Kennedy Space Center in November 1966, with the next five vehicles following in 1967, but it was becoming increasingly evident that this schedule would be difficult to achieve. On 13 September 1965 Shea established the Weight Control Board to enable subsystem managers to meet on a weekly basis and report progress in controlling the weights of the two spacecraft, and when appropriate to create ad hoc task forces to chase up specific issues and report back.

Meanwhile, on 17 February 1965 Shea clarified for North American Aviation the Block I schedule. CSM-009 and CSM-011 were to be configured for unmanned use and fly as AS-201 and AS-202 to test the heat shield. CSM-012 and CSM-014 were to be delivered for manned missions, but be capable of being adapted at the Cape for unmanned flight. The decision for CSM-012 would be made 6 months ahead of the scheduled launch date for AS-204, and if flown unmanned this would be done either to gain additional data on the spacecraft's characteristics or to provide more time for the Marshall Space Flight Center to prepare AS-203 to obtain data on the behaviour of the S-IVB stage in space. North American Aviation was told that CSM-017 and CSM-020, assigned to the early tests of the Saturn V, need not be capable of manned use. The first manned Block II would be CSM-101, which was to fly in conjunction with LEM-2. On 22 March Glynn Lunney, Chief of the Flight Dynamics Branch of the Flight Control Division in Houston, was appointed Assistant Flight Director for AS-201 and AS-202. On 25 June Carroll H. Bolender was made Deputy Director of Mission Operations at the Office of Manned Space Flight, and his first task was to plan these two preliminary missions.

On 10 August 1965 ASPO named LEM-1 to AS-206, LEM-2 to AS-207, LEM-3 to AS-503, LEM-4 to AS-504, LEM-5 to AS-505 and LEM-6 to AS-506. Of the six test articles, LTA-1 was kept by Grumman at Bethpage to resolve issues during the initial fabrication, assembly and checkout procedures, LTA-2 went to the Marshall Space Flight Center for launch vibration tests, LTA-3 and LTA-5 were to be used to assess the structural effects of engine firing, LTA-8 went to the Manned Spacecraft Center for thermal-vacuum environmental testing, and LTA-10 went to the North American Aviation factory in Tulsa, Oklahoma, for fit-checks with the SLA, which was being manufactured there. To cut costs, in July Grumman had been directed to delete LTA-4 (intended for vibration tests), the ascent stage of LTA-5 and the two flight test articles and instead to refurbish two of the test articles for flight once their ground testing role was complete. The company said it would refurbish LTA-10 and LTA-2 in case they were needed for the first two Saturn V test flights. The first three LEMs were to incorporate development flight instrumentation so as to record the dynamic environment at launch. A key requirement was that the differences between LEM-3 and LEM-4 be minimised and that all subsequent production vehicles be identical.

On 21 October 1965 Sam Phillips slipped AS-201 to January 1966 and AS-202 to June 1966 to accommodate the revised delivery dates for CSM-009 and CSM-011, but otherwise preserved the outline schedule which had been drawn up in November 1964. On 2 December 1965 Hugh Dryden died of cancer.[7] Robert Seamans replaced him as Deputy Administrator on 21 December. He retained the duties of Associate Administrator until Homer Newell gained this post in August 1967, and was in turn superseded as Associate Administrator for Space Sciences and Applications by John E. Naugle.

Meanwhile, an operational step toward the chosen Apollo 'mission mode' was achieved when Gemini 6 rendezvoused with Gemini 7 on 15 December 1965. The straightforward manner in which this was done raised the prospect of undertaking the manned test of the LEM without reducing the weights of the CSM and LEM to enable the Saturn IB to lift them both together. On 28 January 1966 Sam Phillips asked ASPO to assess the impact, including the effects on ground support equipment and mission control, of a *dual* AS-207/208 mission as early as the scheduled date for

---

[7] NASA's Flight Research Center at Edwards Air Force Base was renamed in Dryden's honour.

AS-207, which was the Saturn IB that was nominally to have sent them into orbit together. The idea was for near-simultaneous launches of AS-207 with CSM-101 and AS-208 with LEM-2 to facilitate a rendezvous and docking, at which point the mission would unfold as originally planned. On 2 February John P. Mayer, Chief of the Mission Planning and Analysis Division at the Manned Spacecraft Center, informed Chris Kraft, Assistant Director for Flight Operations, that the main constraint would be programming the Real-Time Computer Complex in Houston to plan and support such a mission – in which case the decision on whether it was to be attempted must be taken very soon. Mayer also urged that if the IBM staff who worked on the Gemini 6/7 rendezvous could be spared, they should be reassigned to help to plan the new dual mission. On 4 February John Hodge, Chief of the Flight Control Division, noted that some of the operational issues associated with near-simultaneous launches would be obviated if the interval were extended. On 24 February Mayer's assistant, Howard W. Tindall, recommended that the CSM be launched first and the LEM follow it either 24 hours later or at a recurring daily window. On 1 March Joseph Shea endorsed the concept. On 8 March Sam Phillips directed the Manned Spacecraft Center, Marshall Space Flight Center and Kennedy Space Center to endeavour to launch the dual mission a month later than intended for AS-207 on the previous schedule.

EARLY UNMANNED TESTING

AS-201 was the first in a series of test flights to 'man rate' the Saturn IB and the Apollo spacecraft.[8] It lifted off from Pad 34 at 16:12:01 GMT on 26 February 1966. After the booster cut off, the S-IVB stage separated cleanly and attained the planned suborbital arc. In releasing CSM-009, the stage splayed its four panels to an angle of 45 degrees to allow the service propulsion system engine an unobstructed exit. The spacecraft had neither a guidance and navigation system nor an S-Band transmission system. It was powered by batteries instead of fuel cells, had a 20 per cent propellant load, and an ad hoc electromechanical control sequencer. It began by firing its RCS thrusters for 18 seconds to withdraw from the S-IVB. Upon peaking at an altitude of 226 nautical miles, the spacecraft fired its thrusters again to provide ullage to settle the propellants in their tanks, then fired the service propulsion system. However, 80 seconds into the planned 184-second burn the thrust chamber pressure started to decline owing to inadvertent helium ingestion, and by the time the engine shut down the pressure had declined to 70 per cent. The thrusters were immediately fired for ullage and the engine was reignited for a 10-second burn, during which the chamber pressure oscillated from 70 per cent down to 12 per cent.

---

[8]　At this point, CSM-002 was the only production-line spacecraft to have flown – it was launched on 20 January 1966 at the White Sands Missile Range by a Little Joe II booster as a high-altitude abort test.

Although the manoeuvres on the descending side of the arc were designed to drive the spacecraft into the atmosphere at a speed significantly faster than a normal orbital entry, it was still not as fast as a trajectory returning from the Moon. Several seconds later, the thrusters began a pitch manoeuvre at a rate of 5 degrees per second for 18 seconds to yield a 90-degree change in attitude. On separating, the command module used its own thrusters to continue this pitch rotation for an additional 82.5 degrees and then rolled 180 degrees in order to orient its heat shield for atmospheric entry. The plan was to subject the heat shield to a high heating rate – meaning a high temperature for a comparatively short time – but the velocity at entry was 782 ft/sec slower than the planned 29,000 ft/sec and the flight path was 0.44 degree shallower, with the result that the heating rate was less than that intended. Although the deceleration peaked at 14.3 g rather than 16.0 g, it was still much greater than on an operational mission. A fault in the electrical power system ruled out aerodynamic steering, and the 'rolling' entry which resulted was 40 nautical miles short. Some 37 minutes after launch, the command module splashed into the South Atlantic. It was recovered 2.5 hours later by USS *Boxer*. To allow the time to diagnose and rectify the fault in the service propulsion system, AS-202 was rescheduled to follow AS-203, which, as an S-IVB development flight, would not carry a spacecraft.

The docking by Gemini 8 with its Agena target vehicle on 16 March lent support to the decision to try the AS-207/208 dual mission. On 21 March NASA announced that Gus Grissom was to command the first Apollo mission. He would fly CSM-012 with Ed White and Roger Chaffee. They were to be backed up by James McDivitt, David Scott and Rusty Schweickart respectively. In each case, the commander and senior pilot were Gemini veterans and the third man was a rookie. Deke Slayton earmarked Grissom for this role immediately after the Gemini 3 test flight in March 1965. After commanding Gemini 4 in June 1965, McDivitt was reassigned to back up Grissom. White, who flew with McDivitt on Gemini 4, backed up Gemini 7 in December 1965 and then joined Grissom's crew. Although Slayton was introducing a 'rotation' for Gemini in which a pilot could progress through backup to command a later mission, after flying Gemini 8 Scott was immediately assigned to McDivitt's crew to enable them to obtain early experience of Apollo training prior to attempting the AS-207/208 dual mission. If CSM-011 demonstrated that the problems suffered by CSM-009 had been fixed, then AS-204 would launch CSM-012 in the last quarter of 1966 on an 'open ended' mission of up to 14 days "to demonstrate spacecraft and crew operations and evaluate spacecraft hardware performance in Earth orbit", but if there were significant issues outstanding then CSM-012 would be modified for a third unmanned test.

On 4 April 1966 the Manned Spacecraft Center revised its senior management job titles, replacing 'assistant director for' with 'director of' in order to make explicit the fact that the post had *primary* rather than subordinate responsibility for that activity. Thus, for example, Kraft ceased to be the Assistant Director for Flight Operations and became the Director of Flight Operations. On 12 May NASA deleted the word 'Excursion' from 'LEM', to make the lander the Lunar Module 'LM'. On 25 May, precisely 5 years after President Kennedy made his speech to Congress calling for a lunar landing, a diesel-powered crawler carried the 500-F engineering model of the

Apollo-Saturn V at a maximum speed of 1 mile per hour from the vast cube of the Vehicle Assembly Building a distance of 3.5 miles on a special causeway to Pad 39 on the Merritt Island Launch Area in order to verify the ground facilities and assist in the development of training procedures. It was an awesome demonstration of the 'mobile launcher' concept.

AS-203 lifted off from Pad 37 at 14:53:17 GMT on 5 July 1966 and the S-IVB inserted itself into the desired circular orbit at an altitude of 100 nautical miles. As it did not have a spacecraft, an aerodynamic nose cone was used. At orbit insertion the liquid hydrogen was 'settled' by a combination of tank baffles and deflectors and by ullage induced by venting liquid oxygen. A TV camera in the fuel tank then verified that continuous venting of liquid hydrogen could hold the fluid in this condition during a coasting phase that approximated a flight heading for translunar injection. The fact that the rise in the liquid hydrogen pressure in orbit was greater than predicted gave data on the heat transfer properties of the tank that would be applied in planning Saturn V missions. Radar tracking by ground stations monitored how the parameters of the orbit were changed by the thrusting effect of continuous venting. A simulated restart of the J-2 engine verified the charging of the restart bottles at orbital insertion cutoff, the fuel recirculation chill-down, the fuel antivortex screen, and the liquid oxygen recirculation chill-down. A subcritical cryogenic nitrogen experiment carried in the nose cap successfully maintained pressure control, with a progressive decrease in the fluid quantity indicating that vapour was being uniformly delivered from a two-phase mixture. To save weight, the S-IVB had been designed such that its propellant tanks shared a bulkhead. This sophisticated structure had to cope with the normal difference in pressure between the tanks and also insulate the liquid oxygen at $-172°C$ from the liquid hydrogen at $-253°C$ to preclude the oxygen solidifying. After the ullage trial of the first revolution, the hydrogen valves were closed and the oxygen valves opened to space in order to place an inverse pressure on the common bulkhead and assess its predicted failure point – when this occurred early on the fifth revolution it caused the vehicle to break up.

On 13 July 1966 Deke Slayton and Chris Kraft jointly wrote to Joseph Shea, the Apollo Spacecraft Program Manager: "A comprehensive examination of the Apollo missions leading to the lunar landing indicates there is a considerable discontinuity between the missions AS-205 and AS-207/208. Both missions AS-204 and AS-205 are essentially long-duration system validation flights. AS-207/208 is the first of a series of very complicated missions. A valid operational requirement [therefore] exists to include an optical equi-period rendezvous on AS-205." If this Block I flight were to include a rendezvous with its spent S-IVB, it would offer an opportunity to evaluate the control dynamics, visibility, and piloting techniques for the rendezvous phase of AS-207/208. By this point, every spacecraft on Grumman's production line through to LM-4 was late. The focus, of course, was on LM-1, but late shipments by subcontractors were impeding its assembly. Nevertheless, the 'rate of slippage' was slowing, and on 6 October Shea reported his expectation that the company would be able to deliver LM-1 early in 1967. By the end of 1966 LM-1 and LM-2 were in test stands, and LM-3 through LM-7 were in various stages of assembly, but by the end

As Gemini 11 lifts off from Pad 19 on 12 September 1966, the 500-F model of the Saturn V stands on Pad 39A, known as 'Moonport, USA'.

of January 1967 it was clear that LM-1 would not be able to be shipped on schedule in February.

As its designation suggests, AS-202 was intended to be the second Saturn IB test, but it slipped behind AS-203 as a result of delays involving the spacecraft. CSM-011 was a fully functional Block I spacecraft, minus the crew equipment. But it carried a more sophisticated ad hoc sequencer than on AS-201, a 60 per cent propellant load, a variety of flight qualification instrumentation and four film cameras. It lifted off from Pad 34 at 17:15:32 GMT on 25 August 1966. A key objective was to verify the emergency detection system in closed-loop configuration. At cutoff, the S-IVB was at an altitude of 120 nautical miles and climbing on a ballistic arc. Eleven seconds after separating, the spacecraft fired its service propulsion system in order to place itself on a higher trajectory that would result in entry over the Pacific. As a thermal test, the spacecraft then turned to aim its apex towards the Earth and maintained this attitude through the peak altitude of 618 nautical miles above Africa. On descending over the Indian Ocean it realigned its apex to the velocity vector, then fired its main engine for 89.2 seconds to accelerate for atmospheric entry and concluded by firing it briefly twice more in rapid succession as a demonstration of rapid restart.

In contrast to the 'rolling' entry made by AS-201, this time the command module controlled its attitude to fly a trajectory that 'skipped' off the atmosphere to trace a ballistic arc which led to a second contact and full entry. A similar profile was to be used on returning from the Moon. The double peak in the heating rate was designed to expose the shield to low heat rates with high heat loads – lower temperatures, but applied for longer – than a 'straight in' lunar return. Although the temperature at the base of the shield peaked at 1,482°C, the cabin did not exceed 21°C. After a flight of 93 minutes, the command module splashed into the Pacific and adopted the apex-up flotation attitude. But the flight path angle at entry of −3.53 degrees was steeper than the desired −3.48 degrees and the lift-to-drag ratio of 0.28 (±0.02) was less than the predicted 0.33 (±0.04), causing it to fall short by 205 nautical miles. It was 8 hours before USS *Hornet* recovered the capsule. The planners would have to take into account the lower than expected lift-to-drag ratio of the command module. This qualified the heat shield for Earth orbital missions, but additional tests would be required for a lunar return. Both the Saturn IB and the Block I spacecraft were declared ready for the first manned mission.

As 80 per cent of the objectives specified for CSM-002, CSM-009 and CSM-011 had (between them) been met, AS-204 was released for the manned Apollo 1.

THE GEMINI LEGACY

When NASA decided in 1962 that Apollo would use the lunar orbit rendezvous mission mode, many people doubted that orbital rendezvous would be feasible. The primary objective of the Gemini program was to explore the issues. The ten manned missions flown between March 1965 and November 1966 not only established that rendezvous and docking was feasible, by testing a variety of techniques it gave the Apollo planners the flexibility of options. This inspired a workaround to the fact that the combined Apollo vehicles exceeded the payload capacity of the Saturn IB, in the form of the dual AS-207/208 rendezvous. Gemini also showed that astronauts could endure the space environment for longer than any Apollo mission would require. Given that the longest American space flight at the time of President Kennedy's commitment to Apollo was Al Shepard's 15-minute suborbital arc, on which he was weightless for only a couple of minutes, this was welcome news. The fuel cells that were to power the Apollo spacecraft were tested on Gemini, as were a fully inertial reference platform for guidance and navigation, a spaceborne radar, a state-of-the-art digital computer to process the radar data for rendezvous, and bipropellant ablative thrusters. Gemini established that a spacecraft could be steered through re-entry for recovery at a specific location. This increased confidence in the 'atmospheric skip' manoeuvre that was to be used by an Apollo spacecraft returning from the Moon. By enabling astronauts to learn how to operate outside a spacecraft, Gemini inspired a rescue option for the crew of an Apollo lunar module that was unable to dock with its mothership. And, of course, by training a cadre astronauts and flight controllers Gemini allowed Apollo to get off to a running start.

As Robert Gilruth, Director of the Manned Spacecraft Center, observed: "In order to go to the Moon, we had to learn how to operate in space. We had to learn how to manoeuvre with precision to rendezvous and to dock; to work outside in the hard vacuum of space; to endure long-duration in the weightless environment; and to learn how to make precise landings from orbital flight – that is where the Gemini program came in."

SETBACK AND RECOVERY

On 26 August 1966 the command module of CSM-012 arrived at the Cape in a container prominently labelled 'Apollo One'.

North American Aviation was to have shipped it several weeks earlier, but the failure of a glycol pump in the environmental control system had led to the exchange of this unit with its CSM-014 counterpart. Although the customer acceptance review identified other "eleventh-hour problems" associated with the environmental control system, NASA had taken receipt.

The Office of Manned Space Flight held the AS-204 design certification review on 7 October, and declared that the launch vehicle and the spacecraft "conformed to design requirements" and would be flightworthy once a number of deficiencies had been overcome. Sam Phillips issued a list of these deficiencies to Lee B. James at the Marshall Space Flight Center, Joseph Shea at the Manned Spacecraft Center, and John G. Shinkle, Apollo Program Manager at the Kennedy Space Center, requiring speedy compliance. On 11 October Phillips was informed by Carroll Bolender of a report he had received the previous day from Shinkle detailing increasing delays in the preparation of CSM-012. When the spacecraft was delivered, 164 'engineering orders' had been identified as 'open work' – despite the fact that the accompanying data package had listed only 126 such items. By 24 September the list had grown to 377, and Shinkle ventured that about 150 of the 213 additional orders ought to have been identifiable by the manufacturer prior to the customer acceptance review. The issues included the environmental control system (which had failed again), problems with the reaction control system, a leak in the service propulsion system, and even design deficiencies with the couches that had obliged the company to send engineers to the Cape. On 12 October Phillips wrote to Mark E. Bradley, Vice President of the Garrett Group, whose AiResearch Division had supplied the environmental control system under subcontract to North American Aviation, explaining that its reliability threatened a "major delay" to the AS-204 mission. To Phillips, the problems seemed "to lie in two categories: those arising from inadequate development testing, and those related to poor workmanship". A replacement was delivered on 2 November, and testing resumed as soon as it was in place. However, the unit malfunctioned and had to be returned to the company.

On 25 October the propellant tanks of the service module for CSM-017, assigned to AS-501, failed catastrophically in a test at North American Aviation. The normal operating pressure was 175 psi, but it had failed after 100 minutes at the maximum requirement of 240 psi. The test had been ordered following the discovery of cracks in the tanks of CSM-101, assigned to AS-207. The failure was particularly puzzling because the tanks of CSM-017 had been subjected to 320 psi for several minutes in 'proof testing'. ASPO set up an investigation, which was to report by 4 November. As SM-012 had been through the same test regime, Shea grounded it pending this report. The problem was determined to be stress corrosion in the titanium resulting from the use of methyl alcohol as a test liquid. The point of the test was to verify the integrity of the tanks, and because the hydrazine and nitrogen tetroxide propellants were toxic another fluid had been used – and unfortunately this had caused damage! The remedy was to switch to a fluid that was compatible with titanium, and it was decided to use freon in the oxidiser tank and isopropyl alcohol in the fuel tank, with the additional stipulations that the systems must not have been previously exposed to the actual propellants and that after the tests the system must be purged by gaseous nitrogen. With the issue resolved, the tanks of SM-012 were removed for inspection and confirmed to be free of cracks.

The crew for the CSM-014 mission was announced on 29 September 1966. Wally Schirra would be in command, flying with Donn Eisele and Walt Cunningham. They would be backed up by Frank Borman, Tom Stafford and Michael Collins. Schirra was the only experienced man of the prime crew, but all the backup astronauts were veterans. In fact, Deke Slayton had given Schirra and Borman these assignments in March, on their return from an international 'goodwill tour' after the rendezvous of Gemini 6/7. Stafford and Collins had been assigned following Gemini 9 in June and Gemini 10 in July, respectively. Slayton had actually earmarked the rookies Eisele and Chaffee to Grissom's crew, but in late 1965 Eisele had injured his shoulder in weightlessness training in a KC-135 aircraft and dropped out of training, prompting Slayton to swap Eisele with Ed White, whom Slayton had earmarked for Schirra's crew. This Apollo 2 mission was to be a straightforward rerun of Apollo 1 to further evaluate the spacecraft's systems.

In early December 1966, accepting that Apollo 1 would not fly that year, George Mueller postponed it to February 1967 and also deleted the Block I reflight in order to prevent the slippage of CSM-012 from impacting the Block II missions scheduled for later in 1967. Schirra had hoped to put his crew first in line for the dual mission, but Slayton imposed a rule that the man who would operate the CSM alone while his colleagues flew the LM must be experienced in rendezvous, since if the LM were to become crippled he would have to perform a rescue. Eisele was a rookie but Scott had performed a rendezvous on Gemini 8, so Slayton exchanged Schirra's crew with McDivitt's crew. Schirra was not pleased at being given the backup role, but Slayton had always intended to assign McDivitt the dual mission.

On the new schedule, AS-206 would launch LM-1 for an unmanned test as soon as possible after Apollo 1, and if this was satisfactory McDivitt's crew would fly the dual mission (which was now AS-205/208 because deleting CSM-014 had released AS-205) as the revised Apollo 2 in August. This revision was publicly announced on 22 December, together with the assignment of Tom Stafford, John Young and Gene Cernan to backup McDivitt's crew. Also, if two unmanned tests proved sufficient to 'man rate' the Saturn V, the intention was to launch AS-503 with a CSM and LM. The crew for this mission would be Frank Borman, Michael Collins and Bill Anders, backed up by Pete Conrad, Dick Gordon and Clifton Williams. These assignments had been made after Young flew Gemini 10 in July, Conrad and Gordon flew Gemini 11 in September, and Cernan backed up Gemini 12 in November.

The Gemini missions had demonstrated that for an astronaut on a spacewalk to be able to work effectively he must be provided with mobility and stability aids. On 6 December 1966 Slayton warned Joseph Shea that without handholds and tethering points, a transfer from the forward hatch of the LM to the CSM's hatch would not be feasible. On 26 December Slayton recommended that a spacewalk be scheduled 100 hours into AS-503, after the two firings of the LM's descent propulsion system but prior to the descent stage being jettisoned. One of the two astronauts would egress from the forward hatch and stand on the 'front porch' to assess the environmental control system in the LM during depressurisation, using the hatch, the performance of the life-support backpack, and the egress procedure for the emergency transfer. In addition, whilst outside, the spacewalker was to photograph the exterior of the LM to verify that it had not been damaged during its retrieval from the S-IVB. He would then re-enter the LM and the cabin repressurisation system would be tested, simulating the end of a moonwalk.

On 26 January 1967 Schirra's crew made a 'full up' systems test of CSM-012 on its AS-204 launch vehicle. But the spacecraft drew its power from the pad, and the capsule was not pressurised with pure oxygen. It had not been a very productive day. "Frankly, Gus," Schirra said in the debriefing with Grissom and Shea, "I don't like it. You're going to be in there with full oxygen tomorrow, and if you have the same feeling I do, I suggest you get out." But there was a determination to catch up on the several-times-delayed schedule.

The next day, Friday, 27 January, Grissom's crew attempted the 'plugs out' test in which the spacecraft would be on internal power and pressurised with pure oxygen at 16 psi (i.e. slightly above ambient) for an integrity check. If successful, this would clear the spacecraft for flight. After a simulated countdown, they were to end the day with an emergency egress drill.

In Houston, Flight Director John Hodge was monitoring progress, but the action was at the Cape. Slayton was in the Pad 34 blockhouse talking to Director of Launch Operations, Rocco Petrone. Also present was Stu Roosa, a rookie astronaut serving as the primary communications link with the crew. The Spacecraft Test Conductor, Clarence 'Skip' Chauvin, was in the Automated Checkout Equipment facility of the Manned Spacecraft Operations Building.

"Fire!!" yelled Grissom at 18:31 local time, in a hold at T–10 minutes. "We've got a fire in the cockpit."

In all, there were 25 technicians on Level A8 of Pad 34's service structure, and five more either on the access arm or in the White Room. Henry Rogers, NASA's Inspector of Quality Control, was in the elevator, ascending the service structure. Systems technician L.D. Reece was waiting for the 'Go' to disconnect the spacecraft for the 'plugs out' test, which had been delayed by problems with communications, most notably the whistle from an 'open' microphone that could not be located.

"Get them out of there!" commanded Donald Babbitt, North American Aviation's Pad Leader, on hearing Grissom's call. Mechanical technician James Gleaves was closest, but a spout of flame burst from the capsule before he could react, and he was beaten back by the flame and smoke.

Gary Propst, a technician of the Radio Corporation of America, was on the first level of the pad monitoring a TV camera located in the White Room pointing at the window in the spacecraft's hatch. On hearing Grissom's call, he looked up and saw a brilliant light in the window and gloved hands moving about within.

As soon as Slayton realised what had happened, he sent medics Fred Kelly and Alan Harter to the pad. "You know what I'll find," Kelly observed pointedly. The best that they would be able to do would be to supervise the retrieval of the bodies. On reflection, Slayton decided to accompany them. "We were the first guys from the blockhouse to reach the pad," he later pointed out. Despite the intensity of the fire, Grissom, White and Chaffee had died by asphyxiation as a result of the toxic fumes created by the incomplete combustion of the synthetic materials in the cabin. They had received second and third degree burns, but these in themselves would not have been fatal. After several minutes Slayton left the White Room to call Houston, to report the situation. Shea had just arrived back in Houston and was briefing George Low when the news came through.

The Astronaut Office in Houston was very quiet. All the 'old hands' were absent. With Slayton away, Don Gregory, his assistant, ran the routine Friday staff meeting. The meeting had only just convened when the red phone on Slayton's desk rang. Gregory answered, then reported, "There has been a fire in the spacecraft." Michael Collins was the senior astronaut present. He arranged for Al Bean to track down the wives. In each case, the news had to be broken by an astronaut who was also a close friend of the family. Charles Berry and Marge Slayton went to see Betty Grissom. Pete Conrad was sent to track down Pat White. Gene Cernan would have been ideal for Martha Chaffee because they lived next door, but he was in Downey with Tom Stafford and John Young, so Collins went to give her the bad news himself.

Al Shepard was in Dallas, Texas, about to deliver a speech at a dinner. He was taken aside and told of the fire. Wally Schirra, Donn Eisele and Walt Cunningham were flying home from the Cape, and were told upon touching down at Ellington Air Force Base. Schirra immediately called Slayton at the Cape, who filled him in on the details. James Webb, Robert Seamans, Robert Gilruth, George Mueller, Kurt Debus, Sam Phillips and Wernher von Braun were at the International Club in Washington with corporate officials, including Leland Atwood of North American Aviation, to mark the transition from Gemini to Apollo. Webb immediately ordered Seamans and Phillips to the Cape to investigate. As Webb observed to newsmen shortly thereafter, "Although everyone realised that some day space pilots would die, who would have thought the first tragedy would be on the ground?"

The Board of Inquiry was chaired by Floyd L. Thompson, Director of the Langley Research Center, with Frank Borman as the Astronaut Office's representative. The origin of the fire was near the foot of Grissom's couch, where components of the environmental control system had repeatedly been removed and replaced in testing. Although the investigation did not identify the specific ignition source, it did find physical indications of electrical arcing in a wiring harness. It was concluded that at some time during either manufacturing or testing an unnoticed incidental contact had scraped the insulation from a wire and thereby created the opportunity for a spark. This had ignited nearby flammable material, and in the super-pressurised pure-oxygen situation the result had been a brief but intense 'flash' fire. In fact, there had been some 32 kg of nylon netting, polyurethane foam and velcro – all of it flammable in such conditions. In retrospect, the worst flaw was the inward-opening hatch, which even under ideal conditions took several minutes to open, and would have been *impossible* to open with the internal pressure above ambient. Because neither the launch vehicle nor the spacecraft had been loaded with propellants, the 'plugs out' test had not been judged hazardous. Nevertheless, the launch escape system was directly above the spacecraft, and if the heat from the fire had ignited the solid propellant of this rocket the White Room crew would almost certainly have been killed as well.

In an *Associated Press* interview in December 1966 Grissom had told Howard Benedict: "If we die, we want people to accept it. We are in a risky business and we hope that if anything happens to us it won't delay the program. The conquest of space is worth the risk of life."

In an effort to reduce the risk of a fire during ground testing, it was decided to use an atmosphere comprising 65 per cent oxygen and 35 per cent nitrogen. After liftoff, the nitrogen would be purged and the pressure reduced to the originally planned 100 per cent oxygen at about 5 psi.

Although the investigation into the fire would take months, on 31 January NASA headquarters directed the Manned Spacecraft Center, Marshall Space Flight Center and Kennedy Space Center to proceed as planned with preparations for AS-501 with CSM-017 and LTA-10R, except that the command module was not to be pressurised with oxygen without specific authorisation.[9] On 2 February CSM-014 was delivered

[9]   LTA-10R was a refurbished LM test article serving as a mass-model.

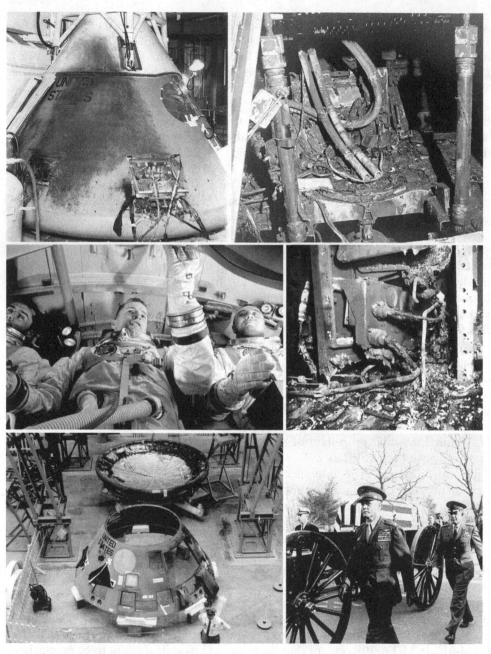

The exterior of the fire-damaged Apollo 1 command module in which Grissom, White and Chaffee died (top left); a view through the hatch; the crew positions, with the hatch above the center couch; the vicinity of the environmental control unit, where the ignition source is believed to have been; and its disassembled outer structures. Glenn, Cooper and Young escort Grissom's coffin.

to the Cape to assist in training the technicians who were to disassemble CM-012 for the investigation. On 3 February George Mueller announced that although manned flights were grounded indefinitely, the unmanned AS-206, AS-501 and AS-502 were to proceed as soon as delivery of the hardware allowed. While the investigation into the fire was underway, Mueller suggested that when the Block II spacecraft became available the CSM-only flight should be deleted and the effort switched to combined testing with the LM, but Robert Gilruth warned that it would not be wise to test *two* new vehicles at once. In March it was decided to fly an 11-day CSM-only mission, in effect to perform Grissom's mission with the upgraded model, and Slayton tipped off Schirra that his crew would fly it, backed up by Stafford's crew. On 21 February, the day that Apollo 1 had been scheduled to launch, Floyd Thompson gave Mueller a preliminary briefing on the investigation's findings, and several days later Robert Seamans sent a memo to James Webb listing Thompson's early recommendations.[10]

On 15 March Deke Slayton proposed that a rendezvous with the S-IVB stage be a primary objective of Schirra's flight, and said that this should occur "after the third period of orbital darkness". On 5 April Sam Phillips told the Manned Spacecraft Center, Marshall Space Flight Center and Kennedy Space Center that the profile for the first manned flight would be based on that developed for Grissom's flight, dated November 1966. As the complexity of the mission was not to exceed that previously planned, and as no rendezvous had been planned, the rendezvous exercise should be assessed in terms of how it would complicate the mission rather than how it would advance the program. As the flight was to focus on evaluating the spacecraft's systems, Chris Kraft pointed out on 18 April that if a problem were to develop that would require the cancellation of the rendezvous, then any manoeuvres which had already been made would complicate the nominal contingency de-orbit procedures. The rendezvous should not be initiated until "after a minimum of one day of orbital flight" and should be "limited to a simple equi-period exercise with a target carried into orbit by the spacecraft". On 2 June Phillips agreed with George Low that there should be a rendezvous but insisted that this should not be listed as a primary objective. The double-hatch of the Block I had been replaced on the Block II by a single 'unified hatch' on a hinge that swung outward. It had a manual release for either internal or external use, could be opened in 60 seconds irrespective of the differential in pressure, and was capable of being opened in order to conduct a spacewalk. But Phillips directed that there "be no additions that require major new commitments such as opening the command module hatch in space or exercising the docking subsystem".

NASA announced on 20 March 1967 that the unmanned LM-1 flight would be transferred from AS-206 to AS-204, which had become available. The rationale for the AS-205/208 dual mission with CSM-101 and LM-2 had been to ensure that testing of the LM would not be held up by the Saturn V development problems. The AS-501 and AS-502 development flights were to carry refurbished LM test articles, but unless

---

[10]   The final report of the Apollo 204 Review Board was submitted on 5 April 1967.

the pace of LM development dramatically picked up, the heavy launcher would become available ahead of the LM, thus rendering the ad hoc dual mission redundant. It was therefore decided that if the LM-1 test flight proved unsatisfactory, AS-206 would launch LM-2 unmanned to address the remaining test objectives. On 25 March George Mueller directed that missions be numbered in the order of their launch, regardless of whether they employed the Saturn IB or Saturn V and whether they were manned or unmanned – previously only the manned missions were to be counted. On the 1966 plan, Apollo 2 was to be CSM-014 (Schirra) and Apollo 3 was to be CSM-101 (McDivitt) flying the dual mission. The cancellation of the Block I reflight advanced CSM-101 to Apollo 2. After the fire, the desire not to reassign the name Apollo 1 had resulted in CSM-101 (Schirra) being seen as Apollo 2. But with paperwork in circulation for a variety of mission plans numbered up to Apollo 3, Mueller precluded the possibility of administrative confusion by directing that the first scheduled mission, AS-501, be named Apollo 4.[11]

On 7 April 1967 Joseph Shea was transferred to Washington as Deputy Associate Administrator for Manned Space Flight, and Low succeeded him as ASPO Manager at the Manned Spacecraft Center. Several days later, Everett Christensen resigned as Director of Mission Operations at headquarters.

A joint meeting of the Manned Spacecraft Center's Flight Operations Directorate and Mission Operations Division announced on 17 April that: (1) successful firings by the descent and ascent stages of an unmanned LM, including a 'fire in the hole' separation of the two stages, should be prerequisites to a manned LM being allotted these functions; (2) a demonstration of EVA transfer should not be a prerequisite to manned independent flight of the LM; (3) the Saturn V should be 'man rated' as rapidly as possible; (4) three manned Earth orbit flights involving both the CSM and the LM should be the *minimum* requirement prior to attempting a lunar landing; and (5) although a lunar orbit mission should *not* be a formal step in the program, this should be planned as a contingency in the event of the CSM achieving lunar-mission capability ahead of the LM.

ASPO sent the Block II Redefinition Task Team, led by Frank Borman, to North American Aviation on 27 April. Having the authority to make on-the-spot decisions which previously would have required referral to the Configuration Control Board, it was to oversee the 'redefinition' of the Block II spacecraft, responding promptly to questions regarding detail design, quality and reliability, test and checkout, baseline specifications, configuration control, and scheduling. Meanwhile, the company had hired William D. Bergen from the Martin Company to supersede Harrison A. Storms as Apollo Project Manager. Bergen brought with him John P. Healy to manage the production of the first Block II at Downey, and Bastian Hello to run the company's operations at the Cape.

---

[11] On 30 March 1967 George Low suggested that the AS-201 and AS-202 test flights be assigned the designations Apollo 2 and Apollo 3 retrospectively in order to fill in the gap, but this was rejected by Mueller on 24 April. AS-203 was not included because it did not carry a spacecraft.

On 8 May 1967 George Low reaffirmed that AS-205 would launch CSM-101 on an open-ended mission of up to 11 days to evaluate its systems. The next day, James Webb told a Senate committee that this mission would be flown by Wally Schirra, Donn Eisele and Walt Cunningham. When Webb canvassed suggestions for how to impress upon Congress that the Apollo program was recovering from the setback of the fire, George Mueller urged that the Saturn V be flown as soon as possible.

16

Apollo picks up the pace

MOON ROCKET

On 19 January 1959 NASA took over the Air Force's contract with Rocketdyne for the development of the F-1 kerosene-burning engine. The prototype was test fired on 10 February 1961. By sustaining 1.55 million pounds of thrust for several seconds, it broke the record for a single-chamber engine by a considerable margin. On 9 April 1961 it was announced that the engine had achieved 1.64 million pounds of thrust. On 26 May 1962 the engine was fired at full power for its intended operating time of 150 seconds. Meanwhile, Rocketdyne began the development of the 200,000-pound-thrust hydrogen-burning J-2 engine that was to power the upper stages of the Saturn launch vehicle. The first full-duration test of this engine was on 27 November 1963. The Douglas Aircraft Corporation fired an S-IVB stage utilising a single J-2 engine at full power for 10 seconds on a static rig at its Sacramento facility on 4 December 1964. But it was a 'battleship' variant (equivalent to a 'boilerplate' for a spacecraft) having tankage made of thick stainless steel instead of the lightweight aluminium of the operational vehicle. On 7 December 1964 the first S-IVB mockup – which was accurate in terms of mass, centre of gravity and structural stiffness, but with models of the engine and other systems – was delivered to the Marshall Space Flight Center for stress testing. On 16 April 1965 the first S-IC stage utilising five F-1 engines was test fired for several seconds at NASA's Mississippi Test Facility. On 24 April the S-II stage utilising five J-2 engines was test fired at Rocketdyne's facility at Santa Susana in California. On 5 August the S-IC made a full-duration test during which it responded to steering commands provided by the blockhouse. On 9 August the S-II made its first full-duration firing. That same day the first production version of the S-IVB was tested, and on 20 August it was fired for 3 minutes, shut down for half an hour and reignited for almost 6 minutes in a simulation of its role on a lunar mission.

Unfortunately, by early 1966 the development of the S-II had slipped. In an effort to recover, North American Aviation hired a new manager, Robert E. Greer, who took a team of engineers to the Mississippi Test Facility. On 23 April 1966 the S-II was successfully fired for 15 seconds, but faulty instrumentation caused premature cutoffs on 10, 11 and 16 May. It fired for 150 and 350 seconds in tests on 17 and 20 May. But fires broke out in two places on the vehicle in a test on 25 May, and as the stage was being removed from the stand three days later its hydrogen tank exploded, damaging the facility and injuring five people. George Mueller in Washington began to send weekly progress reports on the S-II to company president Leland Atwood, at one point advising him that the S-II had an excellent chance of replacing the LM as the 'pacing item' in the program.

But then the fire that killed the Apollo 1 crew during a supposedly routine test of the spacecraft on 27 January 1967 halted the program in its tracks. Nevertheless, the time taken to redesign the CSM provided the opportunity for the development of the Saturn V and the LM to catch up.

On 17 April 1967 the Manned Spacecraft Center proposed a minimum of three manned Saturn V missions involving both the CSM and the LM *prior to* attempting the lunar landing. When George Mueller advocated landing *on* the third mission, Chris Kraft warned George Low that a landing should not be tried "on the first flight which leaves the Earth's gravitational field" because flying to the Moon was such a great step forward in terms of operational capability that this should be demonstrated separately, to enable the landing crew to focus on activities associated with landing. Accepting Kraft's argument, on 20 September Low led a delegation to Washington. Owen E. Maynard, Chief of the Systems Engineering Division in Houston, outlined a step-by-step sequence: (A) Saturn V and unmanned CSM development; (B) Saturn IB and unmanned LM development; (C) Saturn IB and manned CSM evaluation; (D) Saturn V and manned CSM/LM joint development; (E) CSM/LM trials in an Earth orbit involving a 'high' apogee; (F) CSM/LM trials in lunar orbit; (G) the first lunar landing; (H) further 'minimalist' landings; (I) reconnaissance surveys in lunar orbit; and (J) 'enhanced capability' landings.[1] This alphabetically labelled series was not a list of *flights*, as several flights might be required to achieve one *mission*. Two Saturn V development flights were already scheduled as Apollo 4 and Apollo 6, and the LM-1 flight as Apollo 5. Sam Phillips asked whether a second Saturn V test was really necessary, and Wernher von Braun said the second would serve to confirm the data from the first. If the Saturn V development were to prove to be protracted, then the 'D' mission would be done by reinstating the plan in which the CSM and LM would be launched individually by Saturn IBs and rendezvous in orbit. Most of the discussion was devoted to the proposal for a lunar orbital flight "to evaluate the deep space environment and to develop procedures for the entire lunar landing mission short of LM descent, ascent and surface operations". When Mueller argued "Apollo should not go to the Moon to develop procedures", Low said that developing crew operations would not be the

---

[1] The last two categories represented the lunar part of the Apollo Applications Program which was being promoted by George Mueller, and when this fell by the wayside the reconnaissance surveys were deleted and the main program was expanded to include 'enhanced capability' landings.

main reason for the mission; there was actually still a lot to be learned about navigation, thermal control and communications in deep space. Although the meeting left this matter undecided, the alphabetic labels soon became common shorthand.

Sam Phillips confirmed on 2 October 1967 that LM-2 should be configured for an unmanned test flight, and directed that LM-3 be paired with CSM-103 for the first manned mission of the complete Apollo configuration.[2] Grumman's latest schedule called for LM-2 to be delivered in February 1968, LM-3 in April and LM-4 in June. On 4 November George Mueller issued the schedule for 1968: AS-204 with LM-1; then AS-502 as the second unmanned test; AS-503 as the third unmanned test, if this proved necessary; AS-206 with LM-2, if required; AS-205 with CSM-101, manned; and AS-504 with CSM-103 and LM-3, manned. On 15 November George Low said that in the event of AS-503 being unmanned, the payload should be the 'boilerplate' spacecraft BP-30 and lunar module test article LTA-B.

A SPECTACULAR 'ALL UP' TEST

The 1966 schedule had called for the first Saturn V launch early in 1967 but few people believed that this would be feasible owing to problems with the S-II, which had become the 'pacing item'. In fact, the delivery of the first 'live' S-II to the Cape had already slipped from July to October 1966, and on its arrival at the Mississippi Test Facility on 13 August the inspectors found a number of cracks which delayed the start of its acceptance test firings. In November 1966 Sam Phillips revised the schedule to require the S-II for AS-501 to arrive at the Cape on 9 January 1967 for launch in April.

Meanwhile, the S-IC stage had been erected upon a mobile launch platform in the VAB on 27 October. So as not to delay the checkout of the vehicle, a bobbin-shaped 'spacer' was stacked in place of the S-II to support the S-IVB, and on 12 January 1967 the spacecraft comprising CSM-017 and LTA-10R in the SLA was added for its own checkout.

When the S-II arrived on 21 January 1967, several faults were found. By now the launch had slipped into May. On 14 February the spacecraft was transferred to the Operations and Checkout Building for examination as part of the investigation of the Apollo 1 fire, and so many wiring discrepancies were identified that repairs ran into June. In the meantime the S-IVB was de-mated, and on 23 February the spacer was replaced by the S-II. But when factory inspectors discovered cracks in another S-II being prepared for shipment, the S-II was destacked on 24 May for inspection and not restacked until mid-June. Once the S-IVB had been added, the revised spacecraft was installed on 20 June. CSM-017 was a Block I with some Block II modifications for certification, including a heat shield with a simulated unified crew hatch and the

---

[2]   CSM-101 was to fly the Saturn IB and manned CSM evaluation, and CSM-102 was to be retained by North American Aviation as a ground test article.

AS-501 is transported to Pad 39A in readiness for the Apollo 4 mission.

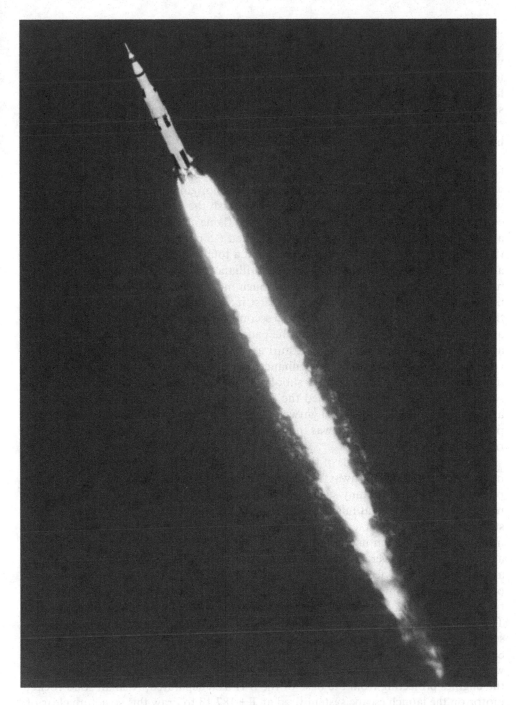

As Apollo 4 soars into the sky it is trailed by a 1,000-foot long plume.

umbilical which crossed the rim of the basal shield from the command module to the service module. The crawler transported AS-501 to Pad 39A on 26 August. The plan was to start the 6-day countdown demonstration test on 20 September but it slipped to 27 September and a variety of issues delayed its completion to 13 October. To be fair, however, this *was* the first Saturn V, it was vastly more complex than any other space vehicle, and the launch operations team was on a learning curve.

The countdown for Apollo 4 began on 6 November, ran smoothly, and the vehicle lifted off on time at 12:00:01 GMT on 9 November 1967.

For the crowds, the first indication that a launch was in progress was a light at the base of the vehicle. A jet of flame passed through a hole in the launch platform to a wedge-shaped deflector, which split and vented it horizontally north and south. The water which was pumped onto the pad to diminish the acoustic reflection from the concrete was vaporised and blasted out with the flame as a roiling white cloud. It was almost inconceivable that a vehicle that weighed 6.5 million pounds could rise from the ground, but the five F-1 engines generated a total of 7.5 million pounds of thrust and as it slowly lifted from the pad the brilliance of the flame rivalled the early morning Sun. Since the Saturn V was so much more powerful than its predecessors, nobody really knew what to expect. At first it was like a silent movie, because the thunderous roar of ignition took fully 15 seconds to reach the facilities for the public and press, which were 3.5 miles from the pad because this had been computed to be as far as an exploding vehicle could hurl a 100-pound fragment. Deafening as this roar was, as the vehicle rose it was enhanced by a staccato pop and crackle that was more felt than heard. The ground shook sufficiently to register on remote seismic sensors. The effect not only rattled the tin roof of the VIP bleacher alarmingly but also threatened to collapse the lightweight booth from which Walter Cronkite was providing TV commentary. It was incredible to think that one day soon men would ride such a rocket.

During the first 10 seconds of the ascent the 363-foot-tall vehicle performed a yaw manoeuvre to 'side step' away from the launch umbilical tower, in order both to preclude a collision with any swing arm that might be tardy in rotating clear and to resist any wind gusts that might otherwise cause it to drift towards the 400-foot-tall structure. Liftoff was on a pad azimuth of 90°E of N, but once clear of the tower it initiated a roll to align its inertial guidance system with the flight azimuth of 72°E and then it pitched over. It achieved Mach 1 at $T + 61.4$ seconds. Then at $T + 78.4$, at an altitude of 37,700 feet and a wind speed of 50 knots, it passed through the region of maximum dynamic pressure. The central engine of the S-IC was shut down by a timer at $T + 135.52$, and the outer engines were cut off by liquid oxygen depletion at $T + 150.77$. The separation proceeded in two phases. Firstly, at $T + 151.43$, the S-IC separated from the interstage, or 'skirt', that extended down over the engines of the S-II; then at $T + 152.12$ the S-II ignited its five J-2 engines and at $T + 181.44$ the skirt was jettisoned. This scheme was designed to ensure that the S-IC could not damage the engines projecting from the base of the S-II as it separated. A small solid rocket motor on the launch escape system fired at $T + 187.13$ to draw this structure clear of the spacecraft. The S-II shut down at $T + 519.76$ and was jettisoned at $T + 520.53$. A pair of solid rockets on the periphery of the S-IVB settled the liquid propellants prior to J-2 ignition and were then jettisoned. The vehicle attained a near-circular parking orbit at an altitude of 100 nautical miles, and the continuous-vent system maintained ullage pressure in the propellant tanks during the coast phase. At 003:11:26.6,[3] after essentially two revolutions with its longitudinal axis in the orbital plane and parallel to the local horizon, it reignited for a simulated translunar injection, although in this case the burn of almost 5 minutes was to create an elliptical atmosphere-intersecting 'waiting orbit' with an apogee of 9,292 nautical miles.

CSM-017 was released at 003:26:28.2, and at 003:28:06.6 the service propulsion system was briefly fired to demonstrate its ability to ignite in the zero-g environment without an ullage impulse to settle its propellants. This burn had the effect of raising the apogee to 9,769 nautical miles. The vehicle then aligned itself with its main axis perpendicular to the Sun and its hatch on the sunny side. It maintained this attitude for about 4.5 hours to induce circumferential thermal stresses and distortions on the command module and its ablator prior to entry. At 005:46:49.5 the vehicle attained its high apogee. During this coast, an automated 70-mm camera took a total of 715 high-resolution pictures of Earth at 10.6-second intervals. At 008:10:54.8 the service propulsion system was reignited to accelerate and set up an atmospheric entry which would subject the heat shield to the most severe operational conditions that a return from the Moon might impose. An inertial velocity of 34,816 ft/sec was intended, but a slight over-burn owing to the manoeuvre being controlled from the ground yielded 35,115 ft/sec.

The command module separated 2 minutes 27 seconds later, and used its thrusters to adopt entry attitude. The entry interface (an altitude 400,000 feet, by definition) occurred at 008:29:28.5 while travelling at an inertial velocity of 36,639 ft/sec and a flight path angle of –6.93 degrees. As a result of the longer than planned final burn of the service propulsion system the conditions at the entry interface were 210 ft/sec faster than nominal and the flight path angle 0.20 degrees shallower, yet still within the desired 'corridor'. Due to the change in the entry conditions, the dynamic load of 7.27 g was less than the predicted 8.3 g. This did not affect the performance of the guidance system in achieving the target, however. The lift-to-drag ratio was 0.365 (±0.015) compared with the predicted 0.350. The command module splashed into the Pacific 10 nautical miles from the aim point at 008:37:09.2, and was soon recovered by USS *Bennington*.

With the spectacular success of this 'all up' demonstration flight, there was a real prospect of achieving Kennedy's challenge of landing a man on the Moon before the decade was out. Indeed, on 20 November 1967 NASA publicly revealed that James McDivitt, David Scott and Rusty Schweickart, who had been trained to fly the dual launch version of the 'D' mission using the Saturn IB, were to be launched on the first manned Saturn V. They were to be backed up by Pete Conrad, Dick Gordon and Al Bean – the latter replacing Clifton Williams, who was killed in an air accident on 5 October 1967. Frank Borman, Michael Collins and Bill Anders, backed up by

---

[3] Times in this hhh:mm:ss format are with reference to the time of launch.

Neil Armstrong, James Lovell and Buzz Aldrin, would still perform the high-apogee 'E' mission, but would ride the second manned Saturn V. If the lunar orbit 'F' mission were deleted, then the next crew would attempt a lunar landing.

THE FIRST LM

In mid-1966 Sam Phillips had hoped that AS-206 would be able to launch LM-1 in April 1967, and Kurt Debus, estimating that it would take 6 months to check out the spacecraft, had asked Grumman to send it to the Cape in September 1966. But it was delayed by manufacturing issues and combustion instabilities in the ascent engine. Nevertheless, AS-206 was erected on Pad 37 in January 1967 in the expectation of launching in April. However, because the AS-204 launch vehicle on Pad 34 had not been damaged by the fire that destroyed the Apollo 1 spacecraft, on 20 March it was reassigned to LM-1. Accordingly, by 11 April AS-206 had been returned to storage and AS-204R – as this was redesignated – erected in its place. In the absence of the spacecraft, Grumman built a plywood mockup on the pad for facilities verification.

On 12 May George Low confided to headquarters that although Grumman had promised to deliver LM-1 in June, he was sceptical. John J. Williams headed a 400-man operations team at the Cape. After the arrival of the ascent and descent stages on 23 June, LM-1 was mated on 27 June. However, the initial examination identified a significant number of departures from specification. On 26 July Carroll Bolender was reassigned to Houston as ASPO's LM Manager. LM-1 was de-mated in August to repair leaks in the ascent stage. After another leak developed in September, it was de-mated and a number of items extracted for return to Grumman. After the testing was finally completed, the spacecraft, minus its legs, was mechanically mated to the launch vehicle on 19 November and a nose cone fitted in place of the absent CSM. The flight readiness tests were finished in late December. The cabin closeout was on 18 January 1968, during the countdown demonstration test. Loading the hypergolic propellants into LM-1 was delayed by procedural issues, but the ensuing tests ended on 19 January.

The terminal countdown began on 21 January, at T–10 hours 30 minutes. The spacecraft went onto internal power at T–42 minutes and, several hours later than planned, AS-206 lifted off as Apollo 5 at 22:48:08 GMT on 22 January 1968.

The S-IVB achieved an orbit ranging between 88 and 120 nautical miles, shed the nose cone and then splayed the four SLA panels. LM-1 was released at 000:53:50. After manoeuvring clear using its attitude control thrusters, the LM adopted a 'cold-soak' orientation, which its guidance system successfully maintained with a minimal engine duty cycle.

The mission plan called for two descent propulsion system manoeuvres, an abort staging, and an ascent propulsion system manoeuvre. The first manoeuvre was to occur on the third revolution and last 38 seconds. It would run at 10 per cent throttle for the first 26 seconds, then be concluded at full throttle. The thrust profile of the second manoeuvre was to be representative of flying a lunar landing, involving five phases over a total of 734 seconds. The abort staging sequence would be initiated at

Installing LM-1 for the Apollo 5 mission.

full throttle, and was to include descent propulsion system shutdown and a 'fire in the hole' 5-second burn by the ascent propulsion system. The final burn was to run to propellant depletion and end the primary mission. The guidance system initiated the first descent propulsion system firing at 003:59:42 but the buildup of thrust did not satisfy the programmed velocity-time criteria, and the guidance system, sensing that the spacecraft was not accelerating as rapidly as expected, aborted the burn after just 4 seconds. In fact, this was a design feature, since on a manned mission it would allow the crew time to analyse the situation and decide whether to restart the engine to continue. In normal circumstances the engine would have fired with full tank pressurisation and achieved the desired thrust in 4 seconds, but in this case the tanks were only partially pressurised and it would have taken 6 seconds to build up thrust. The premature cutoff was merely the result of inadequate coordination between the guidance and propulsion teams, not a problem with the spacecraft. Mission Control sent a command to deactivate the guidance system in order to permit the remainder of the mission to be controlled from Earth using a preplanned sequence which would address the minimum requirements of the mission.

At 006:10:00 the onboard automatic sequencer initiated this program. It began by using the backup control system to control the vehicle's attitude. In performing two burns in this mode, the descent engine gimballed properly and responded smoothly to throttle commands. But the short duration of the three descent propulsion system firings precluded a full evaluation of the thermal aspects of the supercritical helium pressurisation system. In abort staging, all system operations and vehicle dynamics were satisfactory for manned flight. The primary control system was then reselected to control the vehicle's attitudes and rates. However, as this had been off during the abort staging sequence its computer program did not know of the change of mass resulting from this action and its computed thruster firing times were based on the mass of the two-stage vehicle and caused an extremely high rate of propellant usage. The final ascent propulsion system firing started at 007:44:13 and ran to thrust decay at 007:50:03. Since the attitude control system had by that time exhausted its own propellants, this burn was initiated with the thrusters drawing from the tanks of the ascent propulsion system. This continued until the sequencer automatically closed the interconnect valves, whereupon, with the thrusters starved and the ascent engine still firing, the vehicle started to tumble. The rates were soon of such a magnitude as to impede the flow of propellants to the engine, and helium ingestion induced thrust decay prior to propellant depletion. The vehicle had been in a retrograde orientation during the controlled portion of its final manoeuvre, and calculations indicated that it entered the atmosphere over the Pacific Ocean.

On 26 January the LM-2 flight requirements meeting determined that: (1) apart from minor anomalies, LM-1 had achieved all its flight objectives; (2) it should be possible to achieve the objectives for LM-2 either by additional ground testing or on a manned mission; and (3) it was not necessary to undertake additional unmanned flights to 'man rate' the LM. Grumman's own view was that there should be two test flights, but the company relented after the review of the LM-1 data by the Manned Space Flight Management Council on 6 February. On 6 March NASA cancelled the shipment of LM-2 to the Cape. If AS-502 repeated the success of its predecessor, then AS-503 would indeed be manned, and hopefully be launched before the end of the year with CSM-103 and LM-3.

A DARING PLAN

Robert Seamans resigned as NASA's Deputy Administrator on 2 October 1967, and departed 3 months later. On 5 February 1968 the Senate accepted the nomination of Thomas O. Paine, a senior manager of the General Electric Corporation, and he was sworn in as Deputy Administrator on 25 March.

AS-502 was the second in the series of 'A' missions designed to 'man rate' the Saturn V launch vehicle. The payload was LTA-2R and CSM-020, a Block I with some Block II modifications for certification, including a heat shield with an actual unified crew hatch to be tested in lunar return conditions. In view of the fact that the mission would be unmanned, certain systems had been deleted from the command module in order to accommodate an electromechanical control sequencer.

The countdown proceeded without unplanned holds, and Apollo 6 lifted off from Pad 39A at 12:00:01 GMT on 4 April 1968. It contrast to the perfect performance of the first Saturn V, this one suffered a number of problems.

Firstly, between T + 110 and T + 140 seconds it underwent a 'pogo' oscillation, as a longitudinal structural mode frequency coupled with the resonant frequency of the oxidiser lines feeding the S-IC's engines. The greatest disturbance was in the range 5.2 to 5.5 hertz. The oscillations in the engine chamber pressures built up to a peak-to-peak maximum of 8 to 10 psia at T + 125, and the condition was reinforced by the consequent variation in thrust. The ± 0.6-g oscillation measured in the spacecraft exceeded the design criteria, and would have given astronauts a very rough ride. The emergency detection system cast one 'vote' for terminating the mission. Had it gone on to cast a second vote, an abort would have become mandatory and the command module would have been drawn clear by the launch escape system.

Ground-based and airborne cameras noted three small pieces and five or six large pieces separating from the vicinity of the SLA between T + 133.31 and T + 133.68, at which time strain, vibration and acceleration sensors in the S-IVB, IU, SLA, LTA and CSM reported abrupt changes. Subsequent analysis determined that one of the SLA panels had suffered structural failure and shed some of its skin. Fortunately, the supporting elements were able to sustain the loads for the remainder of the powered flight.

The S-IC shut down at T + 148.21 and separated cleanly. The five J-2 engines of the S-II ignited at T + 150 and performed satisfactorily for 169 seconds, but at T + 319 the hydrogen flow rate to engine no. 2 suddenly increased and its thrust declined by 23 psi. The engine ran at this level until T + 412.92, whereupon the temperature in the engine bay suddenly rose and the engine shut down. Engine no. 3, which had shown no sign of distress, cut off 1.26 seconds later.

Despite the loss of 40 per cent of its power, the S-II was able to gimbal the three remaining engines to hold itself stable. However, the controller in the Instrument Unit had been configured to react only with a single-engine-failure contingency and was unable to take into account the loss of the second engine. It naively attempted to recover its trajectory as if it had four good engines. When the IU started to adjust the propellant mixture ratio to optimise consumption leading up to S-II shutdown, it also adopted a mode designed to stabilise the vehicle for separation. If all had been going to plan, at that time the S-II would have been within 5 seconds of cutoff but the loss of two engines reduced the rate of propellant consumption and the other engines had to burn for much longer than planned to trigger the fuel-depletion cutoff. In fact, the burn was 57.81 seconds longer than usual, and because the vehicle held its attitude fixed during this time it climbed higher than intended and the space-fixed velocity at cutoff was 335.52 ft/sec less than nominal.

The inherited trajectory anomaly presented the S-IVB with a serious challenge. It ignited at T + 577.28 and burned for 166.52 seconds, some 28.95 seconds longer than nominal. On finding itself high, slow and short, it pitched down 50 degrees in order to lose altitude, accelerate and gain range. On achieving the desired altitude it raised its nose above the local horizon to overcome the *negative* radial velocity acquired in descending, whilst simultaneously minimising further increasing its horizontal rate. On activating terminal guidance at T + 712.3, the system set the altitude constraints to zero and focused on achieving the desired velocity. Although it ordered the vehicle to pitch up beyond vertical and travel *backwards*, the 1 deg/sec rotation rate meant that when the space-fixed velocity exceeded the objective and the stage shut down at T + 747.04 the angle had reached only 65 degrees. The velocity was 160 ft/sec greater than nominal, the surface range was 269.15 nautical miles longer than nominal and the flight path angle was slightly negative. A circular orbit at 100 nautical miles had been planned, but the actual orbit of 93.49 × 194.63 nautical miles did not preclude continuing with the mission.[4]

Once the S-IVB had realigned its axis with the horizon it initiated a sequence of manoeuvres to be undertaken on the first revolution. First, it rolled 180 degrees and pitched down 20 degrees, and then it pitched up 20 degrees and rolled to resume its original attitude. The only appreciable effect was the sloshing of liquid oxygen at the onset of each change in pitch, but this was rapidly damped out. This qualified such manoeuvres to orient the vehicle to enable astronauts to perform landmark tracking while in parking orbit.

The Apollo 6 plan was to reignite the S-IVB at the end of the second revolution to simulate the translunar injection. Although this was to achieve lunar distance, the apogee was to be away from the Moon in order not to complicate the evaluation of the guidance system in deep space. Immediately after cutoff, the S-IVB was to pitch through 155 degrees and release the CSM in an attitude appropriate for a retrograde burn. A 254-second burn by the service propulsion system would put the spacecraft into an ellipse with an apogee reduced to about 12,000 nautical miles and with an atmosphere-intercepting perigee. The vehicle was to coast in an attitude that would

[4] It is worth noting that the guidance system in the IU performed this magnificent recovery entirely on its own.

'cold soak' the heat shield to approximate lunar return thermal conditions. Late in the descending portion of the ellipse, the spacecraft would fire its engine again to accelerate to 36,500 ft/sec with a flight path angle of –6.5 degrees to simulate lunar return, jettison the service module and orient itself for atmospheric entry leading to splashdown in the Pacific.

Despite the additional propellant consumed in attaining orbit, the S-IVB was still capable of performing the planned program. However, the restart system failed to produce hydraulic pressure. On noting this failure, the IU cancelled the ignition command at 003:13:50.33 and advanced to the next programmed item – just as if it had achieved the burn. At 003:14:10.33 it initiated the 155-degree pitch manoeuvre, and 15 seconds later was still rotating when the ground commanded the spacecraft to separate. The hinges had been designed to rotate the SLA panels to 45 degrees from the vehicle's longitudinal axis within 1.3 seconds, but in this case the S-IVB was still in the process of pitching and the fact that the spacecraft suffered a disturbance in pitch of 1.5 deg/sec during separation implied that one of the panels had nudged the spacecraft. A preplanned alternative mission was selected in which the service propulsion system would fire to obtain the desired apogee of 12,000 nautical miles, although the propellant required to achieve this would rule out the follow-on burn to accelerate to lunar return velocity. As planned, the spacecraft adopted an attitude to 'cold soak' its heat shield for 6 hours. At high altitude, a 70-mm camera snapped 370 colour pictures of Earth in daylight, and instruments monitored how efficiently the command module's wall blocked the charged-particle radiation circulating in the van Allen belts.

On jettisoning the service module, the command module oriented itself for entry, making contact at 009:38:29 at an inertial velocity of 32,830 ft/sec (10 per cent less than intended) on a flight path angle of –5.85 degrees. It splashed down 49 nautical miles short of USS *Okinawa*, on station at the recovery point for the simulated lunar return. For the first time, a capsule adopted the apex-down flotation attitude, but was promptly righted by a set of airbags that inflated on its nose. It was recovered when the ship arrived 6 hours later.

Even while CSM-020 was coasting in space, NASA set up a meeting at the Cape with the contractors to investigate the problems suffered by the upper stages. Within 24 hours a review of the recorded telemetry established that:

- 70 seconds after S-II ignition, sensors in the engine compartment began to report chilling and the flow of liquid hydrogen to engine no. 2 increased;
- at + 110 seconds the thrust of this engine began to decline gradually, and then dropped sharply at + 169 seconds, at which time the load on the mechanism that gimballed the engine suddenly increased to counter a lateral component of thrust;
- at + 263 seconds the thrust fell sharply again, there was a sudden increase in temperature in the engine compartment, the automatic system to shut down an engine if its thrust fell below a specific value intervened; and
- one second later engine no. 3, which had shown no anomalous behaviour, suddenly cut off!

It was soon realised that engine no. 3 had cut off because it received the command intended for its ailing sibling. It was inferred that the control wires for the solenoids of the liquid oxygen prevalves of engines 2 and 3 must have been erroneously cross-connected. A check of the records showed that the wiring had been modified several months previously. Thus the intervention had shut down engine no. 2 by cutting off its fuel supply and shut down engine no. 3 by cutting off its oxidiser.

The telemetry from the S-IVB showed that its J-2 had also behaved anomalously by starting to chill at +68 seconds, the thrust starting to decline at +107 seconds, the rate of decline increasing at +115 seconds, and then the engine compartment temperature increasing at +119 seconds. However, in this case the engine continued until it was shut down nominally at +170 seconds.

Thus two engines on different stages had misbehaved in a strikingly similar way: namely, the onset of chilling in the engine compartment about 70 seconds into the burn, the thrust starting to decay about 40 seconds later, and then a sudden increase in temperature – although at different times.

The chilling implied a propellant leak – and the fact that a temperature colder than liquid oxygen was measured indicated a leak of hydrogen somewhere in the feed to the engine. The way in which the engine compartment of the S-II had been chilled supported the case for a hydrogen leak – that is, it had affected engine no. 5, which was in the centre of the cluster, but the engine on the opposite side of the cluster was unaffected. The structure of the engine and its orientation within the cluster implied that the leak was associated with a stainless steel pipe about 1 inch in diameter that wound its way 'upwards' from the middle of the assembly. This pipe carried liquid hydrogen to the ignition system, a small chamber in the middle of the injector plate in the roof of the combustion chamber. Another pipe delivered liquid oxygen. Spark plugs ignited this mix and issued a jet of flame into the chamber to start and sustain main combustion. The hydrogen pipe incorporated three flexible bellows to permit movement, with one bellows located precisely where the leak was inferred to have occurred. Evidently, the leaking pipe sprayed liquid hydrogen into the engine bay, chilling it. This flow created the lateral component of the thrust which the gimbal counteracted. The reduced flow of hydrogen to the igniter made the flame from the igniter oxygen-rich and turned it into a 'torch' which so eroded the structure as to let blazing gases enter the engine compartment, heating it. Rocketdyne rigged an engine at the Santa Susanna Test Laboratory in California to replicate the leak, and within a month was able to reproduce the behaviour of engine no. 2.

In the case of the S-IVB it was concluded that when the feed pipe broke, it did so completely, and with no hydrogen reaching the igniter there was no 'torch' to erode the structure. However, gas was able to pass from the combustion chamber through the igniter and out of the severed pipe to cause external heating. The cryogenic leak froze the hydraulic fluid, with the result that during the restart attempt both the main and auxiliary hydraulic pumps cavitated and yielded essentially no system pressure. In any case, with no hydrogen reaching the igniter there was no way that the engine could have been ignited even if everything else had worked.

Within a month, therefore, the investigators knew *what* had happened to the J-2s, but not *why*. To discover why the pipe was vulnerable, the telemetry from AS-501 was compared with that from AS-502. It was noted that on AS-502 the power of the engines was greater, and that the liquid hydrogen pressures in the pipes would have been marginally greater. Tests by Rocketdyne established that the bellows section vibrated at the increased flow rate, but even on a prolonged firing remained intact. The entire engine was vibrated prior to ignition to simulate the pogo of the S-IC in case this had been a factor, but the engine fired as before with the bellows vibrating. After several weeks, it was decided to vibrate the engine while passing fluid at the increased rate – but for safety using gaseous nitrogen at room temperature instead of liquid hydrogen. As the flow rate was being adjusted to the value desired, and prior to the vibration being introduced, the bellows failed. The same thing happened to a replacement. An inspection indicated metal fatigue arising from vibration induced by the flow.

Attention then turned to the environmental factors. The flight engine had been in near-vacuum in the upper atmosphere. The firing tests had been at sea level where, with liquid hydrogen at $-253°C$, a coating of ice would have formed on the bellows and tended to dampen out the vibration induced by the flow. Rocketdyne put an engine into an altitude chamber and pumped liquid hydrogen through the pipe, and the bellows failed after 100 seconds. In retrospect, the mystery was not that failures had occurred in flight; it was that the other engines had *not* succumbed to the same problem! Regardless, the fix was to use pipes which incorporated bends capable of absorbing the movements that the flexible bellows had been intended to counter.

Meanwhile, the pogo had been overcome by modifying the prevalves in the liquid oxygen ducts of the S-IC engines to incorporate a cavity containing helium which, by compressing, would dampen pressure fluctuations and thereby maintain a smooth combustion.

An analysis of the structural failure of one of the SLA panels found that this was unrelated to the pogo. Aerodynamic heating had increased the pressure of moisture inside the aluminium honeycomb material sufficiently for part of the facing sheet to puncture and tear off in the slipstream. It was decided to apply a layer of cork to the exterior of the adapter to absorb moisture, and to drill holes to prevent a build up of pressure within the underlying honeycomb.

Sam Phillips said this post-flight investigation was "one of the most aggressive, thorough and determined engineering test and analysis programs I have ever seen".

In fact, AS-502 was a successful *test flight* precisely because it revealed problems which were then fixed.

Notwithstanding the problems suffered by AS-502, on 23 April George Mueller called for AS-503 to be manned. Although Sam Phillips directed the next day that this launch vehicle be prepared with CSM-103 and LM-3 for a manned mission, he also instituted contingency planning to reconfigure it with BP-30 and LTA-B in the event of the decision to undertake a third unmanned test. The Kennedy Space Center replied that, given sufficient notice of the configuration, the 'boilerplates' would be able to be launched in mid-October but the manned mission would not be ready until late November at the earliest. On 26 April James Webb approved this planning for a manned mission, subject to a resolution of the anomalies which afflicted AS-502. In seeking to overcome the pogo, the Marshall Space Flight Center asked whether the

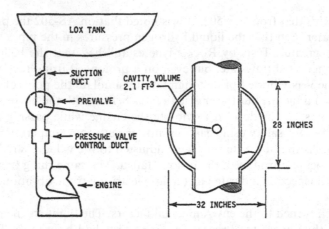

To overcome the 'pogo' suffered by the first stage of Apollo 6, the prevalves that fed liquid oxygen to the F-1 engines were modified to include a cavity of helium which would damp out pressure fluctuations.

emergency detection system could be configured to trigger an abort automatically. When Deke Slayton argued against doing so, George Low ordered the development of a 'pogo abort sensor' with a display in the command module to enable the crew to judge whether to initiate an abort. On 17 August, by which time it was clear that the pogo would be able to be eliminated, Low recommended that work on this sensor be terminated and a week later Phillips concurred.

Meanwhile, when the fuel injector of the ascent propulsion system developed by the Bell Aerospace Company for the LM continued to suffer combustion instability into the summer of 1967, NASA hired Rocketdyne to develop an alternative injector as a contingency measure. In April 1968 Grumman was instructed to coordinate the testing of Rocketdyne's injector in Bell's engine. In May George Low decided that this hybrid should be used, and told Rocketdyne to perform the integration work. By mid-August it was clear from qualification testing that the modified engine was free of instabilities, and there would be no need to mount another test flight. On 13 May Low met Chris Kraft, Deke Slayton and Maxime Faget to consider whether the 'fire in the hole' staging demonstration should be on the 'D' or 'E' mission. A key factor was that LM-3 would be the last to have the development flight instrumentation for monitoring the systems. Faget argued that whilst such data was desirable, it was not essential – it would be sufficient to take photographs of the base of the ascent engine following the rendezvous. In view of this line of argument, and the fact that the 'fire in the hole' staging demonstration would increase the complexity of the 'D' mission, Low postponed making a decision on whether to do it on the 'E' mission until the performance of the engine on LM-1 had been thoroughly analysed. On 17 May Kraft advised Low that the 'E' mission was already a complex affair, and that as further objectives were added the probability of achieving them diminished. In fact, Kraft saw little need for a 'fire in the hole' test. He understood the engineers' desire to test all the systems in space in both normal and backup modes, but the first 'fire in the hole' test at the White Sands Test Facility on 22 December 1967 had achieved all its objectives and further ground testing would provide the data needed to calculate the pressure and temperature transients pertaining to lunar lift off.[5] In parallel with these discussions, the Manned Spacecraft Center was studying extending the apogee of the 'E' mission to almost lunar distance to investigate navigation, communications and thermal control issues in the event of the lunar orbital 'F' mission being deleted, and this alternative mission was labelled 'E-prime'.

On 7 May 1968 CSM-101 passed its final customer acceptance review, and at the end of the month was delivered to the Cape. The inspectors were delighted to find fewer discrepancies than on any previous spacecraft. But Wally Schirra, who would be in command of flying the vehicle, did not accept it as flightworthy until after its altitude chamber tests in June. In contrast, when LM-3 was delivered on 14 June the inspectors found over 100 deficiencies, many of which were classified as major. In July, George White, the Chief of Reliability and Quality Assurance in the Office of Manned Space Flight, briefed George Mueller on the issues the Certification Review Board would require to consider.

On 7 August George Low advised the Manned Space Flight Management Council that the delivery of CSM-103 was imminent, but LM-3 was unlikely to be ready for launch until February 1969.

Low felt that for Apollo to have a chance of achieving a lunar landing in 1969, the first manned Saturn V must be flown in late 1968. By this point, the pogo problem was heading towards resolution and the other issues that marred AS-502 had been fixed. In April 1967 the Manned Spacecraft Center had outlined a contingency for a lunar mission involving only the CSM. On 8 August Low asked Kraft to consider sending CSM-103 to the Moon, and then he flew to the Cape with Carroll Bolender, ASPO's LM Manager, Scott Simpkinson, Chief of ASPO's Test Division, and Owen Morris, ASPO's Chief of Reliability and Quality Assurance, to discuss AS-503 with Sam Phillips, Kurt Debus, Director of the Kennedy Space Center, Rocco Petrone, Director of Launch Operations and Roderick Middleton, the Apollo Manager at the Cape.

At 08:45 local time in Houston on 9 August, Low met Kraft and Robert Gilruth to discuss the CSM-only option. Kraft said that it was feasible and Gilruth was enthusiastic. At 09:30 Deke Slayton was called in, and offered his support. Low then telephoned Sam Phillips, who had remained in Florida, and a meeting was arranged at 14:30 in Huntsville. In attendance were Low, Gilruth, Kraft and Slayton from the Manned Spacecraft Center; Wernher von Braun, Eberhard Rees (his deputy), Lee James (his Saturn V Program Manager) and Ludie Richard from the Marshall Space Flight Center; Kurt Debus and Rocco Petrone from the Kennedy Space Center; and Phillips and George Hage (his deputy) from headquarters. Low opened by saying that if the Apollo 7 evaluation of CSM-101 went well, it would be technically possible to send CSM-103 out to the Moon in December. However, if CSM-101 had

---

[5] In the event, no 'fire in the hole' separation was demonstrated in space.

problems then it would be necessary to confine CSM-103 to Earth orbit to continue the evaluation of the spacecraft's systems. Kraft pointed out that a mere 'loop' around the far-side of the Moon and then back to Earth would be insufficient – the spacecraft would have to enter orbit in order to contribute significantly to the lunar landing mission (which, depending on whether the 'F' mission was undertaken, might be the next mission to venture to the Moon). There was general agreement to initiate the planning for this contingency in secret, pending the decision and public announcement. The meeting broke up at 17:00. On returning to Houston, at 20:30 Low briefed George Abbey (his technical assistant), Kenneth Kleinknecht (his CSM Manager), Carroll Bolender (his LM Manager) and Dale Myers (Apollo Program Director for North American Aviation).

On 10 August Slayton offered this 'new' mission to James McDivitt, who was earmarked to fly AS-503, but McDivitt opted to await LM-3 so as to keep the 'D' mission for which his crew had been training. In contrast, Frank Borman, who had eagerly followed the discussions to extend the apogee of his 'E' mission out to lunar distance, readily accepted, and thereby regained the first manned Saturn V mission.

Kraft told Low on 12 August that a daylight launch would be required to allow an Atlantic recovery after an abort, and this meant the lunar launch window would open on 20 December. Low selected LTA-B as a stand-in for the LM because it had been assigned to fly with BP-30 on the unmanned mission and was already in preparation.

On 14 August the original twelve conferees, minus Rees, were joined at a meeting at NASA headquarters by Deputy Administrator Thomas Paine, Julian Bowman and William Schneider (the latter both of the Office of Manned Space Flight) to make a formal recommendation. With the discussion in progress, George Mueller telephoned from Vienna in Austria, where he and James Webb were at a United Nations Conference on the Exploration and Peaceful Uses of Outer Space. Mueller was sceptical of the proposal, and said he would not be able to discuss it until 22 August. Paine, playing the devil's advocate, pointed out to the conferees that until recently there had been doubts about whether the Saturn V was safe for manned flight, and here they were considering having it send a spacecraft on an impromptu mission to orbit the Moon, then he invited comments. Von Braun pointed out that once it was decided to man AS-503, it did not matter how far the spacecraft went. Hage noted that there were a number of points in the mission where go/no-go decisions could be made, managing the risk. Slayton opined that not to pursue this option would significantly diminish the likelihood of achieving President Kennedy's deadline. Debus had no technical reservations about the launch. Nor did Petrone. Bowman said it would be a 'shot in the arm' for the program. Lee James said it would enhance the safety of later flights. Ludie Richard said it would improve lunar capability. Schneider was fully in favour. Gilruth pointed out that although it was an impromptu mission, it would improve the chance of being able to achieve the overall goal of the program. Kraft reiterated that it should be a lunar orbital rather than a circumlunar mission. Low pointed out that if Apollo 7 succeeded, they could either fly this impromptu mission or await LM-3 and launch in February or March. In view of the deadline for the lunar landing, Low said the decision was obvious: they should send Apollo 8 to the Moon. Paine concurred. Phillips ordered planning to continue.

Phillips and Paine discussed the plan with Mueller and Webb on the telephone the next day. Mueller had warmed to the idea overnight. Webb was "fairly negative" (as Phillips later put it) but asked for information to be sent by telegram. On 16 August Webb called Paine and agreed to the mission planning, with the proviso that there be no public announcement. On 17 August Phillips told Low that although Webb had authorised a manned Saturn V launch in December, there must be no 'leak' that the spacecraft might venture to the Moon – that decision was contingent on the outcome of Apollo 7.

Meanwhile, CSM-103 had arrived at the Cape and a start had been made on the modifications required to send it to the Moon. On 19 August Phillips directed that if AS-503 was manned and did not carry a LM, irrespective of where the spacecraft went the mission was to be designated 'C-prime'. Then McDivitt's crew would fly the 'D' mission riding AS-504 with CSM-104 and LM-3. The 'E' mission had been deleted. That same day, Phillips told the press that if the mission of CSM-101 was a success, AS-503 would be manned and that because the LM would not be ready this would be a CSM-only mission; by not mentioning the option of leaving Earth orbit he readily conveyed the impression of an Earth orbit mission. On 3 September Low directed that if the 'C-prime' mission went to the Moon, it would make ten orbits over a period of 20 hours and then head home. If it was confined to Earth orbit, it would undertake the parking orbit preparations for translunar injection and then fly one of a number of alternative missions, each of which would involve simulating the transposition, docking and extraction of the LM. On 9 September Borman's crew began to train in the simulator at the Cape for the lunar mission. Ten days later, with the AS-502 investigations finished and the remedies implemented, Mueller declared the Saturn V to be 'man rated'. CSM-103 was mated with AS-503 on 7 October, the launch escape system was added on 8 October, and the next day the space vehicle was rolled out and installed on Pad 39A.

On 31 March 1968 Lyndon Johnson announced that he would not seek and would not accept his party's nomination for the presidential election in November. The two main candidates were Hubert H. Humphrey and Richard M. Nixon. Perhaps because James Webb knew that neither would retained him as NASA Administrator, he informed Johnson on 16 September that he would stand down on 6 October, which was his 62nd birthday. Thomas Paine was promoted to Acting Administrator.

Meanwhile, on 20 September CSM-101 passed its flight readiness review, and later that day Wally Schirra announced to reporters that Apollo 7 would be his final mission because he intended to retire from NASA.

BORING HOLES IN THE SKY

AS-205 lifted off from Pad 34 at 15:02:45 GMT on 11 October 1968 to fly the 'C' mission. Flown by Wally Schirra, Donn Eisele and Walt Cunningham, Apollo 7 was

On 11 October 1968 the Saturn IB of Apollo 7 lifts off from Pad 34.

to be open-ended up to 11 days and its purpose was to assess the performance of the Block II spacecraft.

The ascent phase was nominal and the S-IVB achieved a 123×152-nautical mile orbit. Prior to separating from the spent stage, the crew temporarily took command of the Instrument Unit and manually manoeuvred the combined vehicle in pitch, roll, and yaw, then they returned control to the launch vehicle. By the time the spacecraft separated at 002:55:02.40, venting of S-IVB propellants had raised the orbit to 123×170 nautical miles. The spacecraft moved clear, flipped and moved back in as if to retrieve the LM (which was absent). Since one of the four panels of the SLA had not fully deployed, it was decided that in future the panels would be jettisoned. One of the primary objectives was to demonstrate Apollo's rendezvous capability using the spent stage as the target. At Schirra's insistence, one man was awake at all times to monitor the spacecraft's systems, even though the ongoing work made sleeping difficult. The rendezvous rehearsal was successfully achieved on the second day.

Although this was the first US spacecraft to have sufficient habitable volume for a man to leave his couch and move around, the crew suffered no disorientation in the weightless state, despite efforts to induce motion sickness. However, all three men developed head colds early on, making them grumpy, and in-flight TV, which was a secondary objective, provided a focus for their frustration. When the monochrome camera was finally switched on, however, it delivered excellent results and the crew played up to their audience. But it was a long and tedious flight of monitoring the systems to evaluate their performance, always prepared to intervene in the event of a problem. In fact, it was an exercise in would later be derided as "boring holes in the sky".

At 11:11:48 GMT on 22 October the command module splashed in the Atlantic 1.9 nautical miles from the target point. It initially assumed an apex-down attitude, but was soon turned apex-up by the inflatable bags on its nose. The astronauts were retrieved by helicopter and arrived on USS *Essex* an hour later.

The Apollo 7 mission was successful in every respect, with the service propulsion system firing perfectly eight times. Indeed, afterwards Schirra described the flight as a "101 per cent success". In combination with previous missions and ground tests, it certified the CSM for use in Earth orbit and for tests in the cislunar and lunar orbital environments.

MEN ORBIT THE MOON!

On 7 November 1968 George Mueller declared that AS-503 was fit for a mission to the Moon. On 11 November Sam Phillips recommended to the Manned Space Flight Management Council that Apollo 8 enter lunar orbit. Later that day, Mueller told Thomas Paine that he had discussed the mission with the Science and Technology Advisory Committee and with the President's Science Advisory Committee, both of which had endorsed the proposal, and he recommended that it should be undertaken. After speaking to Frank Borman by telephone, who confirmed his willingness to fly the mission, Paine gave the formal go ahead and told Phillips to make the necessary arrangements. The next day, NASA announced that Apollo 8 would be launched on 21 December and attempt a lunar orbital mission. Earlier in the year, Michael Collins had withdrawn from the crew to undergo a surgical procedure, and had been replaced by his backup, James Lovell.

Tom Stafford, John Young and Gene Cernan were announced on 13 November as the prime crew of Apollo 10, backed up by Gordon Cooper, Donn Eisele and Edgar Mitchell. This established the precedent for a crew backing up one mission, skipping two, and becoming the prime crew of the mission after that. It had yet to be decided, however, whether Apollo 10 would fly the 'F' or the 'G' mission.[6]

On 9 October 1968 AS-503, complete with CSM-103 and LTA-B, was rolled out to Pad 39A. The countdown demonstration test was completed on 11 December, and the actual countdown began at 00:00 GMT on 16 December. The launch window ran from 20 to 27 December, and it had been decided to try for 21 December to enable the astronauts to inspect the ALS-1 landing site in eastern Mare Tranquillitatis soon after local sunrise.

Frank Borman, James Lovell and Bill Anders entered the spacecraft with a little under 3 hours on the clock. There were no unplanned holds, and Apollo 8 lifted off at 12:51:00 GMT on 21 December for the 'C-prime' mission.

The ascent was nominal and the deviations from the trajectory when the S-IVB cut off at T + 684.98 seconds were + 1.44 ft/sec in velocity and –0.01 nautical mile in altitude, which was almost perfect. At 002:27:22, after the S-IVB and spacecraft had been thoroughly checked, Collins, serving as the CapCom in Mission Control, made the momentous call, "Apollo 8, you are 'Go' for TLI."

The 317.7-second translunar injection was started at 002:50:37.8 and produced a velocity of 35,505.4 ft/sec. The spacecraft separated 30 minutes later and the four SLA panels were jettisoned. After turning around, the spacecraft's ability at station-keeping with the spent stage was assessed. A 1.1-ft/sec manoeuvre was performed at 003:40:01 using the reaction control system of the service module to move clear of the stage, and a 7.7-ft/sec manoeuvre at 004:45:01 increased the separation rate.

At 004:55:56.0 the S-IVB opened its hydrogen vent valve and at 005:07:55.8 it passed oxygen through the engine. At 005:25:55.8 the auxiliary propulsion system was ignited and burned to depletion. The accumulated velocity increment placed the stage on course to fly by the trailing limb of the Moon at an altitude of 681 nautical miles and pass into solar orbit. The spacecraft's service propulsion system executed a 2.4-second, 20.4-ft/sec midcourse manoeuvre at 010:59:59.2. A 24.8-ft/sec change had been planned, but the engine delivered less thrust than expected and a correction

---

[6]   CSM-101 had flown on Apollo 7, CSM-102 had been retained by North American Aviation for ground testing, CSM-103 had been assigned to the Apollo 8 'C-prime' mission, CSM-104 was to fly the Apollo 9 'D' mission, CSM-105 was for ground testing, and CSM-106, which was delivered to the Cape on 25 November 1968, was assigned to Apollo 10.

was made at 060:59:55.9 to refine the trajectory. These burns served to calibrate the service propulsion system in advance of calculating the orbit insertion manoeuvre.

In contrast to Apollo 7, this time all three crewmen experienced nausea as a result of rapid body movement, with the symptoms lasting up to 24 hours. The first of six TV transmissions started at 031:10:36 and ran for 23 minutes 37 seconds. The wide-angle lens gave an excellent view of the inside of the spacecraft, where Lovell was preparing a meal, but the telephoto lens passed too much light and pictures of Earth were poor. After a procedure was devised to tape a filter of the still camera onto the TV camera, it produced improved pictures of Earth during a transmission starting at 055:02:45. At 055:38:40 the astronauts were alerted that they had become the first people to enter a region where the gravitational attraction of another body exceeded that of Earth. The spacecraft had been slowing as it climbed up from Earth, but now it began to accelerate as it was drawn in by the Moon. However, they were not yet committed. If a reason developed not to brake into lunar orbit, then Apollo 8 would simply continue on its 'free return' trajectory around the back of the Moon and be 'slingshot' back to Earth. Although everything was going well, the translunar coast was frustrating in the sense that at no time were the crew able to see their objective owing to the spacecraft's trajectory in relation to the positions of the Moon and the Sun.

The lunar orbit insertion manoeuvre began at 069:08:20.4 at an altitude of 75.6 nautical miles above the far-side of the Moon, and the 246.9-second burn produced an orbit ranging between 60.0 and 168.5 nautical miles with its high point above the near-side. After the post-burn checklist had been attended to, and while still passing over the far-side, the astronauts had their first opportunity to inspect the surface of the Moon up close. At 071:40:52 they gave a 12-minute TV transmission showing the passing terrain. In contrast to geologists, the astronauts described the surface in terms of "a battlefield", "a sandbox torn up by children", "a volleyball game played on a dirty beach", "plaster of Paris", or (vaguely scientifically) as "pumice". Bright ray craters appeared just as if they had been made by a "pickaxe striking concrete". The colour was varied, sometimes appearing to be black and white, yet other times displaying a distinctly brownish tan. In terms of mood, the surface was "desolate", "bleak" and "forbidding". A 9.6-second burn at 073:35:06.6 circularised the orbit at 60 nautical miles.

As this was the first opportunity for humans to directly observe the Moon at close range, James Sasser of the Apollo Spacecraft Project Office in Houston had served as the 'project scientist' for the mission. He formed an advisory team and this drew up a program of photography and visual observations for the crew to perform using a Maurer 16-mm movie camera and a Hasselblad with a 250-mm lens. In particular, the Manned Spacecraft Center wanted views of the eastern limb to assist in selecting landmarks for a lander's navigational checks prior to the powered descent. Some of this documentation was to be overlapping vertical and oblique pictures which would enable stereoscopic analysis to determine the geographical position and elevation of each feature, but the movie camera was also to be fitted to the spacecraft's sextant to depict the landmarks in context. In addition, some 'scientific' targets were marked on the flight charts as 'targets of opportunity' which were to be inspected if time and circumstances allowed. These were to provide either detailed coverage of specific features or broad coverage of areas which had not been adequately imaged by the Lunar Orbiters. And, of course, the ALS-1 landing site was to be inspected. Most of the scientific observing and photography was assigned to Anders, the LMP without a lunar module. Jack Schmitt, a professional geologist who was hired as an astronaut in 1965, served as the main interface between Sasser's team and the Apollo 8 crew in training, but some briefings were provided by US Geological Survey staff. At the suggestion of Wilmot N. Hess, Director of the Science and Applications Directorate at the Manned Spacecraft Center, SasseTr's team had set up a 'science support' room in Mission Control to listen to the astronauts' commentaries and watch the TV of the lunar landscape passing below the spacecraft.

The astronauts could recognise surface features in shadows lit by Earthshine, and could see detail on sunward-facing slopes which had been 'washed out' in the Lunar Orbiter pictures. In fact, they could perceive detail to within 5 degrees of the 'zero phase' point, which is the line of sight with the Sun directly behind the observer. In planning the lunar landing the lower limit for Sun angle had been set at 6 degrees, but the astronauts could see surface detail at angles as low as 2 degrees. They were able to confirm that the upper limit of 16 degrees provided excellent definition, and their observations suggested that it might be possible to raise the limit to 20 degrees – but no higher than this. This enabled the lighting constraints for the lunar landing to be relaxed.

Of the two candidate landing sites in Mare Tranquillitatis, ALS-1 in the east was brighter; so much so, in fact, that it was debatable whether it was truly mare material or a flattish portion of the adjacent terra. Observing it visually from an altitude of 60 nautical miles, Lovell said it reminded him of an aerial view of Pinacate in Mexico, a volcanic field which he had been shown in training.

Owing to crew fatigue, Frank Borman took the decision at 084:30 to cancel all secondary activities during the final two revolutions, to allow the crew to rest. The only tasks during this period were an alignment of the inertial guidance system and the preparations for transearth injection. But at 085:43:03 they provided the planned 27-minute TV transmission showing the Moon and Earth, and to mark the fact that it was Christmas Eve they recited the first ten verses of the Book of Genesis from the Bible prior to signing off with, "Good night, good luck, a Merry Christmas, and God bless all of you – all of you on the good Earth."

Radio tracking indicated that by the time Apollo 8 was ready to head for home the mascons had perturbed its initially circular orbit into one of 58.6 × 63.6 nautical miles. At 089:19:16.6, after ten revolutions of the Moon, the 203.7-second transearth injection was made on the far-side of the Moon at an altitude of 60.2 nautical miles, which was just about perfect. After returning to the Earth's gravitational influence, the spacecraft progressively accelerated. Only one small midcourse correction was required. It was made at 104:00:00, and the 15.0-second burn by the service module reaction control system imparted a change of 4.8 ft/sec.

On shedding the service module, the command module adopted its entry attitude and at 146:46:12.8 hit the entry interface travelling at 36,221.1 ft/sec. It pursued an automatically guided profile. The ionisation bathed the interior of the cabin in a cold

As the crew of Apollo 8 emerged from behind the Moon for the first time, they were astounded by the sight of Earth rising over the lunar limb.

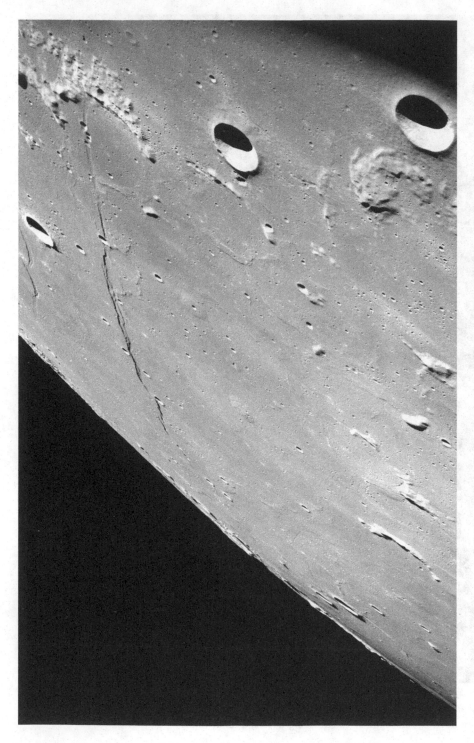

An oblique view by Apollo 8 looking northwest across the eastern part of Mare Tranquillitatis. The crater in the foreground is Taruntius-F, and one of the Cauchy clefts crosses the upper part of the picture. The ALS-1 site is out of frame to the south.

. blue light as bright as daylight. At 180,000 feet, as expected, the lift vector deflected the vehicle to 210,000 feet, then it resumed its downward course. It splashed into the Pacific 1.4 nautical miles from the target at 15:51:42 on 27 December. It adopted an apex-down position, but promptly righted itself. The astronauts were soon recovered and flown by helicopter to USS *Yorktown*.

This audacious mission, described as the "greatest voyage since Columbus", took NASA a giant step towards achieving Kennedy's challenge.

On 6 January 1969 Deke Slayton called Neil Armstrong, Michael Collins and Buzz Aldrin to his office at the Manned Spacecraft Center and told them that they would fly Apollo 11 and should assume their mission would involve a lunar landing.

On 10 January 1969 John Stevenson, Director of Mission Operations at the Office of Manned Space Flight, circulated a revised version of the tentative schedule for the year that was issued early in 1968. This called for launching the delayed 'D' mission on 28 February. As the 'E' mission had been rendered irrelevant by Apollo 8, this meant that if the 'F' mission flew in May and was satisfactory, it should be possible to attempt the lunar landing in July. The rationale for the 'F' mission was to obtain experience of operating in deep space, but after Apollo 8 the issue became whether another test in lunar orbit was required. The decision was postponed until LM-3 had been put through its paces.

THE SPIDER

Apollo 9 was to be the 'D' mission – a lunar module manned flight demonstration in Earth orbit. The payload for the AS-504 launch vehicle was CSM-104 and LM-3. As they were to operate independently, the spacecraft were given radio call-signs. The blue wrapping of the command module for its shipment to the Cape had given it the appearance of a sweet, so it was named 'Gumdrop'. The arachnid-like configuration of the lunar module prompted the name 'Spider'.

The launch was scheduled for 28 February 1969 and the countdown was begun at 03:00:00 GMT on 27 February with 28 hours on the clock, but 30 minutes into the planned 3-hour hold at T–16 hours the clock was recycled to T–42 hours in order to enable the crew of James McDivitt, David Scott and Rusty Schweickart to recover from a mild respiratory infection. The count picked up at 07:30:00 on 1 March and the vehicle lifted off from Pad 39A on time at 16:00:00 GMT on 3 March.

The ascent was nominal and at S-IVB cutoff at T + 664.66 seconds the deviations were + 2.86 ft/sec in velocity and –0.17 nautical mile in altitude, with the result that the initial orbit was almost perfect at 100 nautical miles. At 002:41:16.0 the S-IVB released the CSM, which moved clear, turned end over end to aim its apex at the top of the LM and moved back in. At 003:01:59.3 it docked at the first attempt, marking the first use of this apparatus. Once the tunnel between the two spacecraft had been pressurised, the crew opened the apex hatch of the command module to confirm that all the latches on the docking ring had engaged, and after lines had been connected to supply power to the dormant LM the hatch was reinstalled. On a command issued by the CSM at 004:08:09 the S-IVB released the docked combination.

Preparing the CSM-104 and LM-3 spacecraft for the Apollo 9 mission.

Apollo 9's S-IVB with the Lunar Module 'Spider' exposed.

On his spacewalk from the Lunar Module, Rusty L. Schweickart photographed David R. Scott in the hatch of the Command Module.

After the spacecraft was clear, the S-IVB reignited its engine at 004:45:55.5 to raise an apogee of 1,672 nautical miles. Then, after a period of coasting to allow the engine to cool down, it initiated a final burn at 006:07:19.3 to achieve a velocity of 31,620 ft/sec which would send it into solar orbit.

Meanwhile, at 005:59:01.1 a 5.2-second burn by the service propulsion system raised the spacecraft's orbit to 111 × 128 nautical miles. Three further manoeuvres on the second day in space measured the oscillatory response of the docked vehicles to obtain data designed to improve the autopilot's response in this configuration, and also burned off the CSM's propellant to increase the fidelity of manoeuvres which it would later perform in Earth orbit to rehearse what a mission would do in lunar orbit.

On the third day in space, Schweickart entered the LM to check out its systems. McDivitt joined him 50 minutes later. At about 045:52, shortly after the landing gear was deployed, McDivitt advised Mission Control that Schweickart had twice been sick – this illness would have an impact on the EVA planned for later in the mission. At 046:28 the astronauts made a 5-minute TV transmission from inside the LM. The descent engine was ignited at 049:41:34.5 for a 371.5-second burn in which the autopilot controlled the attitude of the docked vehicles and the astronauts manually throttled the engine to full thrust. The LM was deactivated at 051:00. Several hours later, a service propulsion system burn achieved an almost circular orbit of 125.9 × 131.0 nautical miles in preparation for the rendezvous sequence.

The EVA plan had called for Schweickart to exit the LM's forward hatch, transfer to the command module hatch, and then return. But owing to his bouts of nausea the spacewalk was cut back from 2 hours 15 minutes to just 39 minutes, to be made on a single daylight pass. The LM was depressurised at 072:45, and the hatch opened at 072:46. Schweickart initiated his egress at 72:59:02, feet first and face up, and was completely out by 073:07. He was wearing the Extravehicular Mobility Unit suit and Portable Life Support System backpack which astronauts were to wear on the lunar surface. A 25-foot nylon safety tether precluded him drifting away. For stability, he inserted his feet into a pair of 'golden slippers' on the 'porch' of the descent stage. Meanwhile, at 073:02:00 Scott opened the side hatch of the command module and poked his head and shoulders out to monitor Schweickart. Although the transfer to the command module hatch had been cancelled, Schweickart was able to make an abbreviated study of translation and body-attitude-control using handrails affixed to the upper part of the LM. Before ingressing, Schweickart shot 16-mm movie footage of Scott's activities, and 70-mm Hasselblad pictures of the exterior of both vehicles. Although the EVA was brief and did not involve a period of orbital darkness, it was sufficient to certify the suit and backpack for use on the lunar surface. The LM was repressurised at 073:53, and the CSM several minutes later. After a TV transmission from the LM that started at 074:58:03 and lasted 15 minutes, it was deactivated and McDivitt and Schweickart rejoined Scott.

On the fifth flight day McDivitt and Schweickart were back in the LM by 088:55 in order to prepare that ship for a period of free flight and an active rendezvous. At 092:22 the CSM oriented the pair into the attitude required for undocking. This was attempted at 092:38, but the latches did not fully release until 092:39:36. This was to be the first time that astronauts flew a spacecraft that was incapable of returning to Earth if an emergency were to arise – they relied on Scott to rescue them. Once free, the LM pirouetted while Scott made a visual inspection. At 093:02:54 the CSM used the thrusters of its reaction control system to make a separation manoeuvre. Over the next 6.3 hours, the LM undertook a series of manoeuvres which set up and executed a rendezvous. In the process, the descent propulsion system was fired under different control regimes and with the throttle being varied, after which the descent stage was jettisoned and the rendezvous was performed by the ascent stage. Terminal phase braking began at 098:30:03, and was followed by a period of station-keeping, then formation flying to facilitate mutual photography prior to docking at 099:02:26. McDivitt and Schweickart then transferred back to the CSM. The ascent stage was jettisoned at 101:22:45.0, and half an hour later ignited its main engine and fired it to depletion to enter a 126.6 × 3,760.9-nautical mile orbit.

The remainder of the mission was less hectic, being devoted mainly to conducting multispectral photography to prepare for the Skylab space station. At 169:30:00.4 the service propulsion system was fired in a 24.9-second burn which established the conditions for a nominal de-orbit. Unfavorable weather in the planned recovery area prompted a postponement of the de-orbit by one revolution, and it was performed at 240:31:14.8. The service module was jettisoned a few minutes later. The command module flew the entry profile under the control of its primary guidance system, and splashed into the Atlantic at 17:00:54 on 13 March about 2.7 nautical miles from the target. It settled in the ideal apex-up flotation attitude, and within an hour the crew were onboard USS *Guadalcanal*.

DRESS REHEARSAL

With Apollo 9 having successfully tested the LM in Earth orbit, the next issue was whether to fly the 'F' mission or to push on and attempt the lunar landing. In fact, it would be impossible for LM-4 to attempt the 'G' mission, as the software to conduct the powered descent was still under development. Furthermore, owing to propellant restrictions in the ascent stage of this somewhat overweight LM it would be unable to lift off and rendezvous. Tom Stafford, the Apollo 10 commander, argued against his crew waiting for LM-5 to become available. "There are too many 'unknowns' up there," he noted. "We can't get rid of the risk element for the men who will land on the Moon but we can minimise it; our job is to find out everything we can in order that only a small amount of 'unknown' is left."

On 24 March 1969 NASA stated that Apollo 10 would fly the 'F' mission. The original idea had called for the LM merely to undock, enter a slightly different orbit, rendezvous and redock, but in December 1968 the Mission Planning and Analysis Division at the Manned Spacecraft Center had urged putting the descent propulsion system through a high-fidelity rehearsal in which the LM would lower its perilune sufficiently to test the ability of the landing radar to detect and lock onto the surface. Howard Tindall also proposed that the LM should initiate the powered descent and then execute an early abort by 'fire in the hole' staging, but his colleagues convinced him that this would be too adventurous. One aspect of the decision to go ahead with the 'F' mission was to evaluate the tracking and communications of two vehicles in lunar orbit. In essence, it had been decided to exploit the fortuitous relaxation in schedule pressure and improve on Apollo 8 by performing a rehearsal to the point at which a later LM would initiate its powered descent.

The finally agreed plan called for the LM to separate from the CSM in the circular lunar parking orbit, enter an elliptical orbit having a perilune of about 50,000 feet located just east of the prime landing site, execute a low pass and then jettison the descent stage to make the rendezvous.

In April 1969 the site selectors met to decide the prime target for the first Apollo landing. The photographs of ALS-1 taken by Apollo 8 indicated the presence of a smooth blanket of light-toned material that softened or masked the landscape, and a study of the craters showed that the regolith was quite thick, which in turn implied a considerable age. The fact that the site was *atypical* of the maria made it unattractive for dating the maria, so it was rejected. This left ALS-2 in the southwestern part of Mare Tranquillitatis as the prime target. In early May, Jack Schmitt put it to Tom Stafford that the launch of Apollo 10 be slipped 24 hours from the proposed date so that the low-perilune pass over ALS-2 could be made in illumination matching that of a mission attempting to land there. This would enable high-resolution pictures to be taken of the site and the landmarks on the approach route. Stafford was receptive. Schmitt approached George Low, who asked Chris Kraft, who sought the advice of the flight control specialists – there were issues in favour and against. When the case was put to Sam Phillips he rescheduled the launch.

AS-505 had been installed on Pad 39B on 11 March, and Apollo 10 lifted off on schedule at 16:49:00 GMT on 18 May 1969 with Tom Stafford, John Young and Gene Cernan.

When the S-IVB cutoff at T + 703.76 seconds, the deviations were –0.23 ft/sec in velocity and –0.08 nautical miles in altitude. After translunar injection, CSM-106 'Charlie Brown' separated, turned around and docked with LM-4 'Snoopy', then the pair were released by the stage. The S-IVB then used propulsive venting to adopt a path that would fly past the Moon and enter solar orbit. At 026:32:56.8 the service propulsion system made a 49.2-ft/sec burn to match a July lunar landing trajectory. At 075:55:54.0 the spacecraft entered an initial lunar orbit of 60.2 × 170.0 nautical miles. Two revolutions later, this was refined to 59.2 × 61.0 nautical miles. During a 30-minute colour TV transmission the astronauts showed off the lunar surface. They reported the colour of the surface to be less grey than was described by Apollo 8. In particular, Mare Serenitatis appeared "tan", whereas Mare Tranquillitatis appeared "dark brown".

After undocking at 098:29:20, the vehicles took up station 30 feet apart while Young inspected the LM, and then the CSM moved off. A 27.4-second burn by the descent propulsion system at 099:46:01.6 placed the LM into a descent orbit with its perilune 15 degrees east of ALS-2. The landing radar was tested while passing over that site at an altitude of 47,400 feet an hour later. The pictures taken were of greater resolution than those transmitted by the Lunar Orbiters. Unfortunately, the 16-mm

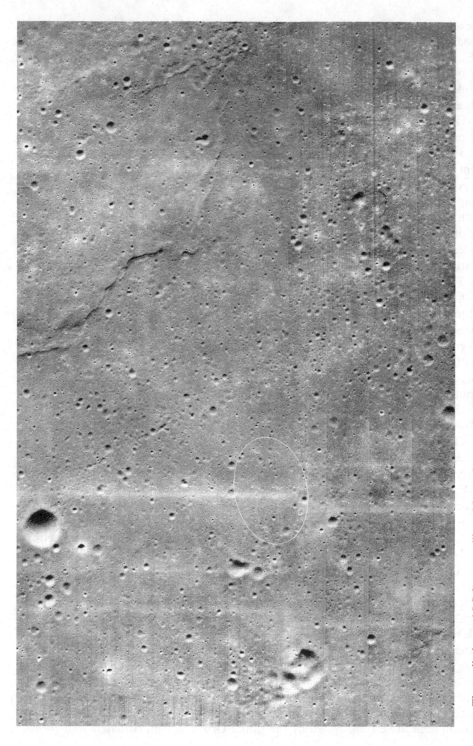

The preliminary ALS-2 target ellipse (shown in an earlier illustration) in the context of part of the Lunar Orbiter II-P-6 mosaic.

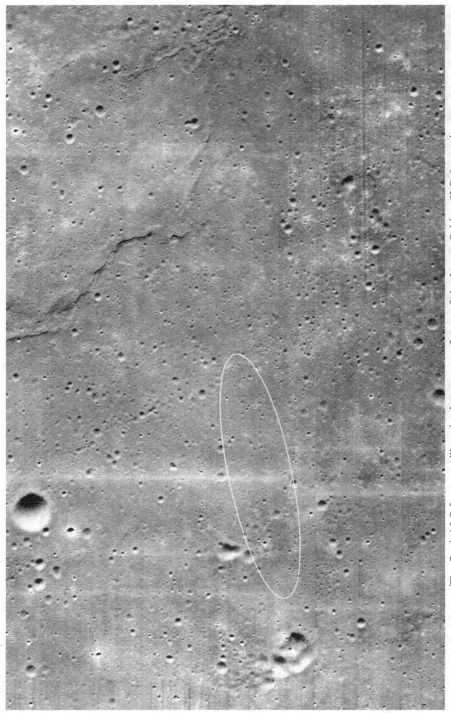

The final ALS-2 target ellipse in the context of part of the Lunar Orbiter II-P-6 mosaic.

This oblique view looking northwest across the crater Maskelyne was taken by the Apollo 10 Lunar Module 'Snoopy' as it flew low over Mare Tranquillitatis towards the ALS-2 target.

This view from 'Snoopy' shows the crater Moltke and Hypatia Rille, located just to the south of ALS-2.

A view of the eastern hemisphere of the Moon taken by Apollo 10 shortly after it departed lunar orbit, heading back to Earth.

movie camera failed. A descent propulsion system burn at 100:58:25.9 put the LM
into an orbit of 12.1 × 190.1 nautical miles to arrange a 'lead angle' equivalent to
that which would occur at cutoff of an ascent from the lunar surface. At 102:44:49,
during preparations to start the rendezvous with the CSM, the LM started to
wallow off slowly in yaw and then stopped, and several seconds later it initiated a
rapid roll accompanied by small pitch and yaw rates. Subsequent analysis revealed
that this anomalous motion was due to human error. The control mode of the
abort guidance system had inadvertently been returned to *AUTO* instead of the
ATTITUDE HOLD mode for staging. In *AUTO*, the abort guidance system steered
the LM to enable the rendezvous radar to acquire the CSM, which at this point was
not in accordance with the plan. The required attitude was re-established by the
commander taking manual control. The descent stage was jettisoned at
102:45:16.9, and 10 minutes later an ascent propulsion system burn achieved an
orbit of 11.0 × 46.5 nautical miles. This matched the insertion orbit for a mission
returning from the surface. The LM had the active role in the rendezvous, and
docked at 106:22:02. Two hours later the ascent stage was jettisoned, and during
the next revolution the ascent propulsion system was fired to depletion in order to
place the vehicle into solar orbit.

At 137:39:13.7, after 31 lunar revolutions, the CSM made the transearth injection.
The aim was so accurate that it required only a 2.2-ft/sec refinement 3 hours prior to
shedding the service module to centre the trajectory in the 'corridor' for atmospheric
entry. The capsule splashed into the Pacific 1.3 nautical miles off target at 16:52:23
on 26 May and adopted the apex-up flotation attitude. The astronauts were aboard
USS *Princeton* within the hour.

While Apollo 10 was in transit to the Moon, AS-506 was rolled out to Pad 39A in
preparation for the Apollo 11 mission. After the pictures taken during the low pass
over ALS-2 were examined, it was confirmed as the prime site for Apollo 11. ALS-3
in Sinus Medii was 2 day's terminator travel westward and would be the backup. If
the launch had to be delayed beyond the date for ALS-3, then the target would be
ALS-5 in Oceanus Procellarum. In the post-flight debriefing, Tom Stafford pointed
out that although the ALS-2 aim point was acceptable, the western end of the ellipse
was much rougher. He advised Neil Armstrong that if he were to find himself at the
far end of the ellipse and did not have the hover time to manoeuvre among the small
craters and boulders to select a spot on which to land, then he would have to "shove
off" – by which Stafford meant abort.

END GAME

A week before Apollo 11 was due to launch, people began to congregate at the Cape
communities of Titusville, Cocoa Beach, Satellite Beach and Melbourne. They came
from all around the world in order to be able to tell their grandchildren they were
present when men set off to try to land on the Moon. By 15 July hotels and motels
allowed late-comers to install camp beds in lounges and lobbies, but most people
spent the night on the beaches and by the roadside, generating the worst congestion

Viewers in the Firing Room of the Launch Control Center watch Apollo 11 set off on its mission to land on the Moon.

in Florida's history. With the notable exception of alarm clocks, which rapidly sold out, shops were able to supply the hoards. As it was to be a dawn launch, the parties ran through the night.

When AS-506 lifted off at 09:32:00 local time on 16 July on a mission to accept President Kennedy's challenge of landing a man on the Moon before the decade was out, it was estimated that there were about a million people present and 1,000 times as many watching on 'live' television.

No-one could be certain that the objective would be achieved, but the way had certainly been well paved.

Index

Printing: Mercedes-Druck, Berlin
Binding: Stein+Lehmann, Berlin